QUANTUM OPTOELECTRONICS
PRINCIPLES AND APPLICATIONS

Kruger Brentt

Publishers

QUANTUM OPTOELECTRONICS
PRINCIPLES AND APPLICATIONS

Kebra Ward

Kruger Brentt
Publishers
2025

Kruger Brentt Publishers UK. LTD.
Company Number 9728962

Regd. Office: 68 St Margarets Road, Edgware, Middlesex HA8 9UU

ISBN: 978-1-78715-384-4

Disclaimer:

Every possible effort has been made to ensure that the information contained in this book is accurate at the time of going to press, and the publisher and author cannot accept responsibility for any errors or omissions, however caused. No responsibility for loss or damage occasioned to any person acting, or refraining from action, as a result of the material in this publication can be accepted by the editor, the publisher or the author. The Publisher is not associated with any product or vendor mentioned in the book. The contents of this work are intended to further general scientific research, understanding and discussion only. Readers should consult with a specialist where appropriate.

Every effort has been made to trace the owners of copyright material used in this book, if any. The author and the publisher will be grateful for any omission brought to their notice for acknowledgement in the future editions of the book.

For information on all our publications visit our website at http://krugerbrentt.com/

PREFACE

"**Quantum Optoelectronics: Principles and Applications**" embarks on an enlightening journey into the interdisciplinary realm of quantum optics and optoelectronics, offering readers a comprehensive understanding of the principles, technologies, and applications that underpin the fusion of quantum mechanics and photonics. This book serves as an essential resource for students, researchers, engineers, and practitioners seeking to explore the frontiers of quantum photonics and its transformative impact on technology and science.

Quantum optoelectronics represents a convergence of quantum mechanics, optics, and electronics, enabling the manipulation and control of light at the quantum level for a wide range of applications, including quantum information processing, quantum communication, sensing, imaging, and metrology. By harnessing the unique properties of photons and quantum systems, quantum optoelectronic devices offer unprecedented capabilities for information processing, communication, and sensing with enhanced performance and security.

The primary objective of this book is to provide readers with a comprehensive introduction to quantum optoelectronics, covering key topics such as quantum optics, quantum information theory, photonic devices, quantum light sources, detectors, and applications in quantum computing, cryptography, and quantum metrology. Each chapter offers insights, theoretical frameworks, practical techniques, and real-world examples to help readers develop a deep understanding of quantum optoelectronic concepts and methods.

Contributions from leading experts in the field explore various aspects of quantum optoelectronics, including techniques for generating, manipulating, and detecting quantum states of light, methods for implementing quantum algorithms and protocols using photonic platforms, and applications in emerging fields such as quantum sensing and imaging. Special emphasis is placed on the interdisciplinary nature of quantum optoelectronics, with discussions on its connections to quantum information science, nanotechnology, and materials science.

As the field of quantum optoelectronics continues to advance and mature, the potential for transformative applications in information technology, telecommunications, healthcare, and beyond is vast. By fostering collaboration and innovation among researchers, engineers, and industry stakeholders, we can harness the power of quantum photonics to address some of the most pressing challenges facing society and pave the way for a new era of technological innovation and scientific discovery.

Kebra Ward

CONTENTS

1 Chapter

STUDY AND DEVELOPMENT OF DEVICES, CIRCUITS, AND SYSTEMS THAT UTILIZE QUANTUM AND OPTOELECTRONIC PRINCIPLES

1. INTRODUCTION: AN OVERVIEW

In recent years, significant scientific breakthroughs have been made in the disciplines of quantum electronics (such as quantum dots and superconducting circuits) as well as optoelectronics. These platforms are composed of quantum and optical systems, as well as electronic and condensed matter systems. They can be used to create various components for applications in quantum computing, optical technology, and information processing, including machine learning. In this Special Issue, we have curated a collection of papers in this field that cover a diverse range of issues, including the applications of quantum dots, quantum information processing, and machine learning for industrial processes. We anticipate that the research outlined in this document will stimulate additional advancements in this field and yield substantial practical uses for both the industry and society. Section 2 provides a concise overview of the papers featured in the Special Issue. In Section 3, we provide a framework for potential future advancements.

2. PUBLISHED ARTICLES: BRIEF DESCRIPTION

In their study, Juárez-Barojas et al. [1] presented a maintenance system that utilizes artificial intelligence to continuously monitor the expansion of the support bed in a 30-L pilot-scale inverse fluidized bed reactor (IFBR) in real-time. Their objective was to acquire a condition-based maintenance method by utilizing a single-level sensor for a biofilm inverse fluidizing bed as a virtual sensor source. To do this, they deployed an artificial neural network on an embedded electrical system, specifically the Raspberry Pi 4.

Moiseev et al. [2] examined the characteristics of the electroluminescence spectra in narrow-gap type II InAs/InSb/InAs heterostructures. These heterostructures included a single layer of InSb quantum dots positioned within a p-n-InAs junction. The researchers employed both forward and reverse bias within the temperature range of 77–300 K to examine the luminescent characteristics of nano-objects embedded in the narrow-gap matrix, while varying the surface density of these objects. Under specific circumstances, researchers noted the inhibition of negative interband luminescence and the prevalence of interface recombination transitions at the InSb/InAs type II heterojunction when the temperature was at room level. Gijare et al. [3] conducted an investigation on the electrochemical measurements of electrodes made from reduced graphene oxide-titanium oxide (rGO/TiO2) with the purpose of developing a glucose sensor. The research examines the responsiveness, durability, and consistency of the sensor electrodes used to measure the glucose levels in the serum. The sensor was employed to detect glucose levels in natural blood serum and exhibited a relative standard deviation of 1.88%, which closely matched the values obtained from the commercial glucose sensor. Venkatesan et al. [4] constructed a flat Si/PEDOT: PSS heterojunction solar cell employing three distinct solvents—ethylene glycol, acetonitrile, and dimethyl sulfoxide—in order to determine the most optimal solvent. The collected samples were analyzed using several techniques, such as diffuse reflectance spectroscopy, scanning electron microscopy, X-ray diffraction, and the current-voltage technique. Balakirev et al. [5] investigated how the optical characteristics of InAs quantum dots are affected by the pressure of arsenic during the low-temperature GaAs overgrowth process. In the spectra of quantum dots grew at a high arsenic pressure, a solitary wide line was detected, which corresponds to the unimodal size distribution of the quantum dots. Additionally, they saw two clear peaks (~1080 and ~1150 nm) at longer wavelengths in the spectra of samples containing quantum dots that were generated with a low arsenic pressure. Cherckesova et al. [6] proposed a quantum protocol modification to enhance the cryptographic strength of the BB84 quantum protocol in traditional quantum cryptography methods. Roslan et al. [7] examined the electrical conduction characteristics of a specific compound called organic vanadyl 3,10,17,24-tetra-tert-butyl-1,8,15,22-tetrakis(dimethylamino)-29H,31H phthalocyanine (VTP). The researchers established that the current-voltage (I-V) profile exhibited rectifying behavior, as evidenced by a single voltage transfer characteristic (VTP). Additionally, they calculated several characteristics of the Schottky diode, such as the ideality factor, barrier height, shunt resistance, and series resistance, along with other investigations. Kotb et al. [8] successfully simulated and proved all fundamental optical logic operations, such as XOR, AND, OR, NOT, NOR, XNOR, and NAND, utilizing small silicon waveguides with a K-shaped design at the 1.55 μm telecommunication wavelength. These waveguides consist of three silicon waveguide strips that are printed on silica. Mohanty et al. [9] investigated the impact of the tungsten nitride (WNx) diffusion barrier layer on the resistive switching operations of aluminum nitride (AlN)-based conductive bridge random access memory. Quapp et al. [10] examined the twist map, specifically its application in the finite Frenkel-Kontorova model, elucidating the significance of tensile forces in some proposed models. Sun et al. [11] introduced a wide reflected angle Ti_3O5/SiO_2 distributed Bragg reflector (WRA-DBR) as a solution to the limitations of the double-stack distributed Bragg reflector in micro-LEDs. This WRA-DBR was specifically designed for AlGaInP-based red and GaN-based green/blue flip-chip micro LEDs, also known as RGB flip-chip micro-LEDs. In their study, Shabbir et al. [12] examined the benefits of utilizing quantum dots in solar cells and quantum dot lasers. They then proceeded to provide a detailed analysis of the various applications of quantum dots in photodetectors. The discussion revolved around several metallic compounds, including lead sulfide and indium arsenide, as well as nonmetallic materials like graphene and carbon nanotubes. A study conducted by Ahmad et al. [13] examined the efficacy of hybrid quantum dots as a theranostic system in several types of cancer. The researchers analyzed the utilization of hybrid quantum dots that were conjugated with different biomolecules for diagnostic and therapeutic purposes in cancer treatment. The discussion also included the biopharmaceutical features of many biomolecules, including folic acid, PEG, and hybrid quantum dots. Parvin et al. [14] examined various methods for creating single-layer and few-/multi-layer graphene. Subsequently, an investigation was conducted to assess the influence of mono-/few-/multi-layer graphene on its quality and properties. The distinctive electrical characteristics of graphene, such as high carrier mobility, ambipolar behavior, and a specific energy band structure, make it suitable for use in field effect transistors (FETs) and various applications including radio frequency (RF) circuits, sensors, memory, and more.

3. PROSPECTS FOR THE FUTURE

The topic of quantum and optoelectronics is anticipated to expand in the future. One promising research direction with potential impact involves integrating this field with machine learning and artificial intelligence. This could potentially enhance the design and regulation of quantum and optoelectronics devices by utilizing machine learning techniques and leveraging physical devices such as neuromorphic architectures, memristors, and similar technologies to achieve more efficient machine learning. Although predicting the future development of a scientific topic might be challenging, it is our belief that there are still several untapped opportunities for industry and society in these domains.

REFERENCES

1. Juárez-Barojas, I.; Posada-Gómez, R.; Alvarado-Lassman, A.; Rodríguez-Jarquín, J.P. Condition-Based Maintenance of an Anaerobic Reactor Using Artificial Intelligence. Electronics 2023, 12, 799. [Google Scholar] [CrossRef]

2. Moiseev, K.; Ivanov, E.; Parkhomenko, Y. Long-Wavelength Luminescence of InSb Quantum Dots in Type II Broken-Gap Heterostructure. Electronics 2023, 12, 609. [Google Scholar] [CrossRef]
3. Gijare, M.; Chaudhari, S.; Ekar, S.; Shaikh, S.F.; Mane, R.S.; Pandit, B.; Siddiqui, M.U.H.; Garje, A. Facile Green Preparation of Reduced Graphene Oxide Using Citrus Limetta-Decorated rGO/TiO2 Nanostructures for Glucose Sensing. Electronics 2023, 12, 294. [Google Scholar] [CrossRef]
4. Venkatesan, R.; Sheik Kadar Maideen, S.M.T.; Chandhiran, S.; Kushvaha, S.S.; Sagadevan, S.; Venkatachalapathy, V.; Mayandi, J. Fabrication and Characterization of Si/PEDOT: PSS-Based Heterojunction Solar Cells. Electronics 2022, 11, 4145. [Google Scholar] [CrossRef]
5. Balakirev, S.; Chernenko, N.; Kryzhanovskaya, N.; Shandyba, N.; Kirichenko, D.; Dragunova, A.; Komarov, S.; Zhukov, A.; Solodovnik, M. Photoluminescence Properties of InAs Quantum Dots Overgrown by a Low-Temperature GaAs Layer under Different Arsenic Pressures. Electronics 2022, 11, 4062. [Google Scholar] [CrossRef]
6. Cherckesova, L.V.; Safaryan, O.A.; Beskopylny, A.N.; Revyakina, E. Development of Quantum Protocol Modification CSLOE–2022, Increasing the Cryptographic Strength of Classical Quantum Protocol BB84. Electronics 2022, 11, 3954. [Google Scholar] [CrossRef]
7. Roslan, N.A.; Supangat, A.; Sagadevan, S. Investigation of Charge Transport Properties in VTP: PC71BM Organic Schottky Diode. Electronics 2022, 11, 3777. [Google Scholar] [CrossRef]
8. Kotb, A.; Zoiros, K.E. K-Shaped Silicon Waveguides for Logic Operations at 1.55 μm. Electronics 2022, 11, 3748. [Google Scholar] [CrossRef]
9. Mohanty, S.K.; Reddy, K.P.K.; Wu, C.-H.; Lee, P.-T.; Chang, K.-M.; Busa, P.; Kuthati, Y. Investigation of Barrier Layer Effect on Switching Uniformity and Synaptic Plasticity of AlN Based Conductive Bridge Random Access Memory. Electronics 2022, 11, 3432. [Google Scholar] [CrossRef]
10. Quapp, W.; Bofill, J.M. An Analysis of Some Properties and the Use of the Twist Map for the Finite Frenkel–Kontorova Model. Electronics 2022, 11, 3295. [Google Scholar] [CrossRef]
11. Sun, Y.; Shi, L.; Du, P.; Zhao, X.; Zhou, S. Rational Distributed Bragg Reflector Design for Improving Performance of Flip-Chip Micro-LEDs. Electronics 2022, 11, 3030. [Google Scholar] [CrossRef]
12. Shabbir, H.; Wojnicki, M. Recent Progress of Non-Cadmium and Organic Quantum Dots for Optoelectronic Applications with a Focus on Photodetector Devices. Electronics 2023, 12, 1327. [Google Scholar] [CrossRef]

13. Ahmad, J.; Garg, A.; Mustafa, G.; Ahmad, M.Z.; Aslam, M.; Mishra, A. Hybrid Quantum Dot as Promising Tools for Theranostic Application in Cancer. Electronics 2023, 12, 972. [Google Scholar] [CrossRef]

14. Parvin, N.; Kumar, V.; Joo, S.W.; Park, S.-S.; Mandal, T.K. Recent Advances in the Characterized Identification of Mono-to-Multi-Layer Graphene and Its Biomedical Applications: A Review. Electronics 2022, 11, 3345. [Google Scholar] [CrossRef]

Chapter 2

UTILIZING ARTIFICIAL INTELLIGENCE FOR THE CONDITION-BASED MAINTENANCE OF AN ANAEROBIC REACTOR

1. INTRODUCTION: AN OVERVIEW

This paper describes the development of a neural network used to create a condition-based maintenance system for an anaerobic reactor. The reactor is a complex process with nonlinear characteristics, and some of the variables involved cannot be measured online using commercially available sensors. Therefore, virtual sensors are implemented to estimate these variables. The study relies on a level sensor implemented in an anaerobic IFBR, as described in [1], where control, diagnosis, and maintenance were not successfully achieved. The demand for reliability and safety in industrial processes, along with economic and environmental constraints, has made online fault detection and isolation (FDI) a significant concern [2]. In the event of a defect, the process's efficiency may be compromised. If this fault is not identified early on, it could result in a catastrophic situation, such as harm to individuals, environmental pollution, and damage to assets and equipment [3]. In modern times, industrial processes are growing more intricate, leading to challenges in promptly and effectively conducting online fault diagnostics [4]. This is mostly due to the nonlinear and interdependent nature of the variables associated with these processes. The necessity to design a suitable problem detection tool for anaerobic reactors is entirely justified due to the practical impossibility of performing online measurement of all the variables involved in the process, particularly in this scenario [5]. Anaerobic reactors pose a challenging scenario for fault detection methods due to the inherent difficulty in monitoring the internal processes of the reactor at any given moment. Additionally, tracking the changes in concentrations of various bacterial groups and organic pollutant load over time is also a complex task [6]. Furthermore, when presented with a collection of indirect signals or measurements obtained from a process, determining the reliability of the information originating from the process can provide a challenge [7]. Moreover, if the process is not functioning properly, determining the specific component responsible for the irregularity can be highly challenging [8]. In order to address this issue, a fault detection and isolation study has been conducted. This study utilizes a differential geometry approach specifically designed for a certain type of nonlinear systems. The observer-based approach to Fault Detection and Isolation (FDI) [9] is considered one of the significant advancements in the field of analytical redundancy-based FDI. Essentially, it involves establishing a function for the system outputs that is only influenced by a specific fault signal or a group of fault signals. The system's redundancy is ensured by comparing the output function in question with its estimate [10]. Reference [11] contains a study on fault detection and isolation (FDI) that utilizes observer-based methods. The majority of the suggested contributions were designed for linear systems; yet, some processes exhibit nonlinearity. It is important to acknowledge that, while the process may function at a specific operating point, the presence of a fault can cause it to deviate from that point, resulting in inaccuracies in the model and the production of false alarms [12]. At time t0, a linear system is considered observable if it is possible to determine the state x(t0) of the system based on the observation of the system output within a finite time interval.

Consider the linear system in continuous time, represented by:

$$\dot{x}(t) = A(t)Bu(t), \qquad t \leq t_0 \qquad x(t_0) = x_0 y(t) = Cx(t) \tag{1}$$

where $x(t) \in \Re^n$ **is the state vector, A** $\in \Re^{nxn}$ **is the state matrix, y(t)** $\in \Re^n$ **is the system matrix and C** $\in \Re^{mxn}$. The observability of the system (Equation (1)) is established according to the observability matrix ϑ:

$$\vartheta = \begin{bmatrix} C \\ CA \\ CA^2 \\ \vdots \\ CA^{(n-1)} \end{bmatrix} \tag{2}$$

The system is observable if and only if the observability matrix has full rank, that is, rank (ϑ)=n.

For nonlinear systems, a concept related to observability is that of detectability or asymptotic observability, which establishes that a system is detectable if the difference between two trajectories with the same output history converges asymptotically to zero [13].

Nonlinear systems can be analyzed for observability by linearizing the model equations around an equilibrium point. This allows for the determination of the observability matrix, which establishes the local observability of the system near the chosen equilibrium point. Another method for assessing the observability of nonlinear systems involves employing differential geometry techniques. This approach involves constructing an observation space under the assumption that the output is only dependent on the state variables [14].

Due to the nonlinear character of biological processes, they are often not fully observable. Therefore, it is more suitable to estimate relationships between parameters rather than estimating them individually [15].

- The estimation of states in biological reactors presents a critical drawback, due to the high uncertainty associated with the following factors [16].
- The variety of biological species present in the reactor makes it difficult to determine the role played by each of them in the process.
- The measurement of the initial total biomass does not correspond to the concentration of the effective biomass.
- The complexity of the system makes it difficult to describe the parameters associated with the kinetics of the biological reactions.
- The growth rates of the bacterial groups are time-varying.

In the work presented by authors of [17], a methodology for determining the global observability and detectability of biological reactors is explored. The method proposed in the article allows formalizing the design of the so-called asymptotic observers.

Recently, many efforts have been focused on the implementation of programmed sensors based on state observers, due to the advantages that the use of these sensors represents, compared to hardware sensors [18]. Some of these advantages are mentioned below:

The maintenance cost of a programmed sensor is much less expensive than that of a hardware sensor.

- Once a programmed sensor has been designed and tuned for a particular application, no periodic maintenance is required to recalibrate it (e.g., pH analyzers, oxygen analyzers, etc.).
- Programmed sensors can be used to predict information that is difficult to obtain using traditional methods, either due to inherent measurement delays or due to process operating conditions.
- Installation costs are cheaper in most cases compared to traditional methods of monitoring systems.

Programmed sensors are used in order to solve problems associated with the lack of sensors for real-time measurement of certain process variables. For example, authors of [19] used a support vector regression algorithm for COD (chemical oxygen demand) estimation in a perfectly stirred anaerobic reactor.

One of the main difficulties in implementing programmed sensors is that the design contemplates previously obtaining a predictive model for a given process. Consequently, the success of a programmed sensor strongly depends on the accuracy of the mathematical model obtained for that process.

Many empirical equations exist in the literature to model the porosity or expansion of an IFBR. Campos-Díaz et al., 2012 [20], proposed the following model to estimate the porosity of the inverse fluidized bed, as a function of the Reynolds particle and Archimedes numbers:

$$1.753e^{-3.807} = Ar\left[0.75\left(24Re_p + C_1 Re_p^2\right)\right]^{-1} \quad (3)$$

where:

$$C_1 = 2.25\left(5.5 / Re_p\right)^{0.34} \quad (4)$$

Authors of [20] developed an empirical correlation for predicting bed expansion as a function of physical and dynamic variables of the system, calculated using the following equation:

$$\frac{H}{H_0} = 0.988 Ar^{-0.2124} Re^{0.4049} \phi_s^{-1.804} \left(D_c / d_p\right)^{0.4613} \quad (5)$$

where ϕ is the particle sphericity.

It is evident that at Re_p **< 0.2, the correlation of the result of** [20] adequately predicts the porosity of the bed, while at Re_p **> 0.7, the correlation of the result of** [21] is more in agreement with experimental data. However, both empirical correlations fail to provide a reasonable overall description at all evaluated operating conditions, and this might be associated with the experimental conditions in which they were fitted.

Therefore, taking into account that there are conductivity sensors placed inside the IFBR, which provide information on the expansion of the fluidized bed, it was decided to use the correlation of the result of [21] as a starting point to train an artificial neural network capable of finding a relationship between pH, influent COD concentration, inflow to the bioreactor and the signal coming from each of the conductivity sensors installed inside the reactor, to estimate the effluent COD concentration, the bioreactor biogas production flow rate (the sum of methane and carbon dioxide production), the methane production flow rate, the carbon dioxide production flow rate and the concentration of volatile fatty acids in the reactor effluent.

The importance of the work presented resides in the considerations made and how the process is treated to ensure the correct functioning of a control strategy and fault diagnosis in the same, to maximize the production of biogas and serve as an aid system to a human operator for the maintenance of the same, through the use of artificial intelligence techniques.

2. MATERIALS AND METHODS

2.1. Support Expansion Monitoring System

Figure 1 shows the block diagram of the bed expansion monitoring system. The expansion of the bed is produced by the flow generated by opening the control valve connected to the distribution tank. The flow is proportional to the valve opening and also dependent on the force of gravity. By circulating the fluid downward, the colonized bed, which has a lower conductivity than the effluent, begins to gradually expand.

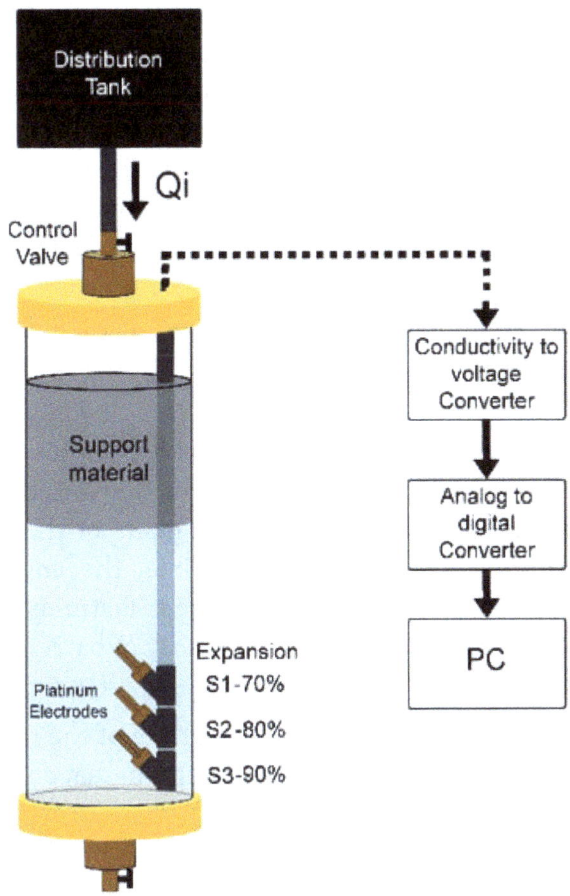

Figure 1. Block diagram of the support expansion monitoring system.

Predevelopment measurements were carried out with a commercial conductivity sensor to validate the above. The values obtained from the measurements using a conductivity electrode, model DFR0300, are shown below in Table 1. All conductivity measurements were performed at 25 °C.

Table 1. Conductivity comparison of different components involved in biodigestion.

Component	Conductivity(mS/cm)
Citric wastewater	2.18
Expanding colonized support	2.08–2.11
Colonized support	0–1.68

The electrode array allows for detecting the substrate change, thus identifying the level of substrate expansion. A conditioning card was developed to obtain the necessary voltage signal, and an acquisition card was used to capture the measured values.

The electrode model used for the array is CON2136-13 with a constant K = 10 and construction material of polysulphone and platinum. It was chosen because the measurement of the variable of interest must be continuous, which represents a problem for many of the metallic materials available, such as copper or tin, since, over time, the environment inside a biodigester is corrosive and can be degraded. Likewise, it is advisable in wastewater treatment to use electrodes with a counter of 10, considering the conductivity levels of the substrates in wastewater treatment. Moreover, the electrode performed adequately in a previous investigation using synthetic wastewater [1], where level measure was achieved, but a control and diagnosis systems were not achieved, conditions that are needed for a condition-based maintenance strategy.

A platinum conductivity probe is proposed at each point in the column. The material was chosen because the measurement of the variable of interest must be continuous, which represents a problem for many of the metallic materials available such as copper or tin, since, over time, the environment inside a biodigester becomes corrosive and can degrade them. Platinum, on the other hand, is immune to oxidation and to the biological activity of bacteria, making it one of the most suitable materials for this application.

The position of the electrodes is set at the bottom of the reactor column with a separation distance of 10 cm from each other and in diagonal position to avoid clogging. The voltage delivered by the conditioning circuit is calculated using Equation (6).

$$V_{out} = (V_{in})(G)(R_f) \tag{6}$$

where:

- V_{out}: is the output voltage of the conditioning circuit.
- V_{in}: is the alternating excitation voltage applied to the electrode.

- G: solution conductivity.
- Rf: feedback resistance of the conditioning circuit.

The value of the proposed feedback resistance in the conditioning circuit is 8.2 KΩ due to the conductivity probe constant and the conductivity values present in the reactor solutions previously presented. Dynamic testing of the system was performed, replicating the expansion of the colonized support. The support material used was colonized (bacteria biofilm around the support media) and noncolonized.

Figure 2 shows the results obtained from the dynamic expansion test of the support bed. This test consisted of introducing colonized support media recirculating in the reactor at a downward velocity of 1.41 m/h. It can be observed how after 650 s (10 min), sensor 1 detects a voltage decrease, which means that it has come into contact with the colonized support media; then, at 950 s (16 min), sensor 2 also comes into contact with the substrate, so its voltage decreases. Finally, sensor 3 comes into contact with the expanded support at 1230 s (20 min). After half an hour of operation, the recirculation is turned off, so the voltage values of the sensors begin to increase, but in reverse order, i.e., the voltage first increases in sensor 3, then in sensor 2, and finally, sensor 1 returns to its initial values.

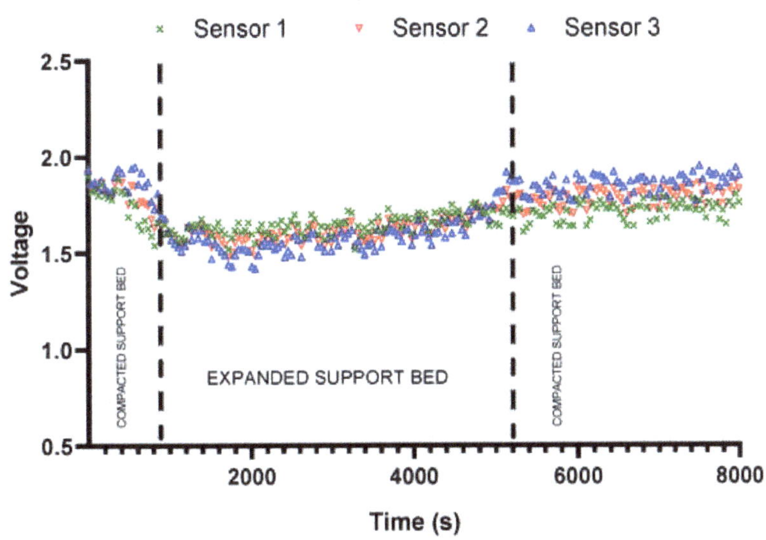

Figure 2. Dynamic test of support bed expansion.

2.2. Description of the Fuzzy Logic PI Type Controller

The fuzzy logic PI-type controller is commonly used to design a robust controller for ensuring suppression of disturbances. In a fuzzy PI-type controller (Figure 3), its nonlinear mapping between the control rules and the fuzzy reasoning can be represented as a function of the error e and the first-time derivative of the error Δe as follows:

$$\Delta u = FLC(e, \Delta e) \tag{7}$$

where Δu is the first-time derivative of the output control action. Δu can be expressed approximately as follows:

$$\frac{\Delta u(t)}{S\Delta u} = \frac{\Delta e(t)}{S\Delta e} + \frac{e(t)}{Se} \tag{8}$$

where $-Se < e(t) < Se$, $-S\Delta e < \Delta e(t) < S\Delta e$, $-S\Delta u < \Delta u(t) < S\Delta u$, or

$$u(t) = k_p e(t) + k_i \int e(t)dt \tag{9}$$

where $kp = S\Delta u / S\Delta e$ and $ki = S\Delta u/Se$.

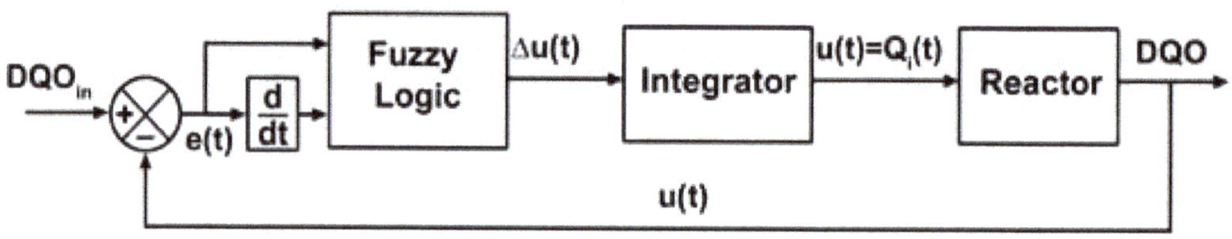

Figure 3. PI control system with fuzzy logic controller.

The membership functions for the inputs e and Δe are defined in the range (-1, 1), see Figure 4a,b, respectively. The fuzzy subsets are defined as follows: NL: negative large; N: negative; Z: zero; P: positive; and PL: positive large, for the input e. N: negative; Z: zero; and P: positive; for the input Δe.

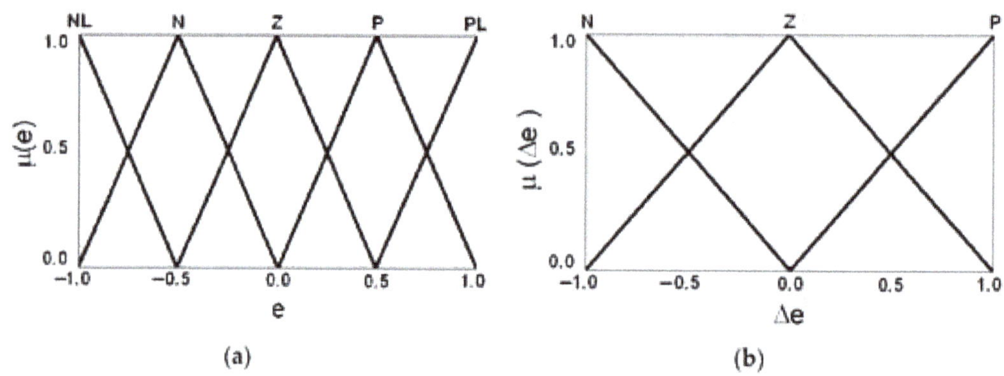

Figure 4. (a) Membership functions for the input e. (b) Membership functions for the input Δe.

The membership functions for the output Δu are defined in the range (-1, 1), see Figure 5. The fuzzy subsets for the output Δu are defined as follows: NL: negative large; NM: negative medium; NS: negative small; Z: zero; PS: positive small; PM: positive medium; PL: positive large; for the output Δu.

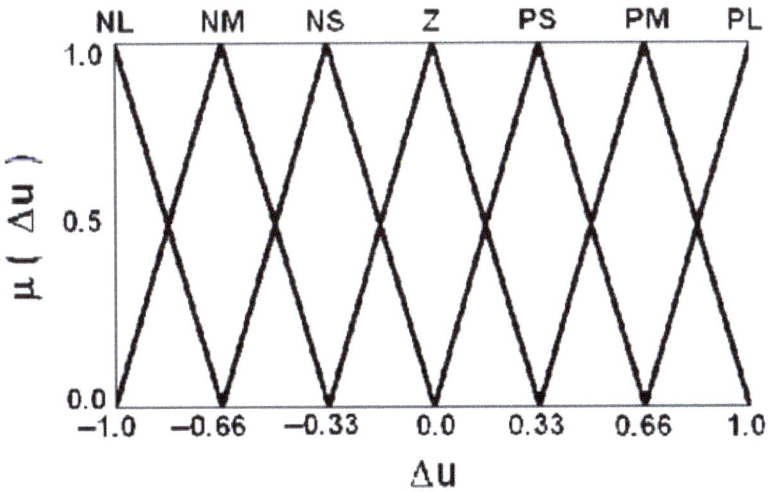

Figure 5. Membership functions for the output Δu.

2.3. Description of Software Sensor Based on Artificial Neural Networks

An artificial neural network was developed to find a relationship between pH, influent COD concentration, bioreactor inflow and the signal coming from each of the conductivity sensors installed inside the reactor to estimate the effluent COD concentration, the biogas production flow rate of the bioreactor (the sum of methane and carbon dioxide production), the methane production flow rate, the carbon dioxide production flow rate and the concentration of volatile fatty acids in the reactor effluent. The neural network was trained from hundreds of experimental tests performed on the bioreactor to relate biogas production to fluidized bed expansion, based on signals from each of the three bioreactor conductivity sensors.

The implementation of the artificial neural network was performed on an embedded electronic system (Raspberry Pi 4), both working together in real-time. This device was selected because it is a low-cost embedded system that does not require a computer to perform its work.

Figure 6 shows how the neural network was trained from the results obtained from the experimental tests performed with the bioreactor. The values of pH, COD concentration of the influent, inflow to the bioreactor and the signal coming from each of the conductivity sensors installed inside the reactor, during each of the experiments performed with the bioreactor, correspond to the input vector (A) of the neural network. This input vector contains 180 data points corresponding to the pH values, COD concentration of the influent, inflow to the bioreactor and the signal coming from each of the conductivity sensors installed inside the reactor during 6 months of experimental tests. There are 180 input neurons (B), which are connected to the neurons of the intermediate layer (C), and these neurons of the intermediate layer are connected to the output neurons (D), which generate the output signals estimated by the neural network, corresponding to the COD concentration of the effluent, the biogas production flow rate of the bioreactor (the sum of methane and carbon dioxide production), the methane production flow rate, the carbon dioxide production flow rate and the concentration of volatile fatty acids in the reactor effluent.

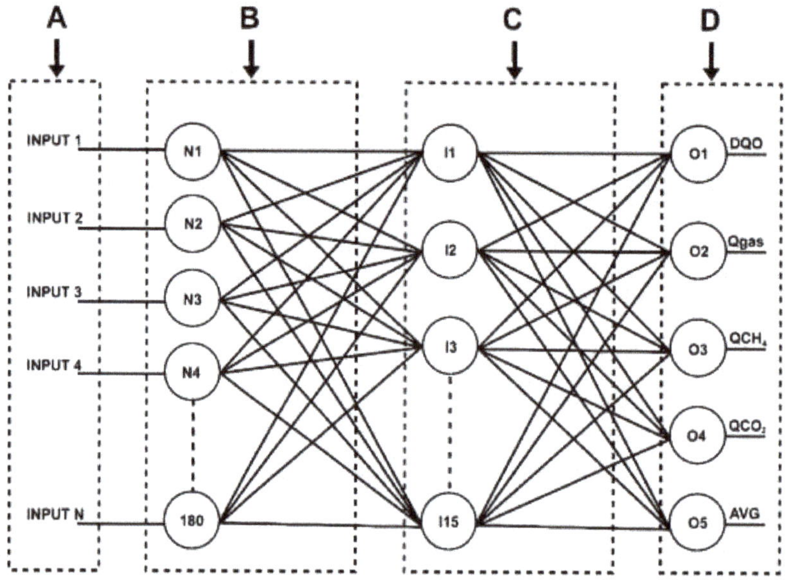

Figure 6. Neural network arrangement used for the determination of pH, Qgas, S1, S2 and S3.

The objective of training the neural network is to make it capable of learning the relationship between the outputs and the input, so that when a new experiment is performed with the bioreactor, a new sample of pH, COD concentration of the effluent, inflow to the bioreactor and the signal coming from each of the conductivity sensors installed inside the reactor will be introduced into the neural network and it will be able, from its learning, to estimate the concentration values of the COD concentration of the effluent, the flow of biogas production from the bioreactor and the signal coming from each of the conductivity sensors installed inside the reactor that is introduced into the neural network and the neural network is able, based on its learning, to estimate the concentration values of the effluent COD concentration, the bioreactor biogas production flow (the sum of methane and carbon dioxide production), the methane production flow, the carbon dioxide production flow and the concentration of volatile fatty acids in the reactor effluent.

One of the primary objectives of this section is to statistically validate the neural networks used to determine the output signals. One of the big questions that always has to be solved when working with neural networks is to know what the number of neurons in the intermediate layer or layers required for its error in the output to be optimal is. To achieve this objective, an automatic algorithm had to be developed to analyze the behavior of the network under different configurations of neurons in hidden layers.

In total, there are 180 data points from each of the 6 input signals to the neural network, which provides us a matrix of 1080 data points, which should generate a correlation with the data obtained in the experimental tests of the anaerobic reactor. Using the 180 data points from each of the 6 input signals to train an artificial neural network can result in an incorrect training of the network, because many of these variables may not have significant relevance. For this reason, a preprocessing of these data was performed to know the variance and correlation of the data in order to reduce the dimensionality of the input vector to the neural network. A PCA is a statistical methodology that performs a linear transformation through the creation of a new coordinate system for the input data, in which, on its axes, it places the data ordered from the highest variance to the lowest variance. In general terms, what a PCA does is create a new system in which it orders the data from most relevant to least relevant, thus reducing the dimensionality of the data.

A PCA was programmed using MATLAB software, using the 1080 data points of the input matrix, which corresponds to 180 dimensions for 6 input signals. According to the results of the PCA, statistically speaking, and looking at its percentage of re-relevance, the first 8 columns of the principal components were selected since they are the most important data. With this dimensional reduction, the training input vector used in the neural network was 8 dimensions.

A neural network was modeled with the 8 input data coming from the dimensional data reduction obtained by the PCA and experimented on with different configurations related to the number of hidden layers and number of neurons in the hidden layer. This neural network delivers 5 outputs as a result, which are the estimation of the COD concentration values of the effluent, the biogas production flow rate of the bioreactor (the sum of methane and carbon dioxide production), the methane production flow rate, the carbon dioxide production flow rate and the concentration of volatile fatty acids in the reactor effluent. Figure 7 and Figure 8 show the structure of the neural network, composed of an input layer and an intermediate layer of neurons.

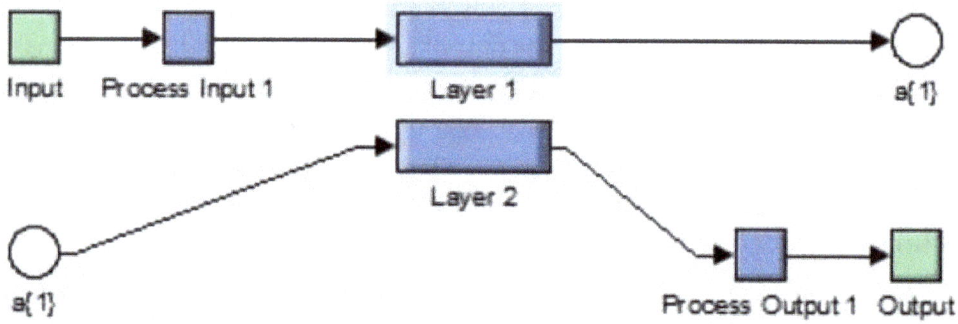

Figure 7. Neural network structure, input layer of neurons and intermediate layer of neurons.

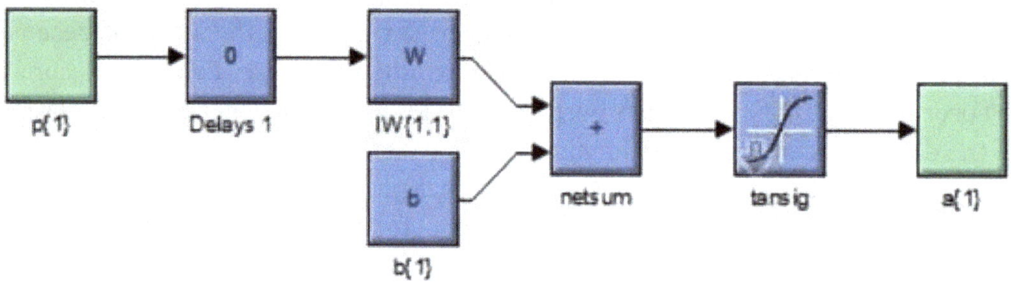

Figure 8. Input layer, composed of the normalization algorithm, the weights and the tansig function.

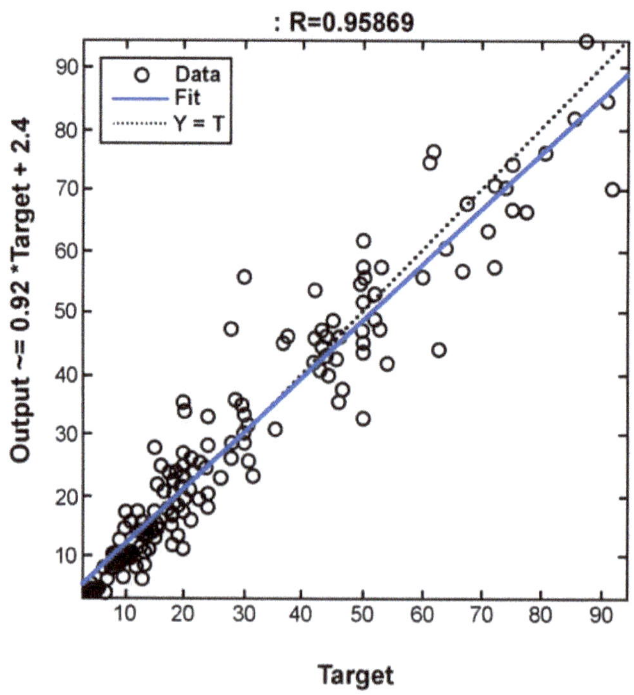

Figure 9. ANN using 5 neurons in the hidden layer.

Figure 9 shows the behavior of a neural network using 5 neurons in the hidden layer of the neural network, obtaining a recognition error of 4.28%. Figure 10 shows the behavior of the neural network using 15 neurons in the hidden layer with an error of 1.45%. The graphs show the target (the blue line) representing the ideal recognition value, and the circles in the graph are the actual recognition values obtained for each target value. It can be seen that the neural network with 15 neurons in the hidden layer has a lower recognition error (Figure 10).

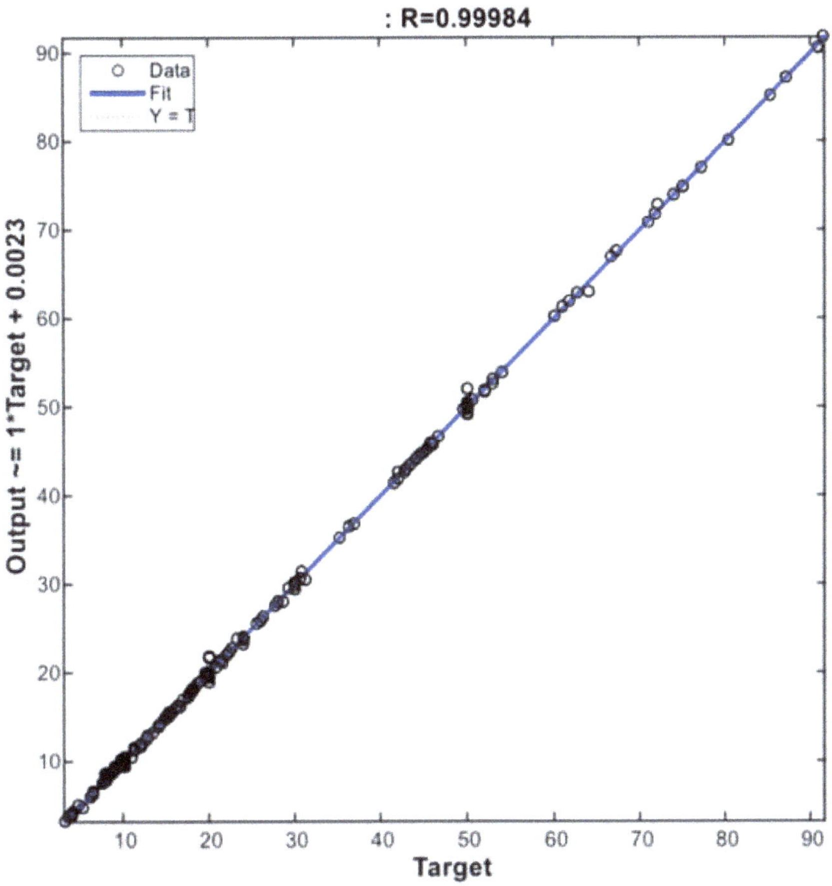

Figure 10. ANN using 15 neurons in the hidden layer.

Figure 11 shows the behavior of the neural network focused on estimating the values of the conductivity sensors, S1, S2 and S3, from the values of pH, COD concentration of the influent and inflow to the bioreactor. In this regard, after evaluating multiple configurations of neurons in the intermediate layer, it was determined that the best performance was obtained with a neural network with 10 neurons in the intermediate layer.

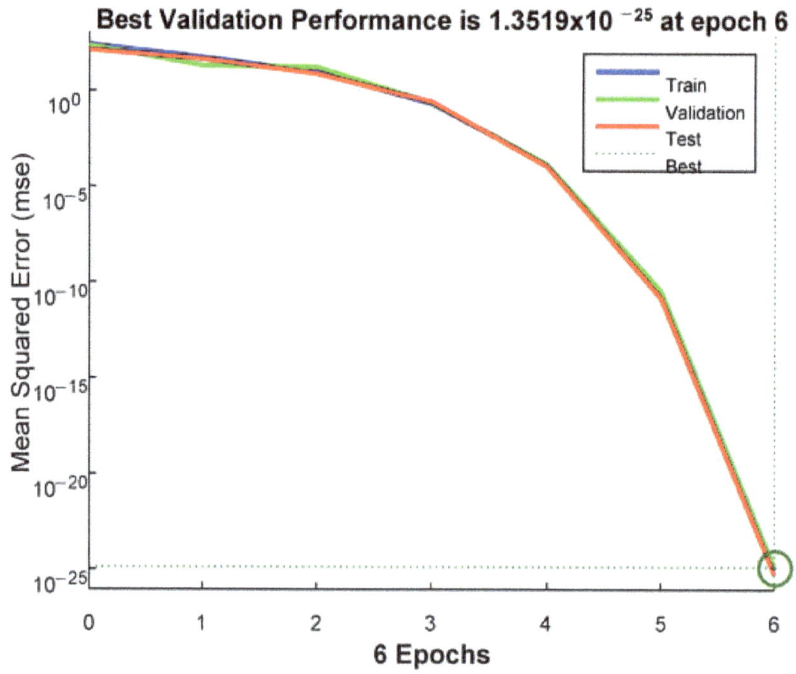

Figure 11. ANN with 10 neurons in the intermediate layer.

After training the neural network and performing the behavioral tests, validation experiments were conducted to determine the margin of error of the neural network in comparison with the hardware sensors. In this regard, the margin of error was less than 2%, which allowed establishing the threshold values for the fault diagnosis scheme developed in this project.

The microcontroller was programmed for real-time execution of the neural networks. Blocks were used to map the input signals from -1 to 1, and a Math Script node was used in which all the weights of the network as well as all the mathematical formulation to run the network were found. This block is connected directly to the outputs of the hardware sensors, which delivers an input vector to the network with the values of pH, COD concentration of the influent, inflow to the bioreactor and the signal coming from each of the conductivity sensors installed inside the reactor.

3. RESULTS

3.1. Experimental Results of Condition-Based Maintenance Scheme

In this work, a fuzzy PI controller was used to regulate the COD concentration in the effluent of the fluidized bed reactor, having the inflow to the Qi bioreactor as a control variable. To validate the performance of the controller, experimental tests were carried out by feeding the reactor with citrus wastewater, which was diluted and conditioned to 6 g COD/L and pH = 7, during the first twenty days of experimentation. Subsequently, the wastewater was diluted to 3 gCOD/L, maintaining the original pH = 7. During the four weeks prior to the start of the experiment, the controller was operating at a set-point value of 1 gCOD/L. On the first day of experimentation, the set-point was changed from 1 gCOD/L to 0.5 gCOD/L, and was maintained unchanged for the forty days of experimentation. The reactor temperature was kept constant at 35 degrees Celsius during the experimentation.

3.2. Operation of the Condition-Based Maintenance Scheme

As mentioned above, the COD sensor is responsible for sending the feedback signal to the fuzzy PI controller that is responsible for regulating the COD concentration in the reactor effluent. Therefore, if the COD sensor fails, then the regulation of COD in the reactor effluent will also fail. To overcome this problem, the COD sensor condition-based maintenance scheme presented in Figure 12 was developed. This maintenance scheme receives the information from the fault diagnosis system (FDI), and when a COD sensor failure occurs, the COD sensor is then disabled and its signal is switched to the COD signal estimated by neural network 1, as long as this neural network is free of failure. However, in case that neural network 1 presents a failure, the feedback signal is then taken from neural network 2, as long as it is fault-free. In the extreme case that both neural network 1 and neural network 2 are faulted, the feedback signal is then taken from neural network 3, as long as it is fault-free.

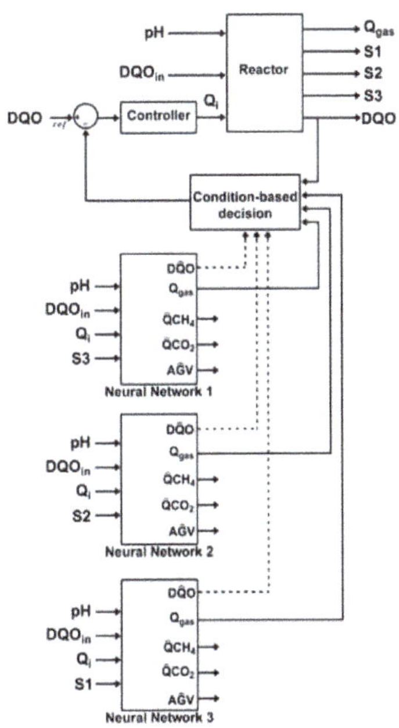

Figure 12. Condition-based decision scheme.

Figure 13 shows the graphs of the COD concentration in the influent, both the experimentally obtained graph and the graphs estimated by asymptotic observers 2 and 3. In this figure, it can be seen that the three graphs are practically identical until day 60 of the experiment, when the sudden and permanent failure of sensor s1 is observed. In Figure 14, a transient is observed on day 60 of the experiment. This transient corresponds to the action of the condition-based maintenance scheme, whereby the signal from sensor s1 is switched by its estimate generated by observer 2, so that the controller continues to operate normally and continues to regulate the COD concentration in the reactor effluent to 0.5 g COD/L.

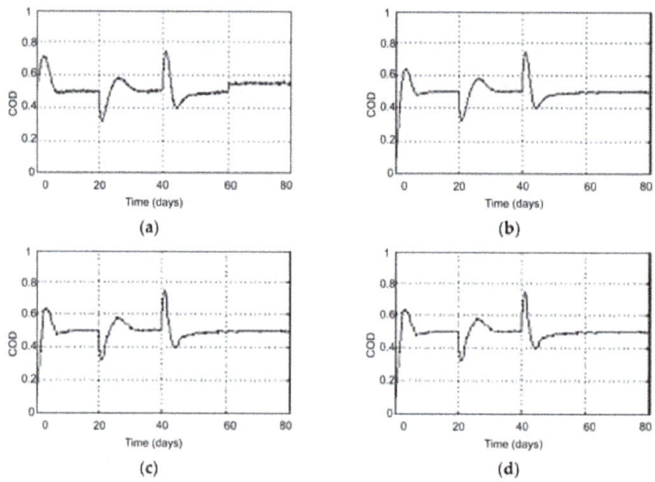

Figure 13. (a) Experimental measurement of COD of the reactor effluent. (b) Estimation of COD of neural network 1. (c) COD estimation of neural network 2. (d) COD estimation of neural network 3.

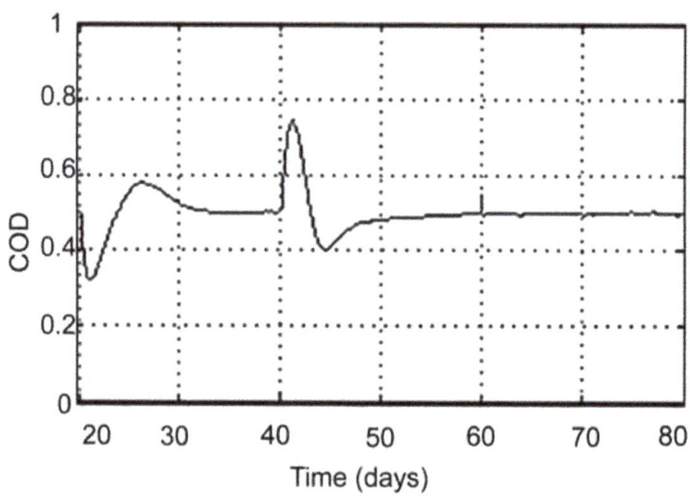

Figure 14. Graph of COD switched by the condition-based maintenance scheme due to a sudden and permanent failure of +50 mgCOD/L on day 60 of experimentation.

The condition-based maintenance scheme effectively enables the system to maintain its control strategy by utilizing the output from an unaffected observer. This allows a human operator to perform corrective maintenance without interrupting the process. Once the fault is rectified, the system detects the proper functioning of the sensor and the control strategy resumes its normal behavior.

4. CONCISE SUMMARY

An empirical investigation was conducted to apply a fuzzy controller to a nonlinear multivariable anaerobic reactor process. This controller proves to be an effective method for detecting faults in situations where the mathematical model is unavailable. Furthermore, a synthetic neural network was created to establish a correlation between pH, influent COD concentration, inflow to the bioreactor, and the signal from each conductivity sensor within the reactor. This network was designed to predict the effluent COD concentration, the flow rate of biogas production in the bioreactor (which includes both methane and carbon dioxide production), the flow rate of methane production, the flow rate of carbon dioxide production, and the concentration of volatile fatty acids in the effluent of the reactor. The neural network was trained using data from numerous experimental tests conducted on the bioreactor. These studies aimed to establish a relationship between biogas output and fluidized bed expansion, using signals from the three bioreactor conductivity sensors. Furthermore, this system can not only identify and separate any malfunctions that may arise in the sensors during the anaerobic digestion process, but the data collected from these sensors can also be utilized as virtual sensors. These virtual sensors are not influenced by specific faults, enabling the controller to remain operational and make adjustments to the physical sensor. In the case of bioreactors, it is crucial to maintain continuous operation due to their complex requirements and conditions. Interrupting the process could lead to a significant decrease in the biomass population responsible for acidogenesis and methanogenesis, which are essential for biogas production. In extreme cases, restarting the bioreactor would be necessary, resulting in months of additional work. This work addresses the issue of the absence of online sensors for measuring variables in anaerobic reactors. Specifically, a sensor for measuring the level of the reverse fluidized bed was considered. The study focuses on a hybrid reactor called reverse fluidized fixed bed, and an innovative design was developed by the Tecnológico Nacional de México Campus Orizaba. Based on this sensor, a control system and fault diagnosis using artificial intelligence were implemented to enable condition-based maintenance. This research employed an artificial intelligence strategy approach to serve as an estimator, predictor, and controller.

At a Glance

This chapter presents a condition-based maintenance system that utilizes artificial intelligence for online monitoring of the support bed expansion in a 30-liter pilot-scale inverse fluidized bed reactor (IFBR). The primary objective is to implement a condition-based maintenance strategy by utilizing a single-level sensor to serve as virtual sensors for a biofilm inverse fluidizing bed. An artificial neural network was implemented on an embedded electronic device, specifically the Raspberry Pi 4. The two components worked in tandem, operating in real time. The neural network's estimated signals are compared to the signals measured by the hardware sensors. If a failure is detected in the physical measurement system, the artificial intelligence-based system utilizes the signal estimated by the artificial neural network to ensure the IFBR operates correctly. This system utilizes an artificial neural network to predict the chemical oxygen demand (COD) concentration of the effluent and the flow rate of biogas produced by a bioreactor. The predictions are based on measurements of pH, the COD concentration of the influent, the inflow rate to the bioreactor, and signals from conductivity sensors installed inside the reactor. These sensors provide information about the expansion of the support media in a pilot-scale inverse fluidized bed reactor. Furthermore, a fuzzy PI controller is introduced, which was applied to a Raspberry Pi circuit board, in order to manage the COD concentration in the effluent of the bioreactor utilized as a case study.

REFERENCES

1. Marín-Peña, O.; Alvarado-Lassman, A.; Vallejo-Cantú, N.A.; Juárez-Barojas, I.; Rodríguez-Jarquín, J.P.; Martínez-Sibaja, A. Electrical conductivity for monitoring the expansion of the support material in an anaerobic biofilm reactor. *Processes* **2020**, *8*, 77. [Google Scholar] [CrossRef]

2. Phoungthong, K. An effective evaluation on fault detection in solar panels. *Energies* **2021**, *14*, 7770. [Google Scholar]

3. Safavi, S.; Safavi, M.A.; Hamid, H.; Fallah, S. Multi-sensor fault detection, identification, isolation and health forecasting for autonomous vehicles. *Sensors* **2021**, *21*, 2547. [Google Scholar] [CrossRef]

4. Ji, C.; Ma, F.; Wang, J.; Wang, J.; Sun, W. Real-time industrial process fault diagnosis based on time delayed mutual information analysis. *Processes* **2021**, *9*, 1027. [Google Scholar] [CrossRef]

5. Kazemi, P.; Bengoa, C.; Steyer, J.P.; Giralt, J. Data-driven techniques for fault detection in anaerobic digestion process. *Process Saf. Environ. Prot.* **2021**, *1*, 905–915. [Google Scholar] [CrossRef]

6. Avilés, J.D.; Torres-Zúñiga, I.; Villa-Leyva, A.; Vargas, A.; Buitrón, G. Experimental validation of an interval observer-based sensor fault detection strategy applied to a biohydrogen production dark fermenter. *J. Process Control* **2022**, *1*, 131–142. [Google Scholar] [CrossRef]

7. Anand, K.; Mittal, A.P.; Kumar, B. ANN-based sensorless adaptive temperature control system to improve methane yield in an anaerobic digester. *Biomass Convers. Biorefinery* **2022**, 1–21. [Google Scholar] [CrossRef]

8. Gupta, R.; Zhang, L.; Hou, J.; Zhang, Z.; Liu, H.; You, S.; Ok, Y.S.; Li, W. Review of Explainable Machine Learning for Anaerobic Digestion. *Bioresour. Technol.* **2022**, *369*, 128468. [Google Scholar] [CrossRef]

9. Wang, J.; Yang, C.; Xia, J.; Wu, Z.G.; Shen, H. Observer-based sliding mode control for networked fuzzy singularly perturbed systems under weighted try-once-discard protocol. *IEEE Trans. Fuzzy Syst.* **2021**, *6*, 1889–1899. [Google Scholar] [CrossRef]

10. Youn, W.; Lim, H.; Choi, H.S.; Rhudy, M.B.; Ryu, H.; Kim, S.; Myung, H. State estimation for HALE UAVs with deep-learning-aided virtual AOA/SSA sensors for analytical redundancy. *IEEE Robot. Autom. Lett.* **2021**, *6*, 5276–5283. [Google Scholar] [CrossRef]

11. Deng, C.; Wen, C.; Huang, J.; Zhang, X.M.; Zou, Y. Distributed observer-based cooperative control approach for uncertain nonlinear MASs under event-triggered communication. *IEEE Trans. Autom. Control* **2021**, *67*, 2669–2676. [Google Scholar] [CrossRef]

12. Lee, D.; Lai, C.W.; Liao, K.K.; Chang, J.W. Artificial intelligence assisted false alarm detection and diagnosis system development for reducing maintenance cost of chillers at the data center. *J. Build. Eng.* **2021**, *1*, 102110. [Google Scholar] [CrossRef]

13. Chen, Z.; Niu, B.; Zhao, X.; Zhang, L.; Xu, N. Model-based adaptive event-triggered control of nonlinear continuous-time systems. *Appl. Math. Comput.* **2021**, *8*, 126330. [Google Scholar] [CrossRef]

14. Touzé, C.; Vizzaccaro, A.; Thomas, O. Model order reduction methods for geometrically nonlinear structures: A review of nonlinear techniques. *Nonlinear Dyn.* **2021**, *105*, 1141–1190. [Google Scholar] [CrossRef]

15. Iida, Y.; Takatani, Y.; Kojima, A.; Ishii, M. Global trends of ocean CO2 sink and ocean acidification: An observation-based reconstruction of surface ocean inorganic carbon variables. *J. Oceanogr.* **2021**, *77*, 323–358. [Google Scholar] [CrossRef]

16. Ortega, R.; Bobtsov, A.; Nikolaev, N.; Schiffer, J.; Dochain, D. Generalized parameter estimation-based observers: Application to power systems and chemical–biological reactors. *Automatica* **2021**, *129*, 109635. [Google Scholar] [CrossRef]
17. Jerono, P.; Schaum, A.; Meurer, T. Observability analysis and robust observer design for a continuous yeast culture. *J. Process Control.* **2021**, *104*, 62–73. [Google Scholar] [CrossRef]
18. Li, C.; Wang, Y.; Liu, F.; Liu, Q.; Buss, M. Model-free incremental adaptive dynamic programming based approximate robust optimal regulation. *Int. J. Robust Nonlinear Control* **2022**, *32*, 2662–2682. [Google Scholar] [CrossRef]
19. Najafzadeh, M.; Niazmardi, S. A novel multiple-kernel support vector regression algorithm for estimation of water quality parameters. *Nat. Resour. Res.* **2021**, *30*, 3761–3775. [Google Scholar] [CrossRef]
20. Campos-Díaz, K.; Bandala-Gonzalez, E.; Limas-Ballesteros, R. Fluid bed porosity mathematical model for an inverse fluidized bed bioreactor with particles growing biofilm. *J. Environ. Manag.* **2012**, *104*, 62–66. [Google Scholar] [CrossRef]
21. Das, B.; Ganguly, U.P.; Bar, N.; Das, S.K. Holdup prediction in inverse fluidization using non-Newtonian pseudoplastic liquids: Empirical correlation and ANN modeling. *Powder Technol.* **2015**, *273*, 83–90. [Google Scholar] [CrossRef]

3 Chapter

THE STUDY FOCUSES ON THE LONG-WAVELENGTH LUMINESCENCE OF INSB QUANTUM DOTS IN A TYPE II BROKEN-GAP HETEROSTRUCTURE

1. INTRODUCTION: AN OVERVIEW

The spectral region between 2.8 and 5.4 μm, known as the mid-infrared spectrum, is crucial for several applications of optical gas sensing. These applications include medical diagnosis through breath analysis and the detection of leaks of poisonous and dangerous gases for environmental protection. Semiconductor heterostructures utilizing the narrow-gap InSb–InAs solid solutions system show great potential for developing light-emitting diodes that operate inside the 4.4–5.4 μm atmospheric transparency window [3,4]. The spectral region mentioned encompasses the strong absorption bands of several nitrogenous chemical compounds such as N2O, NO, (CN)2, HCN, etc. These molecules are highly significant for industrial gas analysis and the identification of explosive and hazardous substances [5]. Quantum dots (QDs) have been found to enhance the quantum efficiency and operating temperature of laser diodes when used in their active region [6]. This has significant possibilities for enhancing the design of optoelectronic devices operating in the mid-infrared range. Unlike extensively researched wide-gap systems like InAs/GaAs and InSb/InP, the energy band map of the narrow-gap InSb/InAs heterostructure displays a type II broken-gap heterojunction at every QD/matrix layer contact [7]. The presence of InSb/InAs nanoheterostructures prompts the inquiry of discerning radiative transitions facilitated by quantum dots. A recent study investigated the photoluminescent characteristics of heterostructures composed of InSb quantum dot (QD) layers embedded in the n-InAs matrix [8]. At low temperatures, recombination transitions involving quantum dot eigenstates resulted in the detection of long-wave radiation in the spectral region of 4–5 μm (0.25–0.3 eV). A study of electroluminescence (EL) was conducted on heterostructures consisting of a single layer of InSb quantum dots (QDs) embedded in the p-InAs matrix [9]. The electroluminescence spectra exhibited luminescence with a positive response when a forward bias was applied, and luminescence with a negative response when a reverse external bias was applied to the structure being investigated within the temperature range of T = 77–300 K. Both biases exhibited interband radiative recombination resulting from the lightly doped p-InAs bulk. The presence of InSb quantum dots (QDs) was only detected through a minor amount of light caused by the contact. In our present research, we describe the findings of our investigation into electroluminescence resulting from radiative transitions that involve the eigenstates of quantum dots. In order to achieve this objective, type II broken-gap heterostructures were fabricated, consisting of a solitary layer of InSb quantum dots (QDs) positioned directly on the interface of the p-InAs/n-InAs junction.

2. MATERIALS AND METHODS

A mixed epitaxial growth process was used to create narrow-gap heterostructures consisting of an InAs matrix with one layer of InSb quantum dots implanted within. The samples under study were grown on p-InAs(001) substrates doped with an acceptor impurity (Mn) to a hole concentration of p_{300}~10^{17} cm^{-3} at T = 300 K. At the first stage of the heterostructure fabrication, the InSb QDs array was deposited on the p-InAs substrate by means of liquid-phase epitaxy (Figure 1a). The technique used for the QDs growth and study on their structural properties in the InAs/InSb system were reported elsewhere [10,11]. Arrays of QDs in the surface density range of 0.5–2 × 10^{10} cm^{-2} with a step of 2 × 10^{9} cm^{-2} were obtained. The layer of InSb QDs was further overgrown with an epitaxial layer of indium arsenide using metal-organic vapor phase epitaxy (MOVPE). The n-InAs capping layer was intentionally undoped. It exhibited n-type conductivity with an estimated residual electron concentration of *n300*~6·10^{16} cm^{-3} at T = 300 K. The presence of the InSb QD arrays buried in the InAs matrix formed was confirmed by transmission electron microscopy investigations (Figure 1b). None of the structures obtained exhibited luminescence from QDs, except for samples with a limiting density of 2 × 10^{10} cm^{-2}. Therefore, the two most typical representatives from the whole series of samples were selected for the EL study (see Table 1). Additionally, the p-InAs/n-InAs structure (without the QD layer) was chosen as a test sample (see sample A in Table 1). It was acquired concurrently with samples B and C through a unified MOVPE method. Undoped compounds were used for epitaxial overgrowth, which effectively prevents the diffusion of the acceptor impurity from the p-InAs matrix to the n-InAs layer during the overgrowth process. Consequently, an interfacial metallurgical p-n junction was created within the InAs matrix, with the InSb QDs layer situated at the interface. Please refer to sample B and C in Table 1 for further details.

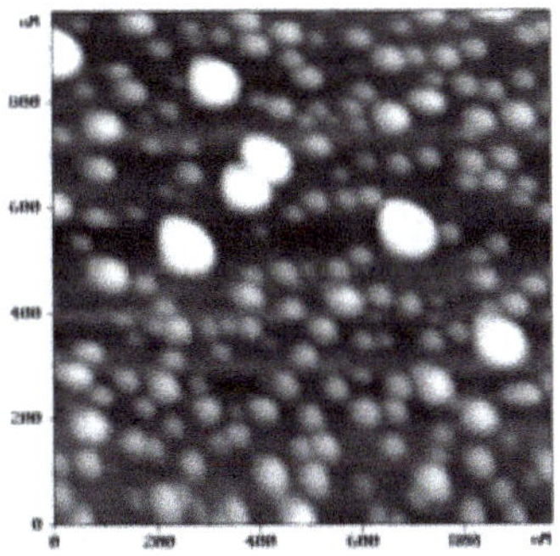

(a)

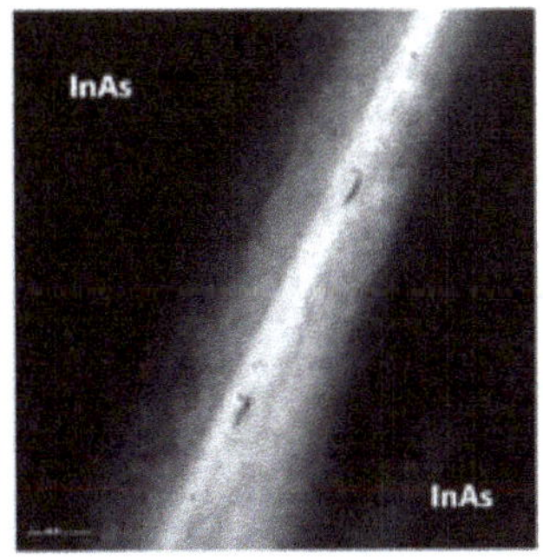

(b)

Figure 1. (a) AFM image of the InSb QD arrays on a surface of the InAs matrix with density *nQD* = 2 × 1010 cm-2. (b) TEM image of a cross-sectional view of the InAs/InSb QDs/InAs heterostructure. Light contrast is the position of the InSb QDs layer.

Table 1. Heterostructures with QDs obtained by combined epitaxial technology.

Sample	Epitaxial Structure	QDs Surface Density(*nQD*)
A	p-InAs/n-InAs	-
BC	p-InAs/InSb/n-InAsp-InAs/InSb/n-InAs	0.7 × 10^{10} cm^{-2}2.0 × 10^{10} cm^{-2}

The EL investigations were conducted using mesa diodes measuring 500 × 500 μm. These diodes had a broad contact on the substrate side and a point contact on the higher epitaxial layer. The mesa, which was created using normal photolithography and selective wet chemical etching, had a diameter of around 300 μm (micrometers). The diameter of the point contact was approximately 50 μm. The mesa diodes under investigation were affixed to the TO-18 package and positioned with the substrate side facing the metal case. In this design, the radiation emitted from the structure was detected from the lateral direction of the capping epitaxial layer. The EL spectra were recorded using an automated setup based on a DK-480 monochromator (CVI Laser Corp., Albuquerque, NM, USA) with a diffraction grating of 150 lines/mm, an InSb photovoltaic detector cooled by liquid nitrogen (Judson Co, Montgomeryville, PA, USA), and an SR-810 lock-in amplifier (Stanford Research Systems Inc., Sunnyvale, CA, USA). The samples were supplied by rectangular current pulses with a duty cycle of η = 50% and a frequency of f = 512 Hz, both with forward and reverse external bias. The forward bias corresponds to the case where the positive potential was applied to the p-InAs substrate; whereas, the negative one was applied to the upper n-InAs layer. Therefore, reverse bias has the opposite configuration of polarities. The pulse amplitude of the driving current (i) was varied from 20 mA to 200 mA.

To analyze the obtained EL spectra, preliminary calculations of the band energy diagrams of the heterostructures studied in the experiment were carried out. Estimation of the energy spectrum of quantum-size structures was performed using available resources [12]. For instance, a single InSb QD can be represented as a narrow one-dimensional potential well with a set of localized levels (Figure 2a). This makes it possible to determine the energy levels for electrons and holes depending on the well width, material compositions and parameters of adjacent semiconductor compounds. It can be seen from the figure that the ground hole level of the InSb QD (μ_h) is situated in energy above the bottom of the conduction band of the InAs matrix layer due to the peculiarity of the type II broken-gap heterojunction in the InSb-InAs system. As was shown in [13], structural crystalline defects in indium arsenide can create acceptor states (IS) on the surface of a semiconductor.

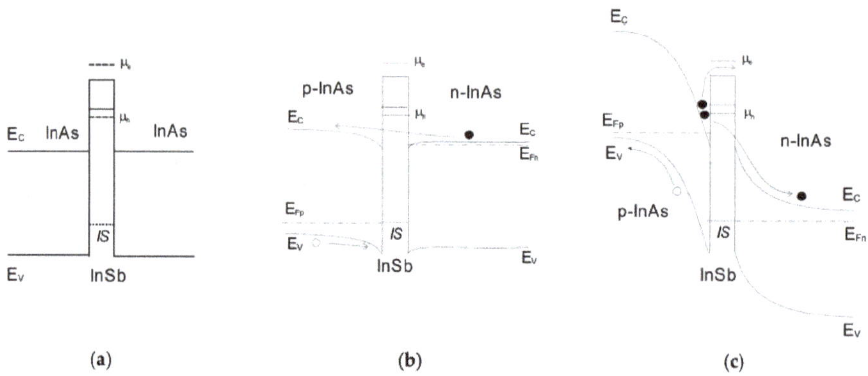

Figure 2. (a) Schematic energy band diagram of the type II InAs/InSb/InAs heterostructure with a layer of QDs. p-InAs/InSb QDs/n-InAs heterostructure under external forward (b) and reverse (c) bias. *IS*—the level of surface states at the interface InAs/InSb; μe and μh are the ground electron and hole states of the InSb QD, respectively.

3. RESULTS

The electroluminescence spectra at room temperature exhibited comparable characteristics for all the structures investigated when a forward bias was applied (Figure 3a). The typical EL spectrum contained one pronounced emission band with an intensity maximum near $h\nu_{300}(+)$ = 0.36 eV and FWHM = 48 meV, which corresponded to interband radiative transitions in the bulk of indium arsenide [14]. Additional emission bands can be distinguished near $h\nu_{IS}$ = 0.31 eV as a shoulder on the low-energy slope of the main EL band. At forward bias (Figure 2b), potential barriers near the InAs/InSb interface appear on the moving path of electrons in an external electric field from the populated conduction band of n-InAs to the unfilled conduction band of p-InAs. The presence of these barriers is caused by the energy band structure of the quantum dots and the interface states at the border between the matrix layers. It is important to highlight that we did not observe any other radiative recombination transitions involving impurity states with an activation energy lower than 30 meV. This can be explained by the fact that shallow levels are depleted due to their thermal warming at a temperature of 300 K.

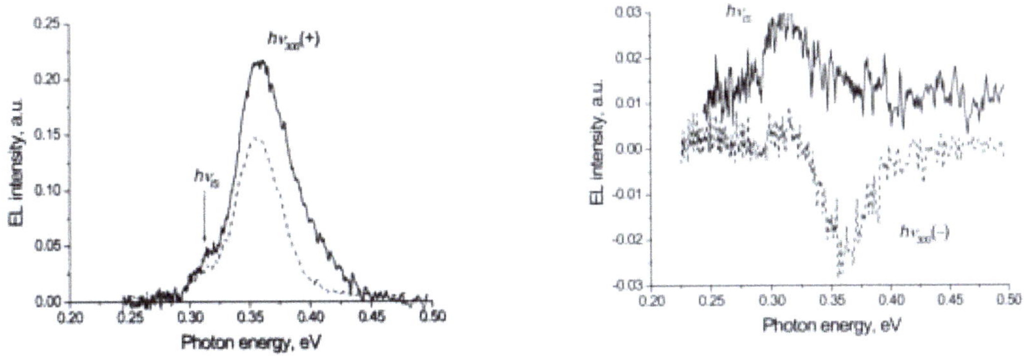

(a) (b)

Figure 3. EL spectra obtained for *p*-InAs/*n*-InAs (dashed line) and *p*-InAs/InSb/*n*-InAs (solid line) heterostructures. The measurements were performed at forward (a) and reverse (b) biases, *T* = 300 K, and driving current *i* = 50 mA.

Under reverse bias conditions (Figure 2c), the emission spectra of the electroluminescence (EL) at room temperature exhibited distinct differences between samples with and without quantum dots (QDs) (Figure 3b). The EL spectrum for the p-InAs/n-InAs structure showed an emission band in the vicinity of $h\nu_{300}(-)$ = 0.36 eV, reversed in phase by 180° relative to the zero-intensity level, i.e., negative luminescence was observed in the bulk of p-InAs [9,15]. Together with this band, a weak peak of positive EL was found at photon energy *hνIS* = 0.31 eV. Contrary to sample A, the heterostructures with InSb QDs (p-InAs/InSb/n-InAs) demonstrated exclusively positive luminescence. Note that these EL spectra contain only one emission band and exactly with an intensity maximum in the vicinity of the photon energy of 0.31 eV. Radiation in the range of photon energy close to the InAs band gap was not detected for samples B and C at all. The space charge region is shunted by the ohmic contact of the type II broken-gap heterojunction due to the fact that the hole levels of the InSb QD are situated in energy above the conduction band of the InAs matrix layers. The main drop in external bias can be expressed as a linear slope in the region of the flat energy bands in the InAs layer that provides conditions for the free movement of electrons to the unfilled conduction band of the n-InAs and further to a metal contact without any potential barriers in their path. In this way, the process of extraction of charge carriers from the depleted regions can be significantly suppressed. A similar effect of negative luminescence suppression was recently observed for heterostructures with an ultrathin oxide layer separating the surfaces of two contacting semiconductors, which are forming a type II broken-gap heterojunction under conservation of the levels of localized states on the surface of indium arsenide [16]. Thus, radiative recombination transitions are realized with the participation of the interface level (IS) for any direction of the injection current (both forward and reverse biases) that is expressed in the presence of an emission band with unchanged polarity in the EL spectrum. Therefore, the $h\nu_{IS}$ emission band observed at room temperature can be ascribed to radiative transitions involving the surface states at the p-InAs/n-InAs heterointerface which are placed 50 meV above the top of the InAs valence band.

As the temperature is lowered to T = 77 K, the contribution of radiative recombination transitions with the participation of impurity states in p-InAs, which have an activation energy greater than the energy of thermal level smearing (kT_{77}~6 meV), becomes noticeable. The EL spectra for all the studied samples contained two distinct emission bands $h\nu_{77}$ = 0.408 eV and $h\nu_{77}A$ = 0.376 eV upon application of forward external bias (Figure 4). The high-energy band ($h\nu_{77}$) can be attributed to interband radiative transitions in InAs [14]. The energy distance of the low-energy band $h\nu^{77}A$ from the high-energy band was 32 meV, which points to radiative recombination transitions associated with impurity acceptor states of Mn situated in the forbidden gap of InAs [17]. Furthermore, the presence of an additional emission band is assumed, which can be revealed as a shoulder on the low-energy slope of the EL spectrum with an emission maximum near 0.35 eV. We suppose that it can be ascribed to radiative transitions involving interface states ($\Delta h\nu_{77}(IS)$~58 meV).

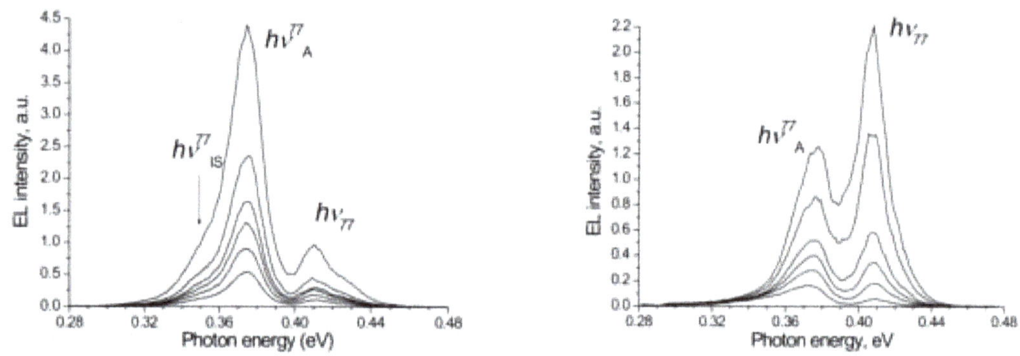

(a) (b)

Figure 4. EL spectra obtained for different values of the driving current (i = 20, 30, 40, 50, 75 and 100 mA) at forward bias applied at T = 77 K: (a) for samples A and B; (b) for sample C.

The EL measurements revealed different behavior in intensity of the emission bands with increasing driving current for the samples under study depending on the surface density of the InSb QDs at the *p-n* junction interface. For sample A (without QDs) and sample B (with a low concentration of QDs), the dominance of the emission band $hv^{77}A$ was retained over the entire range of driving currents (Figure 4a). For sample C (with a high concentration of QDs), a redistribution of intensity between the EL bands was observed, when the dominance of the low-energy band *hv77A* at low injection levels ($i < 50$ mA) was replaced by the advantage of the high-energy band *hv77* at higher injection levels (Figure 4b). In this case, the current dependence of EL intensity for the *hv77* band that occurred by interband radiative transitions in indium arsenide was changing within equal range for all samples. In other words, an increase in the surface density of nano-objects at the interface led to a decrease in the contribution of radiative transitions involving impurity acceptor states of p-InAs to the total EL spectrum. According to the energy diagram shown in Figure 1b, the appearance of a space charge region on the p-InAs side near the heterointerface prevents electrons from penetrating into the region of flat bands, where radiative recombination with the participation of impurity levels takes place.

When reverse bias was applied to the heterostructures under study, the EL spectra at $T = 77$ K showed no emission band corresponding to interband transitions in InAs (Figure 5), as well as at room temperature performance. This characteristic feature of EL behavior at low temperatures was previously observed for heterostructures with the InSb QDs layer at the type II broken-gap InAs/InSb heterointerface in the p-InAs matrix, where interband transitions began to manifest themselves as a negative luminescence at high pump levels only [9]. The main emission band ($hv_{IS+Channel}$) in the EL spectra recorded at reverse bias had an asymmetric shape with a sharp high-energy edge and exhibited a "blue" shift; i.e., the spectral position of the EL peak was shifting towards higher photon energies with an increase in driving current (Figure 6).

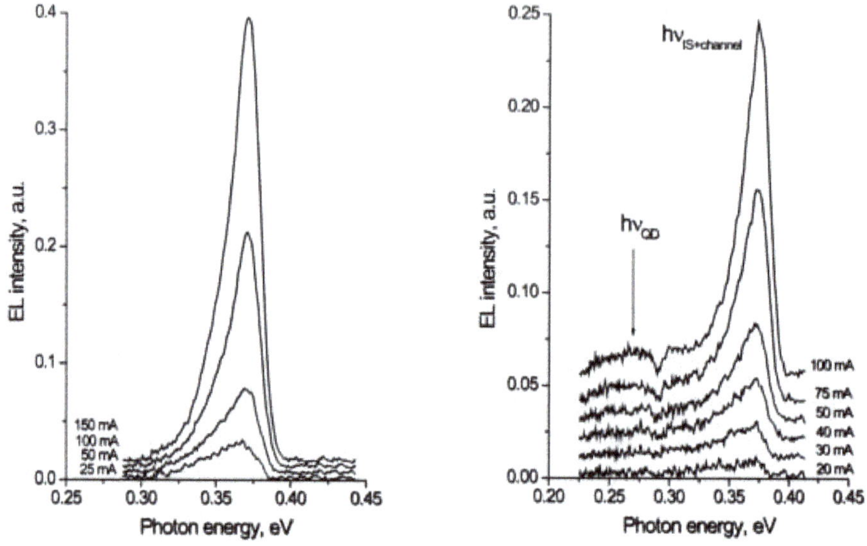

(a) (b)

Figure 5. EL spectra of *p*-InAs/InSb/*n*-InAs heterostructures with InSb QDs ((a) sample B and (b) sample C), measured at reverse bias at T = 77 K.

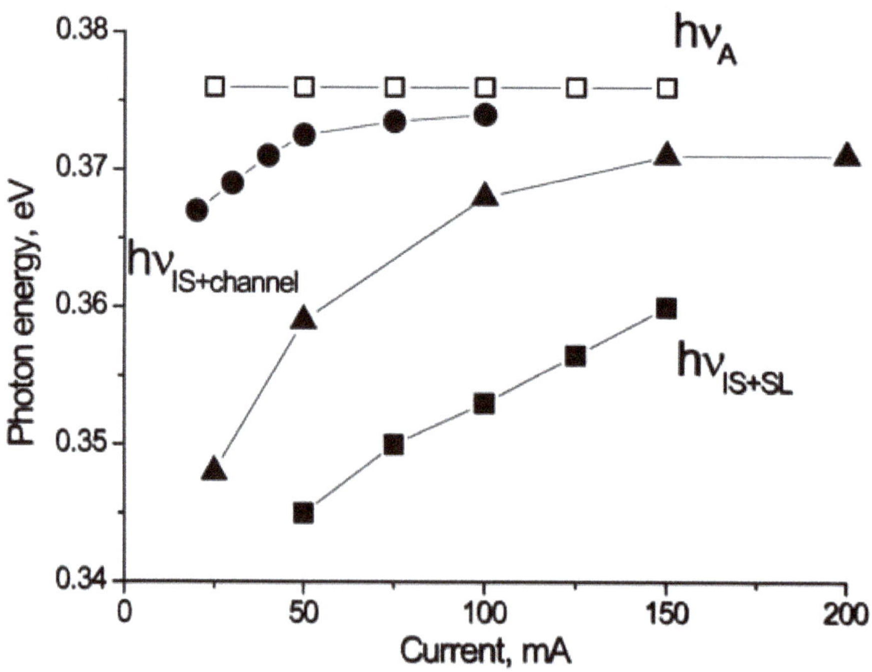

Figure 6. Dependence of the spectral position of the EL bands on the driving current at the reverse bias applied at T = 77 K (filled symbols). For comparison, the dependence of the band $h\nu A$ is shown when the forward bias has been applied (open squares).

Different behavior should be distinguished between the "blue" shift for each dependence of the EL emission band on the current. A linear dependence of the hv_{IS+SL}(i) band for sample A was observed in the photon energy range from 0.345 eV to 0.358 eV in the current interval i = 50–150 mA (Figure 6, filled squares). Comparing the energy band diagram of sample A at reverse bias (not shown) and the photon energy range in which the shift of the band h_{vIS+SL} was observed, one can estimate the energy position of the interface level IS at low temperature. It turns out that the linear "blue" shift observed in the EL spectra for sample A is the result of radiative transitions from shallow impurity levels (SL) formed by natural structural defects in the n-InAs capping layer to the interface level *IS*. Then, an increase in the reverse bias will cause a gradual depletion of the donor levels in n-InAs near the heterojunction. According to our estimates, the level *IS* was approximately 50 meV above the top of the InAs valence band at a low temperature (T = 77 K), which is in good agreement with the experimental data obtained at room temperature (see above).

Note also that both dependences $hv_{IS+channel}$(i), which describe the "blue" shift for samples with QDs, demonstrated non-linear behavior and saturation at high pumping. The specific shape of these dependences was determined by the surface density of nano-objects at the InAs/InSb heterointerface. For sample C, with a high concentration of the InSb QDs at the interface ($n_{QD2} > 10^{10}$ cm^{-2}), luminescence began to appear at a photon energy $h_{IS+channel}$ = 0.366 eV (Figure 6, filled circles). In this case, the range of motion of the given EL band from the high-energy side was limited by the value of the radiative transition energy hv_A = 0.376 eV (Figure 6, open squares). In contrast to samples A and C, the curve ascribed to sample B, with a low concentration of QDs at the interface ($n_{QD} < 10^{10}$ cm^{-2}), occupies an intermediate position, representing the result of a superposition of the two extreme cases considered (Figure 6, filled triangles). For this sample, the onset of luminescence was recorded at *hIS+channel* = 0.348 eV, and the "blue" shift was saturated at a **photon** energy of 0.37 eV.

4. DISCUSSION

This section will examine the genesis of the band hvIS + channel and analyze its conduct. Applying reverse bias to the heterostructures containing InSb quantum dots (samples B and C) caused a potential well to form on the p-InAs side at the InAs/InSb interface. This was a result of the conduction band bending, as shown in Figure 2c. When the external electric field is increased, electrons fill the potential well at the p-InAs/InSb heterojunction, creating a 2D-electron channel [18]. Additionally, there is the potential for radiative recombination transitions to occur from the quasi-Fermi level (EFp) states in the electron channel to the interface states IS [19].

The quantum energy of such transitions can be described by the following expression

$$h\nu_{IS+channel} = E_{IS} + E_{Fp}, \quad (1)$$

where E_{IS} and E_{Fp} are the energy of the interface level and Fermi level position relative to the bottom of the InAs conduction band, respectively. With a further increase in the external bias and, consequently, the filling of the electron channel, the E_{Fp} level is shifted upward in energy that will be expressed in the "blue" shift of the EL band corresponding to these transitions. The observed shift in the driving current is saturated due to the continuous filling of the electron channel until electrons start to leak past the potential barrier created by the band gap of the InSb QD. A supplementary pathway for electrical current is already established through the involvement of the quantum dot electron states. Consequently, samples containing varied concentrations of quantum dots (QD) at the interface exhibit distinct levels of saturation in the "blue" shift of the electroluminescence (EL) bands (refer to Figure 6).

The appearance of the additional current flow channel upon application of a reverse bias enables filling of the eigenstates in the InSb QDs. Therefore, the EL spectra at low temperatures can contain the emission band corresponding to recombination transitions between localization levels in a quantum dot. The photon energy of such transitions can be estimated as

$$h\nu_{QD} = E_{QD} + \delta E_{QD} + \mu_h + \mu_e, \quad (2)$$

where E_{QD} is the band gap of the InSb binary compound, δE_{QD} is the deviation in the band gap of InSb due to internal compression strain in the InSb QD, μ_e and μ_h are the energies of the ground electron and hole levels of the InSb QD, respectively. The EL spectra of sample C exhibited the emission band in the region $h\nu_{QD}$ = 0.270 eV with a half-width of about 60 MeV (Figure 5b). Note that a sharp dip at the low-energy edge of the EL spectrum near a photon energy of 0.29 eV is associated with the absorption of optical radiation by CO_2 molecules (λ = 4.27–4.29 μm) present in the laboratory atmosphere [3,5].

The spectral position of the $h\nu_{QD}$ band agrees well with the experimental data obtained by measuring the PL spectra at T = 77 K for structures with InSb QDs in an InAs matrix grown by molecular beam epitaxy [8]. It should be emphasized that the mentioned PL band was recorded only for structures with a surface density of nano-objects of $n_{QD} > 2\cdot10^{10}$ cm^{-2}. Because the intensity of EL radiation induced by transitions between localized levels in QDs also depends on the concentration of nano-islands, the band $h\nu_{QD}$ was not detected for samples with lower surface density of the InSb QDs (Figure 5a). In the EL spectra of sample C recorded at T = 300 K, the corresponding EL band was not observed either (Figure 3b). The reason for this was that the observations were conducted using an InSb photovoltaic detector that had a long-wavelength cutoff at $\lambda = 5.3$ μm. This cutoff prevented the detection of luminescence with a photon energy lower than 0.23 eV. Other scientific groups have made efforts to achieve electroluminescence in heterostructures containing InSb quantum dots (QDs) in an InAs-based matrix. However, the limited number of nano-objects on the surface hinders the interpretation of the observed emission band, which has a photon energy of approximately 0.37 eV, as being a result of radiative recombination involving the states of InSb QDs [20]. The presence of surface levels at the InAs/InSb heterointerface is the most probable cause for the observed EL band.

5. CONCISE SUMMARY

Therefore, the electroluminescence spectra, which involved the inherent states of InSb quantum dots positioned in a low-energy band matrix composed of InAs compounds, were discovered for the first time. A mixed epitaxial growth process was used to create narrow-gap heterostructures that consist of a single layer of InSb quantum dots inserted in the p-InAs/n-InAs junction. At forward external bias, Type II broken-gap p-InAs/InSb/n-InAs heterostructures showed radiative recombination in the bulk of indium arsenide. Additionally, there was luminescence at the InAs/InSb heterointerface involving surface states. This phenomenon occurred throughout a wide temperature range. Under the application of reverse bias, it is feasible to observe electroluminescence (EL) using the eigenstates of InSb quantum dots (QDs). This occurs when the QD states are filled from the 2D-electron channel at the type II broken-gap heterointerface.

At a Glance

The study focused on analyzing the characteristics of the electroluminescence spectra of narrow-gap type II InAs/InSb/InAs heterostructures. These structures included a single layer of InSb quantum dots positioned within the p-n-InAs junction. The study examined the luminous characteristics of the heterostructures when subjected to both forward and reverse bias within a temperature range of 77–300 K. The investigation focused on how the surface density of nano-objects embedded in the narrow-gap matrix affected these features. Applying reverse bias to the heterostructures being examined resulted in the inhibition of negative interband luminescence and the prevalence of interface recombination transitions at the InSb/InAs type II heterojunction, even at room temperature. The emission resulting from recombination transitions involving localized electron-hole states of the InSb quantum dots was observed and recorded at low temperatures.

REFERENCES

1. Schiff, H.I.; Mackay, G.I.; Bechara, J. *Air Monitoring by Spectroscopy Techniques*; Sigrist, M.W., Ed.; John Wiley and Sons: New York, NY, USA, 1994; p. 560. [Google Scholar]
2. Wang, C.; Sahay, P. Breath Analysis Using Laser Spectroscopic Techniques: Breath Biomarkers, Spectral Fingerprints, and Detection Limits. *Sensors* **2009**, *9*, 8230–8262. [Google Scholar] [CrossRef] [PubMed]
3. Hudson, R.D. *Infrared System Engineering*; John Wiley and Sons: New York, NY, USA; London, UK; Sydney, Australia, 1969; p. 642. [Google Scholar]
4. Romanov, V.V.; Ivanov, E.V.; Pivovarova, A.A.; Moiseev, K.D.; Yakovlev, Y.P. Long-wavelength LEDs in the atmosphere transparency window of 4.6–5.3 μm. *Semicond.* **2020**, *54*, 253–257. [Google Scholar] [CrossRef]
5. Pierson, R.H.; Fletcher, A.N.; Gantz, E.C., St. Catalog of Infrared Spectra for Qualitative Analysis of Gases. *Analyt. Chem.* **1956**, *28*, 1218–1239. [Google Scholar] [CrossRef]
6. Zhukov, A.E.; Maksimov, M.V.; Kovsh, A.R. Device characteristics of long-wavelength lasers based on self-organizing quantum dots. *Semiconductors* **2012**, *46*, 1249–1273. [Google Scholar]
7. Wei, S.H.; Zunger, A. InAsSb/InAs: A type-I or a type-II band alignment. *Phys. Rev. B* **1995**, *52*, 12039–12044. [Google Scholar] [CrossRef] [PubMed] [Green Version]
8. Karim, A.; Gustafsson, O.; Hussain, L.; Wang, Q.; Noharet, B.; Hammar, M.; Anderson, J.; Song, J. Characterization of InSb QDs grown on InAs (100) substrate by MBE and MOVPE. *Proc. SPIE* **2012**, *8439*, 84391J. [Google Scholar]
9. Parkhomenko, Y.A.; Ivanov, E.V.; Moiseev, K.D. Specific features of electroluminescence in heterostructures with InSb quantum dots in an InAs matrix. *Semiconductors* **2013**, *47*, 1523–1527. [Google Scholar] [CrossRef]
10. Bert, N.A.; Nevedomsky, V.N.; Dement'ev, P.A.; Moiseev, K.D. TEM-study of free-standing self-assembled InSb quantum dots grown on InAs-rich surface. *Appl. Surf. Sci.* **2013**, *267*, 77–80. [Google Scholar] [CrossRef]
11. Moiseev, K.D.; Parkhomenko, Y.A.; Nevedomsky, V.N. Uniform InSb quantum dots buried in narrow-gap InAs(Sb,P) matrix. *Thin Solid Films* **2013**, *543*, 74–77. [Google Scholar] [CrossRef]
12. Bunin, V.A.; Matveev, Y.A. Electronic structure of an anion vacancy in indium arsenide. *Sov. Phys. Semicond.* **1985**, *19*, 2018–2022. [Google Scholar]

13. Landolt-Börnstein. *Numerical Data and Functional Relationships in Science and Technology*; Madelung, O., Ed.; Springer: New York, NY, USA, 1982; Volume 17, p. 642. [Google Scholar]

14. Grigoryev, M.M.; Alekseev, P.A.; Ivanov, E.V.; Moiseev, K.D. Two-colour luminescence in a single type II InAsSbP/InAs heterostructure. *Semiconductors* **2013**, *47*, 28–32. [Google Scholar] [CrossRef]

15. Moiseev, K.; Ivanov, E.; Romanov, V.; Mikhailova, M.; Yakovlev, Y.; Hulicius, E.; Hospodková, A.; Pangrác, J.; Šimeček, T. Intense interface luminescence in type II narrow-gap InAs-based heterostructures at room temperature. *Phys. Procedia* **2010**, *3*, 1189–1193. [Google Scholar] [CrossRef] [Green Version]

16. Georghitse, E.I.; Postolaki, I.T.; Smirnov, V.A.; Untila, P.G. Photoluminescence of p-InAs<Mn>. *Sov. Phys. Semicond.* **1989**, *23*, 745–746. [Google Scholar]

17. Moiseev, K.D.; Berezovets, V.A.; Golenitskii, K.Y.; Averkiev, N.S. Band structure of a hybridized electron-hole system at a single broken-gap type II heterointerface. *Low Temp. Phys.* **2019**, *45*, 153–158. [Google Scholar] [CrossRef]

18. Mikhailova, M.P.; Moiseev, K.D.; Yakovlev, Y.P. Interface-induced optical and transport phenomena in type II broken-gap single heterostructures. *Semicond. Sci. Technol.* **2004**, *19*, R109–R128. [Google Scholar] [CrossRef]

19. Zhuang, Q.; Carrington, P.J.; Krier, A. Growth optimization of self-organized InSb/InAs quantum dots. *J. Phys. D Appl. Phys.* **2008**, *41*, 232003. [Google Scholar] [CrossRef]

Chapter 4

EFFORTLESS GREEN SYNTHESIS OF REDUCED GRAPHENE OXIDE UTILIZING CITRUS LIMETTA-DECORATED RGO/TIO2 NANOSTRUCTURES FOR GLUCOSE SENSING

1. INTRODUCTION: AN OVERVIEW

According to the International Diabetes Federation, diabetes, a metabolic condition, has emerged as a significant global concern, impacting millions of individuals. This epidemic poses a serious risk to human health and is one of the major obstacles to long-term social development and economic growth in the present time. Precise monitoring and control of individuals' blood glucose levels are essential for diabetic patients to receive correct diagnosis and appropriate therapy. In order to prevent diabetes emergencies such as hypoglycemia (low blood sugar, ≤3 mM), it is necessary to regularly monitor the physiological glucose level [1]. The detection of glucose has gained major importance in the battle against diabetes and in mitigating the substantial financial burden associated with diabetes and its severe effects. Consequently, throughout the past few decades, there has been significant social and economic significance in the development of dependable and affordable glucose sensors that possess exceptional sensitivity and selectivity. Glucose biosensors have received substantial financial support to monitor glucose levels in individuals with diabetes [2]. A multitude of researchers are currently involved in the advancement of novel biosensors for diverse applications across different fields, following the initial proposal of enzyme electrodes by Clark and Lyons in 1962. Enzymatic sensors exhibited satisfactory sensitivity and stability, nevertheless, they also presented several limitations such as inadequate reproducibility and intricate mobility [3,4]. Despite the advantages of enzymatic sensors, their commercial uses are limited due to their high manufacturing cost [5]. Non-enzymatic glucose sensors were developed as a reaction to these restrictions [6]. Various candidate materials, such as metal oxides, carbon materials, polymers, graphene metals, and graphene oxide materials, have demonstrated their excellence as electrocatalysts in glucose detection [7]. Graphene is a 2D substance composed of a solitary layer of carbon atoms arranged in a hexagonal pattern.

It has garnered significant attention because of its atypical physical, optical, and electrical characteristics. It is the most slender and resilient substance in the world, possessing outstanding electrical and thermal conductivity. Graphene is produced using several processes, including as micromechanical cleavage, chemical vapor deposition, epitaxial growth on silicon carbide, liquid-phase exfoliation, reduction of graphene oxide using different reducing agents, and graphite intercalation [8]. The remarkable mechanical [9], thermal [10], and electrical [11] qualities of this object have garnered significant attention. Commonly, reducers like hydrazine or sodium borohydride are used in the chemical synthesis of GO [12,13]. These reducing agents include inherent drawbacks, such being costly and highly poisonous. Furthermore, the Van der Waals interactions between the graphene layers cause a permanent clustering of GO [14]. When it comes to minimizing the use of harmful chemicals, it is important to do a study on appropriate phytochemicals that can be used as reducing agents. In addition, a green reduction is a straightforward, environmentally friendly, and economical approach. Given this motivation, the suggested study aims to achieve an eco-friendly reduction of GO (graphene oxide) by utilizing Citrus limetta (sweet lime/Mosambi fruit waste) and TiO2 electrodes for electrochemical glucose sensing, without the need for enzymes. To mitigate the negative impacts of dangerous compounds, green reduction and bio-synthesis methods are utilized. These methods involve the use of phytoextracts or bacterium microorganisms, which serve as environmentally benign agents for reduction and as capping agents [15,16,17]. An aseptic atmosphere is essential for cultivating microorganisms, and competence is required for this procedure. GO reduction can be achieved by employing citrus plant extract, which exhibits a reduced reaction rate to regulate the formation of crystals. Previous research has investigated the utilization of several plant and fruit extracts for the eco-friendly production of reduced graphene oxide (rGO) [18]. Plant extracts and fruit wastes provide desirable qualities such as being green, cost-effective, and locally accessible, which make them excellent for the production of reduced graphene oxide. Citrus limetta is the main natural provider of vitamin C and a stimulant of energy [19]. The fruit peels of sweet lime are considered trash in the juice industry, but they are actually a valuable source of phytochemicals such as flavonoids, pectins, carotenoids, hesperidin, terpineol alkaloids, essential oils, d-limonene, and powerful antioxidants. These peels also possess antibacterial capabilities [20]. Graphene produced through the environmentally friendly reduction process can be modified with metal oxides for a range of applications, such as electrochemical supercapacitors, batteries, fuel cells, magnetic data storage, water splitting, chemical sensors, water filters, dye removal, and dye-sensitized solar cells, among others [21,22,23,24]. The current study selected citrus limetta as a reducing agent for the synthesis of reduced graphene oxide (rGO) due to its little environmental impact and absence of any detrimental biological responses. The soxhlet extraction process is favored for its simplicity and efficacy in obtaining delicious lime peel extract. Graphene that has been modified with metallic oxide, resulting in exceptional sensing capabilities, has been well-known for a considerable period of time [25]. Luo et al. utilized an electrochemical deposition approach to fabricate graphene and CuO nanocubes [26]. Yang et al. successfully produced reduced graphene oxide (rGO) via a chemical reduction process. The rGO showed exceptional sensitivity and selectivity within linear ranges of 4 mM and 7.5 mM, as reported in reference [27]. Komathi et al. demonstrated the creation of electrochemical biosensors by modifying titanium dioxide (TiO2) nanowires/graphene as nano-stocks [28]. These findings indicate that the graphene-based metal/metal oxide nanoparticles possess high sensitivity, a broad linear range, and a low detection limit, while also exhibiting good selectivity. Therefore, the rGO/TiO2 composite has been chosen as the sensor electrode in the proposed study and utilized as an enzyme-free biosensor for efficient detection of glucose concentration using cyclic voltammetry (CV) testing. Furthermore, there is a scarcity of literature regarding the green synthesis of biosensors based on rGO/TiO2. The glucose biosensor, in its original state, is utilized to examine glucose levels in the presence of obstructive compounds in actual blood serum.

2. MATERIALS AND METHODS

2.1. Materials

All the chemicals used were of the analytical grade, meaning they were of the highest purity. Graphite powder (98%), $KMnO_4$ (99%), $NaNO_3$ (98%), H_2O_2 (30%), H_2SO_4 (99.99%), HCl (30%), H_3PO_4 (85%), KCl (99.99%) and NaCl (99%) TiO_2 (99.9%), etc., were obtained from Finar chemicals (India). Qualigens fine chemicals in India provided polyvinyl alcohol (PVA, 99%), ethanol, Fluorine-doped Tin Oxide (FTO) substrates, D (+) glucose, dopamine, L-ascorbic acid, D (-) fructose, and lactose, phosphate buffer solution (PBS) was prepared according to AAT Bioquest, Inc. [**29**] and sweet limes were bought from a local market.

2.2. Preparation of GO

A modified Hummer method was used to synthesize GO [**30**]. Briefly, the concentrated H_2SO_4 and H_3PO_4 were mixed in a 9:1 proportion in a round bottom flask, followed by graphite powder and $NaNO_3$ in a 2:1 proportion. To achieve the desired result, it was necessary to stir the mixture at 100 °C continuously for an hour before adding the $KMnO_4$. Adding H_2O_2 to the mixture served the purpose of oxidation, resulting in the formation of a yellow-colored product. SO_4^{2-} is completely removed from the sample by repeatedly washing with 5% HCl and deionized (DI) water, followed by centrifugation (4000 rpm). After nearly 24 h of drying at 60 °C, the resultant product (GO) was obtained which was collected for the next processes.

2.3. Preparation of Aqueous Phytoextract

Fresh sweet limes were purchased from a local market and cleaned using DI water. The citrus limetta fruit waste, such as peel, was cut into small pieces with a domestic cutter. In soxhlet extraction, 15 gm of finely cut peels were placed in a thimble of soxhlet tube, and the DI water was used as the solvent. During condensation, solutes are moved to the reservoir for repeated cycles. The system was cooled for 10-12 h and then the collected extract was preserved at 4 °C for further use.

2.4. Mechanism of Reduction of GO by Citrus Limetta Phytoextract

Citrus Limetta phytoextract can reduce GO by utilizing the oxidized products found in sweet lime fruit peel. These oxidized products include flavonoids, polyphenols, tannis, terpenoids, and saponins, among others. These phytochemicals show high binding linkage towards O_2- containing functionalities to produce their respective oxides and water molecules. Flavonoid compounds are reduced to the phenolic hydroxyl groups. While reacting with the O_2- containing groups, the phenolic hydroxyl group may release water molecules on GO sheets. GO contains reactive, hydroxyl, epoxide, and reactive carbonyl species. As shown in **Figure 1**, acidic polyphenols in the aqueous GO and Citrus Limetta phytoextract were combined, ultra-sonicated, and refluxed at 95 °C. Through a nucleophilic reaction, epoxide fraction and phytoextracts could react with each other. This caused the carbon-oxygen bond to break. Carbonyl and hydroxyl groups in polyphenols could be attacked by nucleophiles in the same way, which would destroy water molecules. GO is changed into rGO in this process, which makes graphene, which is a stable black compound.

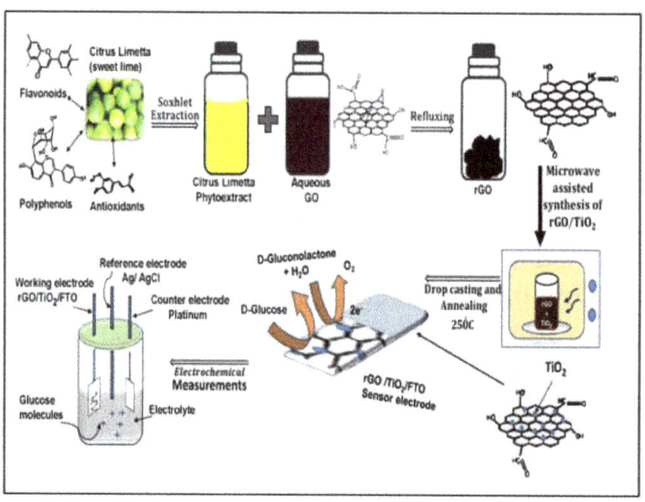

Figure 1. Mechanism of green reduction of GO and rGO/TiO$_2$/FTO for glucose sensing application.

2.5. Reduction of GO Using Aqueous Citrus Limetta Peel Extract

Briefly, 100 mg GO (0.1 mg/mL) was disseminated in ultrapure water by probe sonication for 40 min. The GO solution was mixed with 5 wt.% of 10 mL of aqueous sweet lime fruit extract. The solution's composition was refluxed at 98 °C for 8 h until it turned brown-black. The supernatant (rGO) was repetitively washed with DI water which was centrifuged at 4500 rpm, followed by drying at 100 °C in the oven.

2.6. Synthesis of rGO/TiO_2 Nanocomposite

The rGO/TiO_2 composite was prepared using microwave treatment. A fine powder of rGO and TiO_2 in a 2:1 ratio was dispersed in DI water. After sonication, the mixture was stirred. The compound was kept at 300 °C for 15 min in a microwave (500 W). Finally, the rGO/TiO_2 nanocomposite dry powder was collected. First, 10 mg of rGO/TiO_2 nanocomposites were mixed with 5 mL of DI water and 5 μL of PVA and sonicated in a bath for 30 min. Then, 10 μL of the mixture was drop-cast onto a clean conducting FTO substrate. The working area of the electrode was 1 cm^2. The as-prepared electrode was annealed to remove the binder at a temperature of 250 °C for 1 h. A three-electrode setup consisting of a working electrode made of rGO/TiO_2/FTO, a reference electrode made of Ag/AgCl, and a counter electrode made of platinum was used to conduct electrochemical measurements for glucose detection.

2.7. Glucose Sensing through rGO/TiO_2

For rGO/TiO_2 nanocomposite preparation, in-situ thermal reductions of rGO and TiO_2 were developed. The epitaxial growth process was used to prepare TiO_2 nanobelts. Using Phosphate-Buffered Saline (PBS) as an electrolyte, the glucose sensing performance of the rGO/TiO_2/FTO modified electrode was evaluated. When glucose dissolves in PBS, D-gluconolactone and hydrogen peroxide are produced (H_2O_2). Electro-oxidation of glucose at TiO_2 grain boundaries also produces D-gluconate with H^+ ions. As a result, gluconolactone is the primary product of the oxidation process that hydrolyzes gluconic acid. By releasing electrons, lowering the barrier potential between successive grains, and increasing electrical conductivity, H^+ ions reduce pre-absorbed oxygen [31].

$$\text{Glucose} + O_2 \rightarrow H_2O_2 + \text{gluconic acid} \tag{1}$$

2.8. Characterization Techniques

The as-synthesized GO, rGO, and rGO/TiO_2 sample products were analyzed using a Bruker D8-Advanced Diffractometer with Cu Kα radiation (λ = 0.154 nm) from 5° to 85° at a scanning rate of 2°/min for structural elucidation. Field Emission Scanning Electron Microscopes (FE-SEM): FEI Nova Nano SEM 450 was used to examine the surface morphology and elemental composition of GO, rGO, and rGO/TiO_2 samples. A Micro-Raman spectrometer (Jobin-Yuon HR 800 UV) with a He–Ne (633 nm) laser excitation source was used to conduct the Raman analysis. The multifunctional X-ray photoelectron spectroscopy (XPS) was used to perform XPS analysis on GO and rGO (PHI ulvac probe III Scanning Microprobe). The Fourier-transform infrared spectroscopy (FTIR) study was conducted using an FTIR-6100 spectrometer (JASCO) in transmission mode in the wavenumber range 4000–400 cm^{-1}.

3. RESULTS AND DISCUSSION

3.1. Structure Elucidation Confirmation

The crystal structures of the GO, rGO, and rGO/TiO_2 were characterized and confirmed using X-ray diffraction spectra, respectively. As shown in **Figure 2**a, the reflection peak obtained at 11.8° indicates the presence of oxygen functionalities on the GO surface, which has an interlayer basal spacing of 7.4 Å, suggesting the presence of oxygen functionalities on the GO surface (0 0 2). The spacing between the graphite grains was 3.36 Å. There could be several oxygen-containing groups on the edges of each layer in the first one, which eventually would have increased the distance between them. A broad peak around 25.17° was assigned to the (0 0 2) basal plane of rGO with a separation of 0.37 nm. The fall in basal space and broad peak at 24.7° suggested that GO was transformed to rGO. The prior XRD patterns of rGO/TiO_2 confirmed peaks at 25.3°, 37.8°, 48.0°, 53.8°, 55.0°, 62.6°, and 68.8°, for the (1 0 1), (0 0 4), (2 0 0), (1, 0 5), (2 1 1), (2 0 4), and (1 1 6) reflection planes), respectively of anatase TiO_2 (JCPDS#21-1272 [32]. The narrow and sharp peaks confirmed a high degree of crystallinity of rGO/TiO_2, supporting an easy electron transfer in an electrochemical reaction. Nevertheless, there were no broad peaks corresponding to rGO in the rGO/TiO_2 nanocomposites. The rGO surface can be extensively coated on the TiO_2 nanostructures, which has the effect of suppressing XRD signals that are caused by the stacking of rGO layers [**33**]. The average crystallite size (D) calculated using the Debye-Scherrer formula is as follows,

$$D = (0.9 \times \lambda)/(\beta \times \cos\theta) \tag{2}$$

where, λ is a wavelength of Cu Kα radiation of 1.54 Å, and β is the full width of the diffraction peak at the half maximum at Bragg diffraction angle 2θ. **Table 1** shows the XRD results of the rGO/TiO_2 nanocomposite. The average crystallite size for TiO_2 was 35 ± 2 nm.

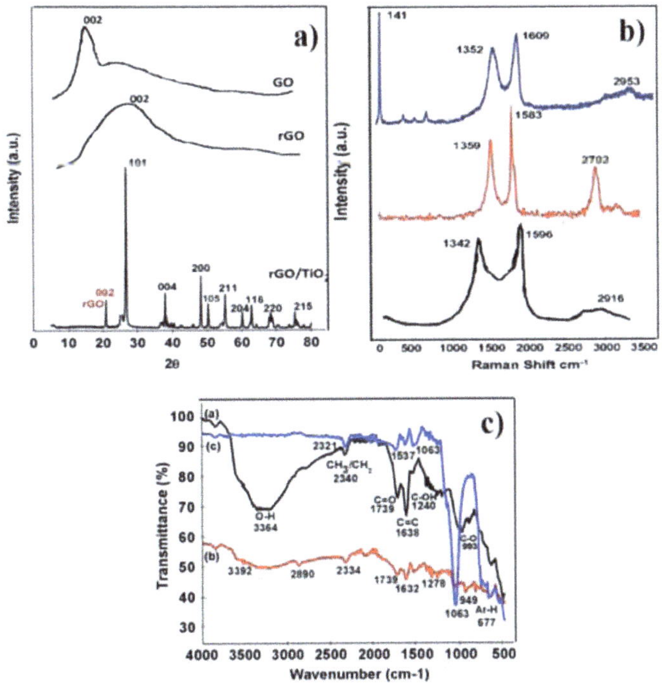

Figure 2. (a) X-ray diffraction, (b) Raman shift, and (c) FTIR analysis spectra of GO, rGO, and rGO/TiO_2.

Table 1. XRD analysis for rGO and TiO_2 products.

No	Two Theta (deg.)	(h k l)	Phase	Crystallite Size (D) ± 2 (nm)
1	20.9	002	rGO	19
2	26.7	101	TiO_2	81
3	37.89	004	TiO_2	14
4	48.14	200	TiO_2	41
5	49.89	105	TiO_2	46
6	55.13	211	TiO_2	41
7	60.05	204	TiO_2	22
8	62.77	116	TiO_2	22
9	68.31	220	TiO_2	23
10	75.15	215	TiO_2	47

The D-bands of GO (black color), rGO (red color), and rGO/TiO_2 (blue color) at 1345 cm^{-1}, 1349 cm^{-1}, and 1353 cm^{-1} were respectively responsible for the obtained peaks (**Figure 2**b). They correspond to the structural defects developed by attaching various functional assemblies with carbon. The peaks at 1593 cm^{-1}, 1598 cm^{-1}, and 1604 cm^{-1} reflect the G band, which appears due to the primitive scattering of the E_{2g} phonon of *sp*^{2} carbon atoms in the same experiment. The wide 2D band of GO with a higher wavenumber was found at 2917 cm^{-1}, confirming the multilayered nature of GO. The reduction of GO to rGO using citrus limetta phytoextract, which causes rGO sheets to stack, was predicted by the two D bands of rGO. Furthermore, the ID/IG ratio was used to measure structural disorder, which increased slightly from 0.85 to 0.87 as GO was reduced to rGO. The higher ID/IG ratio could be assigned to higher structural defects. As a result, defects form in addition to removing oxide functional groups attached to the GO surface. Despite this, the ID/IG ratio for the rGO/TiO_2 nanocomposite was 0.6. This was assigned to the microwave heating process during the reduction of sp^3 carbon to sp^2 carbon, causing the sp^2 domain size of carbon atoms to become decreased [34]. In rGO/TiO_2 spectrum, the peaks below 1000 cm^{-1} may be attributed to the A_g and B_g modes of TiO_2. The crystalline size of the rGO/TiO_2 nanocomposite was estimated using Tuinstra- Koening relationship as [**35,36**],

$$L_a = \left(2.4 \times 10^{-4}\right) \times \left(\frac{I_G}{I_D}\right) \times \lambda^4 \tag{3}$$

L = crystallite size, λ = wavelength of the laser source, and I_D/I_G is the Raman peak intensity ratio of D and G bands, respectively. The nanocomposite‘s crystallite size was found to be 34 ± 2 nm, which agrees well with XRD analysis. The O_2 functional groups in GO were confirmed by the broad O-H stretching vibrational band, C=O, and epoxy group at 3364 cm^{-1}, 1739 cm^{-1}, and below 1000 cm^{-1}, respectively (**Figure 2**c). A few absorption peaks at 2340 cm^{-1}, 1638 cm^{-1}, and 993 cm^{-1} of C-H stretching, bending vibrations of C=C, and aromatic =C-H around, respectively, were obtained. For rGO (curve b of **Figure 2**c), the peak at 3340 cm^{-1} changed to broad and less intense than GO, revealing the significant removal of the hydroxyl group. Also, it was noticed that the other peaks became less intense compared to the GO at the same location. It could be due to the incomplete removal of O_2 functionalities while synthesizing GO using citrus limetta phytoextract. In short, the polar functional groups support hydrophilic properties and high dispersibility of synthesized rGO in DI [37]. The peak noted at 677 cm^{-1} was credited to Ti-O-Ti and Ti-O-C in rGO/TiO_2 composite (curve c).

3.2. Morphological Study

The FE-SEM surface micrograph of GO (**Figure 3**a), synthesized using an improved Hummer method, clearly demonstrated a two-dimensional sheet-like structure. As GO has a multiple laminar layer structure the edges of individual sheets are distinguished from one another in the FE-SEM images. The films are stacked one on top of the other, with some wrinkles. As per reported literature [**38**], the thickness of GO sheet was significantly greater than the thickness of the single-layer graphene. The increase in thickness results from the oxygen-containing functional groups that were introduced. It can also be noted that the GO sheets were thicker at the edges [**38**]. Since the oxygen-containing functional groups could primarily be combined at the edges of the GO. It is clear from the FE-SEM images that the GO sheets are firmly suspended in their frames. The FE-SEM image of rGO (**Figure 3**b) obtained from GO demonstrates nanosheets' agglomeration and the irregular arrangement of stacked crumpled sheets. This could be attributed to the nucleophilic reaction of flavonoids and polyphenols, which occurred during the synthesis of rGO and involved the removal of water molecules from carbonyl and hydroxyl groups. The difference in morphological characteristics between GO's leafy, stacked structure and rGO wrinkled structure suggests that the green reduction process has prime importance in GO to rGO conversion.

Figure 3. FE-SEM images of; (a) GO, (b) rGO, (c) rGO/TiO_2, and (d) EDS spectrum of rGO/TiO_2.

The granular shape of TiO_2 nanoparticles displayed in **Figure 3**c is attributed to the aggregation of nanoparticles during the evaporation process due to microwave heating [39]. The elemental compositions of rGO and rGO/TiO_2 were studied using an energy-dispersive spectrum. **Figure 3**d depicts carbon (C) and oxygen (O) contents in rGO and titanium (Ti) in rGO/TiO_2. From the EDS analysis, it is evident that the rGO contains around 84.88 at. % C, 15.12 at. % O whereas rGO/ TiO_2 avails about 65.99 at.% C, 28.95 at.% O and 28.95 at.% Ti, confirming the formation of TiO_2 nanostructures on the rGO surface.

3.3. X-ray Photoelectron Spectroscopy Analysis

Figure 4a–c displays the XPS spectrum of the C1s spectrum of GO, rGO, and rGO/TiO_2. The peak in **Figure 4**a that corresponds to the C1s bands in deconvoluted

GO into C-C, C=O, and C(O)OH bonds, respectively, has binding energies of 284.8, 286.5, and 288.5 eV [**40**]. When compared to GO, the C=O and C(O)OH bands of the rGO C1s spectrum show a marked decrease in intensity, indicating that rGO has undergone deoxygenation. However, as shown in **Figure 4**b, after being reduced, the peak oxygen intensity with functionalities is sharply reduced. These results have demonstrated the decrease in GO. The C-C, C=O, and C(O)OH bonds, respectively, are ascribed to the peaks with binding energies of 284.5, 285.4, and 283.7 eV that correspond to the C1s bands in deconvoluted rGO. The variation in the chemical surrounding of carbon atoms after reduction may cause a shift in binding energy [41]. Additionally, the surface C/O atomic ratios increased from 2.12 to 5.57, validating the successful removal of O_2 groups from the GO surface using aqueous peel extract of citrus limetta. In **Figure 4**c, three peaks found in the C1s deconvoluted spectrum of rGO/TiO_2 are for C-C/C-H sp^3 hybridized carbon bonds, C-O bonds, and the sp^2 carbon bonds of graphene (c), respectively.

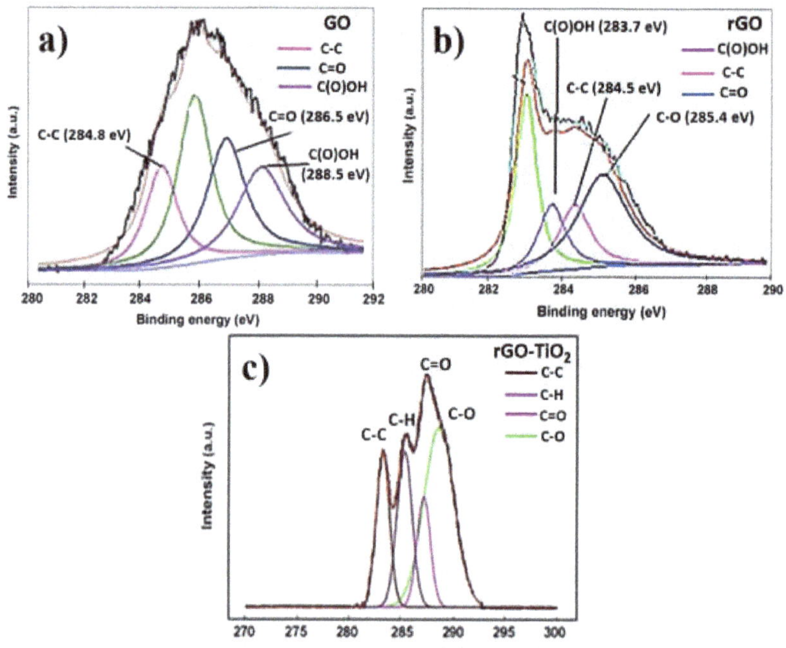

Figure 4. XPS binding energy spectra of C1s spectrum of (a) GO, (b) rGO, and (c) rGO/TiO_2.

3.4. Electrochemical Non-Enzymatic Detection

The electrochemical characteristics of pristine FTO, TiO_2/FTO, rGO/FTO, and rGO/TiO_2/FTO samples were investigated using CV scans in a PBS (7.4 pH) solution containing 2 mM D (+) glucose at a scan rate of 50 mV/s and a working area of 1.0 cm^2 (**Figure 5**a). For FTO, it was found that a faradic peak does not exist. It was noticed that rGO/FTO had a clearly defined peak for glucose oxidation that occurred at -0.25 V. At -0.25 V, the TiO_2/FTO and rGO/TiO_2/FTO oxidation peaks were significantly enhanced. The current rGO/TiO_2/FTO reaction was practically double in comparison to the rGO/FTO response when the voltage was set to -0.25 V.

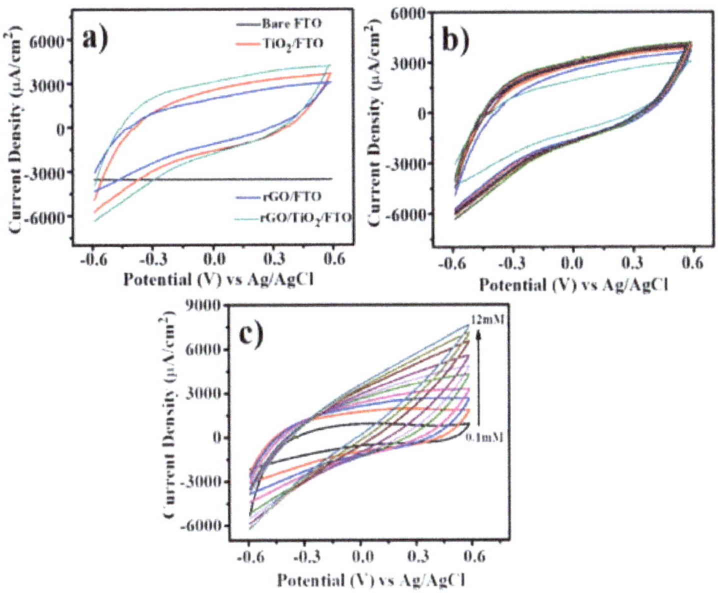

Figure 5. (a) CV response of various electrodes, (b) Effect of variable scan rate on rGO/TiO_2/FTO electrode, and (c) Effect of variable glucose concentration on rGO/TiO_2/FTO electrode.

Consequently, it is confirmed that the rGO/TiO_2/FTO composite is essential for obtaining increased sensitivity and electrocatalytic activity. A further investigation into the electrochemical effect on rGO/TiO_2/FTO with variable scan rate was carried out using CV which is illustrated in **Figure** 5b. The electrode was scanned between 10 and 300 mV/s in an electrolyte containing 2 mM D (+) glucose using a CV between -0.6 V and +0.6 V. Regarding current flow, the oxidation that took place at -0.25 V is the most notable. that the rGO/TiO_2 nanocomposite undergoes an electrochemical reaction, demonstrating the enzyme detection capability. Furthermore, electrochemical measurements were carried out at the optimal potential of -0.25 V vs. Ag/AgCl, which was optimal.

Figure 5c displays the I-V responses of the modified sensor electrode at a scan rate of 50 mv/s for a variable glucose concentration in the range of 0.1 to 12 mM in a PBS electrolyte solution (7.4 pH) (c). The sensor's current density increases linearly as the glucose concentrations increase. The rGO/TiO_2 composite's high electrical conductivity, electron mobility, and electrocatalytic activity are in charge of enhancing the electron transport process between the sensor electrode and the PBS solution during a redox reaction.

3.5. Calibration of the Sensor

The calibration curves for various glucose concentration rage are presented in **Figure 6.** A calibration curve for glucose concentrations in 0.1 to 12 mM is presented in **Figure 6**a. This curve can be divided into two linear glucose concentration ranges: 0.1–1 mM and 2–12 mM, as presented in **Figure 6**b,c. For the calibration curve (0.1-1 mM), the equivalent linear regression equations are Ip = 1488.1c + 930.4 and R^2 = 0.9862 (N = 4). The sensor's sensitivity determined from the slope of the graph is 1425 μA/cm² (S/N < 3). For the calibration curve (2-12 mM) the equations are Ip = 498.5c + 1503.7 and R^2 = 0.9878 (N = 6). The sensor's sensitivity determined from the slope of the graph is 491.75 μA/cm² (S/N < 3).

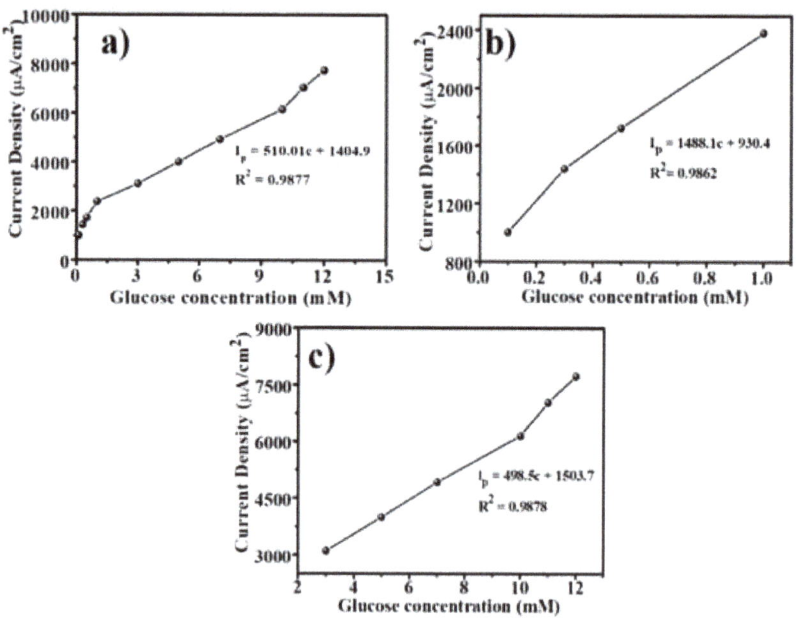

Figure 6. Calibration curve for glucose concentration determination using the Glucose liquid color kit (a) 0.1–12 mM (b) 0.1–1 mM (c) 3–12 mM for different glucose concentration.

3.6. Selectivity and Stability of the Sensor

The proposed biosensor was examined for interfering species in blood serum, including ascorbic acid, dopamine, fructose, and lactose. The selectivity of the rGO/TiO_2/FTO sensor electrode is shown in **Figure 7**a. The sample's stability (aging effect) tested periodically for 30 days with a 1 mM glucose concentration, yielding an RSD of 1.14%, is presented in **Figure** 7b. For ten cycles, the sensor electrode's reproducibility was investigated. To observe the current response of each sensor, 2 mM of glucose concentration was used. The RSD of 1.3% confirms the significant reproducibility.

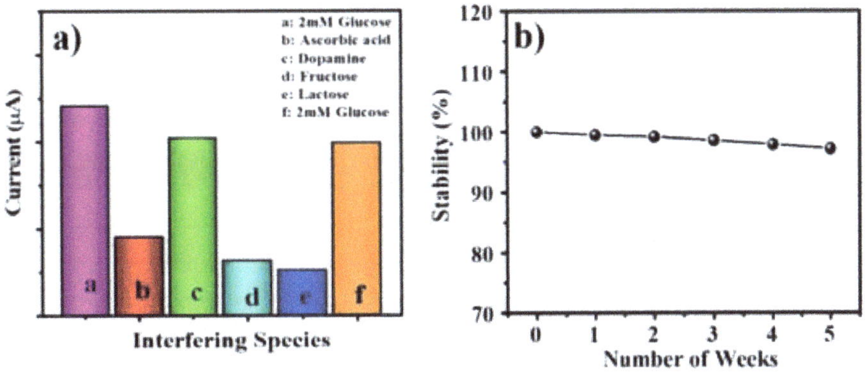

Figure 7. (a) Selectivity and (b) stability of rGO/TiO_2/FTO sensor electrode.

3.7. Repeatability of the Sensor

The reproducibility of the proposed sensors was assessed by doing five measurements of the current response over the course of one week. The performance of the suggested sensor is evaluated by comparing it to previously documented values, and the findings are presented in Table 2. The sensor demonstrates excellent recovery with a Relative Standard Deviation (RSD) of 1.34%.

Table 2. Comparative analysis on sensitivity, detection limit, and linear range of rGO/TiO_2/FTO glucose sensor.

Electrode	Detection Potential (V)	SensitivityµA/ mM1cm2	Linear Range/ mM	Detection Limit / µM	Reference
rGO/TiO_2/FTO	-0.25	1425	0.1–12	0.32	This work
Nafion/GO_x/ rGO-6/ZnONRs/ Au/PET	-0.8 to +0.8	2.26	11.5	37.5	[42]
Ni Ps/ATP/r-GO	0.0 to -1.5	1414.4 µA	1–710 µM	0.37	[43]
$NiCo_2O_4$-rGO	--	548.9	1-25	0.35	[44]
rGO/NiO	-0.6	6.2	1–15	19.35	[45]
GO_x/n-TiO_2/ PANI/GCE	--	6.31	0.02–6	18	[46]
3D Ag-GR-TiO_2	--	12	--	--	[47]
TiO_2-r-GO-PANI	--	--	10–180 µM	7.46	[48]

3.8. Investigation of Glucose Level in Blood Serum

The glucose concentration in blood serum samples was measured using the sensor electrode that was produced. The blood serum samples were obtained from individuals aged 40 to 50 years, and were taken from a certified pathology laboratory. The samples were exposed to analysis after being multiplied by a factor of 100 using the conventional dilution procedure. The objective of the investigation was to assess the findings of the suggested sensor by comparing them with the certified values received from the pathology laboratory. Table 3 displays the empirical data regarding the concentration of glucose in blood serum.

Table 3. Investigation of glucose in blood serum and its recovery.

Serum Sample	Glucose Added (mM)	Glucose Obtained (mM)	Recovery (%)	RSD (%)
1	2.00	1.97	98.5	1.5
2	4.00	3.92	98.8	1.2
3	6.00	5.87	97.5	2.2
4	8.00	7.89	98.6	1.4
5	10.00	9.71	97.1	2.9

The proposed sensor for serum glucose biosensor application demonstrates its reliability with an average recovery rate of 98.2% (RSD 1.88%).

4. CONCISE SUMMARY

The sensor electrode made of rGO/TiO2 demonstrated a remarkable sensitivity of 1425 μA/mM cm2 within the range of 0.1 mM to 12 mM. This sensitivity is considerably higher than what is reported in existing literature. The data demonstrates a rapid response time of 5 seconds, a correlation coefficient (R2) of around 0.98, and a limit of detection (LOD) of 0.32 μM.

Additionally, it exhibits a high level of selectivity towards the interfering species. The sensor has exceptional stability, repeatability, and reproducibility, as evidenced by the RSD values of 1.14%, 1.34%, and 1.3% respectively. This demonstrates the utilization of reduced graphene oxide (rGO)/titanium dioxide (TiO2) electrodes in the context of glucose sensing applications. The sensor exhibits a Relative Standard Deviation (RSD) of 1.88% when quantifying glucose concentrations in unaltered blood serum. The measured results obtained using a commercially available glucose sensor exhibit a high level of agreement. Furthermore, the procedure employed for the synthesis of reduced graphene oxide (rGO) and rGO/TiO2 composite is uncomplicated, swift, economical, and environmentally friendly, hence demonstrating promising prospects for commercialization.

At a Glance

The present study reports the crucial electrochemical measurements of reduced graphene oxide-titanium oxide (rGO/TiO2) electrodes for the purpose of developing a glucose sensor. One of the unique characteristics of this study is the examination of the sensitivity, stability, and reproducibility of sensor electrodes that were created and utilized to assess the concentration of glucose in the serum. This study demonstrates the utilization of waste from citrus limetta (sweet lime) fruit peel to synthesis a sustainable reduction of graphene oxide (rGO). The structural and morphological properties of the rGO/TiO2 composite produced through microwave heating are assessed using several methods. An electrode for an enzymeless glucose sensor is modified using a substrate made of conducting fluorine-tin oxide. Cyclic voltammetry is employed to conduct electrochemical experiments on rGO/TiO2 sensor electrodes. The rGO/TiO_2 sensor electrode exhibits a high sensitivity of 1425 μA/mM cm^2 towards glucose concentration in the range of 0.1 to 12 mM. The sensor was found to be extremely stable and repeatable with a response time of 5 s along with a minimum detection limit of 0.32 μM of glucose. The rGO/TiO_2 sensor shows relative standard deviation (RSD) of 1.14%, 1.34%, and 1.3% which reveals its excellent stability, repeatability, and reproducibility respectively. The sensor was employed to detect glucose levels in natural blood serum and exhibits a Relative Standard Deviation (RSD) of 1.88%. which is consistent with the values obtained from the commercial glucose sensor.

REFERENCES

1. Zhu, H.; Li, L.; Zhou, W.; Shao, Z.; Chen, X. Advances in non-enzymatic glucose sensors based on metal oxides. *J. Mater. Chem. B* **2016**, *4*, 7333–7349. **[Google Scholar] [CrossRef]**
2. Bruen, D.; Delaney, C.; Florea, L.; Diamond, D. Glucose Sensing for Diabetes Monitoring: Recent Developments. *Sensors* **2017**, *17*, 1866. **[Google Scholar] [CrossRef] [Green Version]**
3. Sai-Anand, G.; Gopalan, A.-I.; Kang, S.-W.; Komathi, S.; Lee, K.-P. One Pot Synthesis of New Gold Nanoparticles Dispersed Poly(2-aminophenyl boronic acid) Composites for Fabricating an Affinity Based Electrochemical Detection of Glucose. *Sci. Adv. Mater.* **2014**, 6, 1356–1364. **[Google Scholar] [CrossRef]**
4. Gopalan, A.; Muthuchamy, N.; Lee, K. A novel bismuth oxychloride-graphene hybrid nanosheets based non-enzymatic photoelectrochemical glucose sensing platform for high performances. *Biosens. Bioelectron.* **2017**, *89*, 352–360. **[Google Scholar] [CrossRef] [PubMed]**
5. Kumar, G.G.; Amala, G.; Gowtham, S.M. Recent advancements, key challenges and solutions in non-enzymatic electrochemical glucose sensors based on graphene platforms. *RSC Adv.* **2017**, 7, 36949–36976. **[Google Scholar] [CrossRef] [Green Version]**
6. Gopalan, A.; Muthuchamy, N.; Komathi, S.; Lee, K.-P. A novel multicomponent redox polymer nanobead based high performance non-enzymatic glucose sensor. *Biosens. Bioelectron.* **2016**, *84*, 53–63. **[Google Scholar] [CrossRef] [PubMed]**
7. Dong, Q.; Ryu, H.; Lei, Y. Metal oxide based non-enzymatic electrochemical sensors for glucose detection. *Electrochim. Acta* **2021**, *370*, 137744. **[Google Scholar] [CrossRef]**

8. Manthey, J.A.; Grohmann, K. Phenols in Citrus Peel Byproducts. Concentrations of Hydroxycinnamates and Polymethoxylated Flavones in Citrus Peel Molasses. *J. Agric. Food Chem.* **2001**, *49*, 3268–3273. **[Google Scholar] [CrossRef]**

9. Ghule, B.G.; Shinde, N.M.; Raut, S.D.; Gore, S.K.; Shaikh, S.F.; Ekar, S.U.; Ubaidullah, M.; Pak, J.J.; Mane, R.S. Self-assembled α-Fe_2O_3-GO nanocomposites: Studies on physical, magnetic and ammonia sensing properties. *Mater. Chem. Phys.* **2022**, *278*, 125617. **[Google Scholar] [CrossRef]**

10. Balandin, A.A. Thermal properties of graphene and nanostructured carbon materials. *Nat. Mater.* **2011**, *10*, 569–581. **[Google Scholar] [CrossRef]**

11. Wang, Z.; Chen, Y.; Li, P.; Zhou, J.; He, J.; Zhang, W.; Guo, Z.; Li, Y.; Dong, M. Modulation of N-bonding configurations and their influence on the electrical properties of nitrogen-doped graphene. *RSC Adv.* **2016**, *6*, 92682–92687. **[Google Scholar] [CrossRef]**

12. Shin, H.-J.; Kim, K.K.; Benayad, A.; Yoon, S.-M.; Park, H.K.; Jung, I.-S.; Jin, M.H.; Jeong, H.-K.; Kim, J.M.; Choi, J.-Y.; et al. Efficient Reduction of Graphite Oxide by Sodium Borohydride and Its Effect on Electrical Conductance. *Adv. Funct. Mater.* **2009**, *19*, 1987–1992. **[Google Scholar] [CrossRef]**

13. Ren, P.-G.; Yan, D.-X.; Ji, X.; Chen, T.; Li, Z.-M. Temperature dependence of graphene oxide reduced by hydrazine hydrate. *Nanotechnology* **2010**, *22*, 055705. **[Google Scholar] [CrossRef]**

14. Wang, Y.; Zhang, P.; Liu, C.F.; Zhan, L.; Li, Y.F.; Huang, C.Z. Green and easy synthesis of biocompatible graphene for use as an anticoagulant. *RSC Adv.* **2012**, *2*, 2322–2328. **[Google Scholar] [CrossRef]**

15. Mhamane, D.; Ramadan, W.; Fawzy, M.; Rana, A.; Dubey, M.; Rode, C.; Lefez, B.; Hannoyer, B.; Ogale, S. From graphite oxide to highly water dispersible functionalized graphene by single step plant extract-induced deoxygenation. *Green Chem.* **2011**, *13*, 1990–1996. **[Google Scholar] [CrossRef]**

16. Salas, E.C.; Sun, Z.; Lüttge, A.; Tour, J.M. Reduction of Graphene Oxide *via* Bacterial Respiration. *ACS Nano* **2010**, *4*, 4852–4856. **[Google Scholar] [CrossRef] [PubMed]**

17. Nasrollahzadeh, M.; Maham, M.; Rostami-Vartooni, A.; Bagherzadeh, M.; Sajadi, S.M. Barberry fruit extract assisted in situ green synthesis of Cu nanoparticles supported on a reduced graphene oxide–Fe_3O_4 nanocomposite as a magnetically separable and reusable catalyst for the O-arylation of phenols with aryl halides under ligand-free conditions. *RSC Adv.* **2015**, *5*, 64769–64780. **[Google Scholar] [CrossRef]**

18. Arias, B.; Ramón-Laca, L. Pharmacological properties of citrus and their ancient and medieval uses in the Mediterranean region. *J. Ethnopharmacol.* **2005**, *97*, 89–95. **[Google Scholar] [CrossRef] [PubMed]**

19. Manthey, J.A.; Grohmann, K. Concentrations of Hesperidin and Other Orange Peel Flavonoids in Citrus Processing Byproducts. *J. Agric. Food Chem.* **1996**, *44*, 811–814. **[Google Scholar] [CrossRef]**

20. Majhi, S.; Mirzaei, A.; Kim, H.; Kim, S. Reduced Graphene Oxide (rGO)-Loaded Metal-Oxide Nanofiber Gas Sensors: An Overview. *Sensors* **2021**, *21*, 1352. **[Google Scholar] [CrossRef]**

21. Pradhan, S.; Konwar, K.; Ghosh, T.; Mondal, B.; Sarkar, S.; Deb, P. Multifunctional Iron oxide embedded reduced graphene oxide as a versatile adsorbent candidate for effectual arsenic and dye removal. *Colloid Interface Sci. Commun.* **2020**, *39*, 100319. **[Google Scholar] [CrossRef]**

22. Molak, A.; Paluch, M.; Pawlus, S.; Klimontko, J.; Ujma, Z.; Gruszka, I. Electric modulus approach to the analysis of electric relaxation in highly conducting $(Na_{0.75}Bi_{0.25})(Mn_{0.25}Nb_{0.75})O_3$ ceramics. *J. Phys. D Appl. Phys.* **2005**, *38*, 1450–1460. **[Google Scholar] [CrossRef]**

23. Pradhan, S.S.; Ghosh, T.N.; Marik, A.; Raul, K.K.; Sarkar, S.K. Magnetodielectric effects in three reduced graphene oxide–polymer nanocomposites. *Bull. Mater. Sci.* **2020**, *43*, 208. **[Google Scholar] [CrossRef]**

24. Ghule, B.G.; Shinde, N.M.; Raut, S.D.; Shaikh, S.F.; Al-Enizi, A.M.; Kim, K.H.; Mane, R.S. Porous metal-graphene oxide nanocomposite sensors with high ammonia detectability. *J. Colloid Interface Sci.* **2021**, *589*, 401–410. **[Google Scholar] [CrossRef] [PubMed]**

25. Liu, S.; Sun, H.; Liu, S.; Wang, S. Graphene facilitated visible light photodegradation of methylene blue over titanium dioxide photocatalysts. *Chem. Eng. J.* **2013**, *214*, 298–303. **[Google Scholar] [CrossRef]**

26. Luo, L.; Zhu, L.; Wang, Z. Nonenzymatic amperometric determination of glucose by CuO nanocubes–graphene nanocomposite modified electrode. *Bioelectrochemistry* **2012**, *88*, 156–163. **[Google Scholar] [CrossRef] [PubMed]**

27. Yang, Z.; Hao, X.; Chen, S.; Ma, Z.; Wang, W.; Wang, C.; Yue, L.; Sun, H.; Shao, Q.; Murugadoss, V.; et al. Long-term antibacterial stable reduced graphene oxide nanocomposites loaded with cuprous oxide nanoparticles. *J. Colloid Interface Sci.* **2019**, *533*, 13–23. **[Google Scholar] [CrossRef] [PubMed]**

28. AAT Bioquest, Inc. Quest Calculate™ PBS (Phosphate Buffered Saline) (1X, pH 7.4) Preparation and Recipe. AAT Bioquest. Available online: **https://www.aatbio.com/resources/buffer-preparations-and-recipes/pbs-phosphate-buffered-saline** (accessed on 7 December 2022).

29. Komathi, S.; Muthuchamy, N.; Lee, K.-P.; Gopalan, A.-I. Fabrication of a novel dual mode cholesterol biosensor using titanium dioxide nanowire bridged 3D graphene nanostacks. *Biosens. Bioelectron.* **2016**, *84*, 64–71. **[Google Scholar] [CrossRef]**

30. Zaaba, N.I.; Foo, K.L.; Hashim, U.; Tan, S.J.; Liu, W.W.; Voon, C.H. Synthesis of Graphene Oxide using Modified Hummers Method: Solvent Influence. *Procedia Eng.* **2017**, *184*, 469–477. **[Google Scholar] [CrossRef]**

31. Zhang, L.; Hai, X.; Xia, C.; Chen, X.-W.; Wang, J.-H. Growth of CuO nanoneedles on graphene quantum dots as peroxidase mimics for sensitive colorimetric detection of hydrogen peroxide and glucose. *Sens. Actuators B Chem.* **2017**, *248*, 374–384. **[Google Scholar] [CrossRef]**

32. Shaikh, S.F.; Mane, R.S.; Min, B.K.; Hwang, Y.J.; Joo, O.-S. D-sorbitol-induced phase control of TiO_2 nanoparticles and its application for dye-sensitized solar cells. *Sci. Rep.* **2016**, *6*, 20103. **[Google Scholar] [CrossRef] [Green Version]**

33. Nipane, S.V.; Lee, S.-W.; Gokavi, G.S.; Kadam, A.N. In situ one pot synthesis of nanoscale TiO_2-anchored reduced graphene oxide (RGO) for improved photodegradation of 5-fluorouracil drug. *J. Mater. Sci. Mater. Electron.* **2018**, *29*, 16553–16564. **[Google Scholar] [CrossRef]**

34. Atchudan, R.; Edison, T.N.J.I.; Perumal, S.; Karthikeyan, D.; Lee, Y.R. Effective photocatalytic degradation of anthropogenic dyes using graphene oxide grafting titanium dioxide nanoparticles under UV-light irradiation. *J. Photochem. Photobiol. A Chem.* **2017**, *333*, 92–104. **[Google Scholar] [CrossRef]**

35. Ghosh, T.N.; Pradhan, S.S.; Sarkar, S.K.; Bhunia, A.K. On the incorporation of the various reduced graphene oxide into poly(vinyl alcohol) nano-compositions: Comparative study of the optical, structural properties and magnetodielectric effect. *J. Mater. Sci. Mater. Electron.* **2021**, *32*, 19157–19178. **[Google Scholar] [CrossRef]**

36. Kołodziej, A.; Długoń, E.; Świętek, M.; Ziąbka, M.; Dawiec, E.; Gubernat, M.; Michalec, M.; Wesełucha-Birczyńska, A. A Raman Spectroscopic Analysis of Polymer Membranes with Graphene Oxide and Reduced Graphene Oxide. *J. Compos. Sci.* **2021**, *5*, 20. **[Google Scholar] [CrossRef]**

37. Zhu, J.; Zeng, G.; Nie, F.; Xu, X.; Chen, S.; Han, Q.; Wang, X. Decorating graphene oxide with CuO nanoparticles in a water–isopropanol system. *Nanoscale* **2010**, *2*, 988–994. **[Google Scholar] [CrossRef]**

38. Song, J.; Wang, X.; Chang, C.-T. Preparation and Characterization of Graphene Oxide. *J. Nanomater.* **2014**, *2014*, 276143. **[Google Scholar] [CrossRef]**

39. Qianqian, Z.; Tang, B.; Guoxin, H. High photoactive and visible-light responsive graphene/titanate nanotubes photocatalysts: Preparation and characterization. *J. Hazard. Mater.* **2011**, *198*, 78–86. **[Google Scholar] [CrossRef]**

40. Stankovich, S.; Dikin, D.A.; Piner, R.D.; Kohlhaas, K.A.; Kleinhammes, A.; Jia, Y.; Wu, Y.; Nguyen, S.T.; Ruoff, R.S. Synthesis of graphene-based nanosheets via chemical reduction of exfoliated graphite oxide. *Carbon* **2007**, *45*, 1558–1565. **[Google Scholar] [CrossRef]**

41. Jana, M.; Saha, S.; Khanra, P.; Murmu, N.C.; Srivastava, S.K.; Kuila, T.; Lee, J.H. Bio-reduction of graphene oxide using drained water from soaked mung beans (*Phaseolus aureus* L.) and its application as energy storage electrode material. *Mater. Sci. Eng. B* **2014**, *186*, 33–40. **[Google Scholar] [CrossRef]**

42. Mao, Q.; Jing, W.; Zhou, F.; Liu, S.; Gao, W.; Wei, Z.; Jiang, Z. Depositing reduced graphene oxide on ZnO nanorods to improve the performance of enzymatic glucose sensors. *Mater. Sci. Semicond. Process.* **2021**, *121*, 105391. **[Google Scholar] [CrossRef]**

43. Shen, Z.; Gao, W.; Li, P.; Wang, X.; Zheng, Q.; Wu, H.; Ma, Y.; Guan, W.; Wu, S.; Yu, Y.; et al. Highly sensitive nonenzymatic glucose sensor based on nickel nanoparticle–attapulgite-reduced graphene oxide-modified glassy carbon electrode. *Talanta* **2016**, *159*, 194–199. **[Google Scholar] [CrossRef] [PubMed]**

44. Ni, Y.; Xu, J.; Liu, H.; Shao, S. Fabrication of RGO-$NiCo_2O_4$ nanorods composite from deep eutectic solvents for nonenzymatic amperometric sensing of glucose. *Talanta* **2018**, *185*, 335–343. **[Google Scholar] [CrossRef] [PubMed]**

45. Alghazzawi, W.; Danish, E.; Alnahdi, H.; Salam, M.A. Rapid microwave-assisted hydrothermal green synthesis of rGO/NiO nanocomposite for glucose detection in diabetes. *Synth. Met.* **2020**, *267*, 116401. **[Google Scholar] [CrossRef]**

46. Tang, W.; Li, L.; Zeng, X. A glucose biosensor based on the synergistic action of nanometer-sized TiO_2 and polyaniline. *Talanta* **2015**, *131*, 417–423. **[Google Scholar] [CrossRef]**

47. Jang, H.D.; Kim, S.K.; Chang, H.; Jo, E.H.; Roh, K.M.; Choi, J.-H.; Choi, J.-W. Synthesis of 3D Silver-Graphene-Titanium Dioxide Composite via Aerosol Spray Pyrolysis for Sensitive Glucose Biosensor. *Aerosol Sci. Technol.* **2015**, *49*, 538–546. **[Google Scholar] [CrossRef] [Green Version]**

48. Du, J.; Tao, Y.; Xiong, Z.; Yu, X.; Xie, A.; Luo, S.; Li, X.; Yao, C. Titanium Dioxide–Graphene–Polyaniline Hybrid for Nonenzymatic Detection of Glucose. *Nano* **2019**, *14*, 1950093. **[Google Scholar] [CrossRef]**

5

Chapter

PRODUCTION AND ANALYSIS OF SI/PEDOT: PSS-BASED HETEROJUNCTION SOLAR CELLS

1. INTRODUCTION: AN OVERVIEW

Solar energy is the one viable renewable energy option that can rival conventional low-cost thermal power sources. Silicon currently holds a significant majority in the solar energy business, representing about 90% of the market share [1]. Nevertheless, the exorbitant expenses, energy-intensive manufacturing methods, and limited return on investment continue to hinder their progress compared to alternative energy sources. A study has been conducted to decrease the expenses and amount of materials used, while also enhancing its effectiveness. The low efficiency of silicon solar cells can be attributed to its high reflectance throughout the surface, losses in the sub-bandgap spectrum, and a high recombination rate [2]. Two distinct variations of Si, namely pure Si and amorphous Si, were utilized in the construction of the cells. Nevertheless, the utilization of photovoltaic cells has been restricted due to their exorbitant manufacturing expenses. Thin silicon films have been produced using high-purity single-crystal materials and the absence of efficient mass-production methods. Despite being the most efficient and modern type of solar cells, single-crystalline silicon cells are challenging to adopt due to their high cost [3]. Therefore, there are substitutes for silicon in the shape of thin-film substances as cadmium telluride [4], copper indium: diselenide (CIS), dye-sensitized solar cells (DSSC) [5,6], perovskite solar cells [7], and polymer solar cells. Nevertheless, these technologies are in their early stages of development and necessitate additional research. The use of nanotechnology has been the primary focus of research in the advancement of silicon processing, particularly in the alteration of traditional silicon solar cells. Organic heterojunction solar cells offer a viable solution for achieving both high efficiency and cheap cost in comparison to silicon solar cells. The generation of energy-intensive emitters can be minimized in hybrid solar cells (HSC) that are based on nanostructures. The emitter junctions can undergo solution processing, hence eliminating the need for extra antireflection layers. Initially, the production of planar bulk silicon organic solar cells was carried out [8,9,10,11]. Silicon organic heterojunction (SOH) solar cells employ several organic materials, including poly-(3-hexythiophene) (P3HT) [12], Spiro-OmeTAD [13], poly-(Si-cyclooctatetrocene) [14], polyaniline [15], 4-Tricyanovinyl-N, N-diethylaniline [16], and PEDOT: PSS [17,18,19].

Lin et al. first demonstrated the possibility of using PEDOT:PSS as a hole-transport layer in a Si/organic hybrid solar cell and achieved a PCE of 0.008% in a planar configuration [20]. PEDOT: PSS was considered as the hole transport layer owing to its superior properties such as high transmittance (>90%), high conductivity, good absorption in the visible spectrum, and a favorable bandgap of ~1.6 eV [**21**]. The HOMO (highest occupied molecular orbital) of PEDOT: PSS and valence band (VB) of Silicon is nearly similar to ~5.1 eV [20]. He et al. investigated the effect of Si surface termination conditions on the device performance. Their work proved that oxide-terminated (SiOx-Si) cells exhibit superior performance compared to the hydrogen-terminated (H-Si) cells. Hydrogen-terminated cells exhibited 0.002% while oxide-terminated cells showed 10.6% [22] Zhao et al. improved the efficiency of planar Si/PEDOT: PSS cells by 12.70% by growing a SiO_2 layer at the interface of Si and PEDOT: PSS, resulting in a hydrophilic surface allowing PEDOT: PSS to adhere easily on Si [9]. Avasthi et al. demonstrated an 11.70% efficient planar Si/ PEDOT: PSS HSC by spin-coating a PEDOT: PSS layer at the temperatures below 100 °C [11]. Leung et al. demonstrated the effect of a solvent on the conductivity of the PEDOT: PSS layer. Ethylene glycol (EG) proved to be more efficient in improving the conductivity [18].

Planar-Si cells with fluorosurfactant (FS) as the surfactant and EG as the co-solvent exhibited a record with the efficiency of 13.3%. The co-solvent-optimized PEDOT:PSS also acts as an antireflection coating [18]. Despite the simple fabrication, planar Si/PEDOT: PSS cells have a high surface recombination rate because the separation of the electron–hole pair occurs only at the junction, whereas electron–hole pairs in PEDOT: PSS and bulk silicon are recombined. Moreover, they have a very small PN-junction surface area. To reduce the material cost, Wang et al. fabricated a planar Si/PEDOT: PSS Solar cell on thin-film with Si deposited using a thermal chemical vapor deposition reactor (CVD) and achieved an efficiency of 8.2% [**10**]. Recently, nanostructures have been incorporated into Si wafers to enhance light trapping and reduce reflection. Theoretically, SiNW/PEDOT: PSS solar cells can surpass the efficiencies of commercial crystalline silicon solar cells when a radial p–n junction is associated between PEDOT: PSS and Si Nanowires [23].

Therefore, the present work aims to fabricate a hybrid solar cell using Si/ PEDOT: PSS, and then analyze the various solvents for dissolving PEDOT: PSS to choose the best one. We optimized the formation of Si/PEDOT: PSS and analyzed the influence of the spin-coating speed and annealing temperature of PEDOT: PSS to achieve a radial p–n junction between Si and PEDOT: PSS to attain the enhanced efficiency.

2. MATERIALS AND METHODS

2.1. Materials

3,4 Poly (3,4-ethylene dioxythiophene) poly (styrene sulfonate) (PEDOT: PSS) was purchased from Sigma Aldrich; acetone (AR grade) was purchased from MERCK; and hydrofluoric acid (51%), dimethyl sulphoxide (DMSO), ethylene glycol, acetonitrile, Triton X-100 was purchased from CDH. All the chemicals were purchased in India.

2.2. Sample Preparation

In this study, an n-type electronic-grade silicon wafer (100 μm) with a thickness of 525 μm was used. Before the etching process, the silicon wafers were made to cut to the desired size of 2 × 2 cm and cleaned using DI water under ultrasonication for 10 min. Furthermore, the oxide layer of the silicon wafers were extracted using H_2O: HF in a volume ratio of 10:1 for 3 min at room temperature. Three different polymeric solutions were prepared using various solvents: (i) PEDOT: PSS solution (1.3% in H_2O) was prepared using 7% ethylene glycol (EG) and 0.25% Triton-X. (ii) A PEDOT: PSS solution (1.3% in H_2O) was prepared using 5% acetonitrile and 0.25% Triton-X. (iii) PEDOT: PSS solution (1.3% in H_2O) was prepared using 5% DMSO and 0.25% Triton-X. The cleaned and oxide-removed wafers were dipped into the acetone solution for 5 min to induce adhesiveness. Subsequently, the wafers were dried and spin-coated with 50 μL PEDOT: PSS at 3000 rpm for 60 s. The coated wafers were then annealed at 150 °C for 30 min. To improve the adhesive nature of the silicon wafer, a carbon coating was applied to the wafer. The black camphor smoke was subjected to the falling onto silicon wafers. The silicon wafers were maintained far above the camphor smoke to maintain the thickness of the carbon layer. The contacts from the silicon wafers were made with silver paste using the screen-printing method.

2.3. Instrumentation

The prepared samples were characterized using scanning electron microscopy (SEM) using a VEGA TESCAN 3 model with a tungsten filament. The diffuse and specular reflectance properties of the silicon nanostructures were measured in the range from 200 nm to 800 nm using an integrating sphere with 8° incidence angle (JASCO V-650 UV–VIS spectrophotometer) to determine the reflection properties. X-ray diffraction was performed using a RIGAKU SmartLab X-ray diffractometer with a CuKα radiation source (λ = 1.5406 Å) operated at 30 kV. All the samples were scanned from 20° to 80° at a scan rate of 0.02°/s at the 2θ position at room temperature. The current-voltage characterization was performed using an AAA-grade PET solar simulator with a Keithley 4200 source meter with an air mass of 1.5 (AM 1.5, 100 mW/cm^2).

3. RESULTS AND DISCUSSION

3.1. Optical Properties

The optical properties of the Si-coated PEDOT: PSS were studied using a JASCO (V 650) UV–VIS Spectrometer by DRS method to obtain the surface reflectance and absorbance. **Figure 1**a, b shows the reflectance and absorption spectra of PEDOT: PSS-coated N-type silicon wafers using different solvents. **Figure 2**a, b shows the reflectance and absorption spectra of carbon-coated Si-PEDOT: PSS using different solvents (DMSO, EG, acetonitrile) and compared with a standard N-type silicon wafer as a reference. The obtained surface absorption and reflectance (300–800 nm) of Si-PEDOT: PSS fabricated using acetonitrile, DMSO, and ethylene glycol are listed in **Table 1**. The results show that the surface reflectance was reduced significantly from a bare silicon wafer to a PEDOT: PSS-coated silicon wafer [18]. The absorbance spectrum shows the Si-coated PEDOT: PSS has significantly more than a 50% increase in the UV region and for the EG sample there is a 20% increase in the entire visible spectrum [24]. Similarly, the surface reflectance of the carbon-coated Si-PEDOT: PSS also decreased from that of the bare silicon wafer, as shown in **Table 1**.

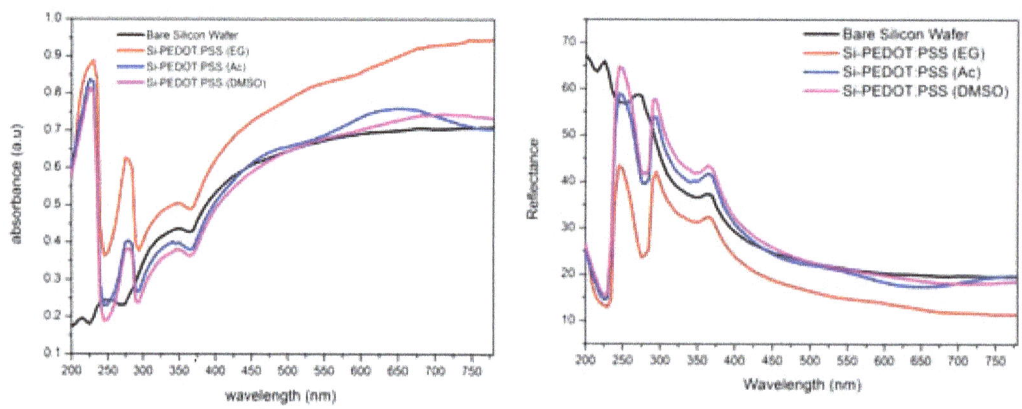

(a) (b)

Figure 1. (a) Absorption and (b) reflectance spectra N-type Si wafer with PEDOT: PSS prepared in different solvents (EG) ethylene glycol, (AC) acetonitrile, and (DMSO) dimethyl sulphoxide with bare silicon as reference.

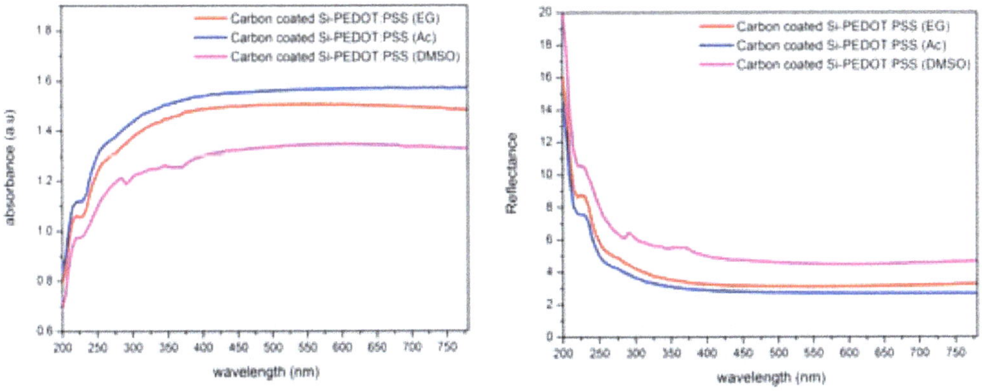

(a) (b)

Figure 2. (a) Absorption and (b) reflectance spectra carbon coated N-type Si wafer with PEDOT: PSS prepared in different solvents (EG) ethylene glycol, (AC) acetonitrile, and (DMSO) dimethyl sulphoxide.

Table 1. Optical properties of Si-PEDOT: PSS and carbon coated Si-PEDOT: PSS.

Samples	Optical Mode	Bare Silicon	EG	Acetonitrile	DMSO
Si/PEDOT: PSS	Reflectance	67	43	58	62
	Absorbance	0.24	0.87	0.84	0.81
Carbon coated—Si/PEDOT: PSS	Reflectance	67	8.7	7.5	10.2
	Absorbance	0.24	1.06	1.11	0.99

3.2. Surface Morphology Analysis

The morphologies of the bare and carbon-coated Si/PEDOT: PSS wafers were analyzed using a scanning electron microscope. The morphology of the PEDOT: PSS deposited on the Si wafer surface is crucial for device performance. The type of PN junction formed between Si and PEDOT: PSS and the lateral or radial PN junctions was investigated by SEM. **Figure 3** displays the top view (a–c) and cross-sectional view (d) of PEDOT: PSS-coated Si. The SEM image reveals that PEDOT: PSS on the surface of Si results in the formation of a lateral PN junction with a low PN junction surface area. This improper coverage (agglomeration and island formation) of PEDOT: PSS might be due to the smaller spacing of the highly dense Si, surface tension, and hydrophobicity of the wafer. PEDOT: PSS was deposited and agglomerated on the surface of silicon [8]. **Figure 3**d shows a cross-sectional view of the PEDOT: PSS-coated silicon wafer.

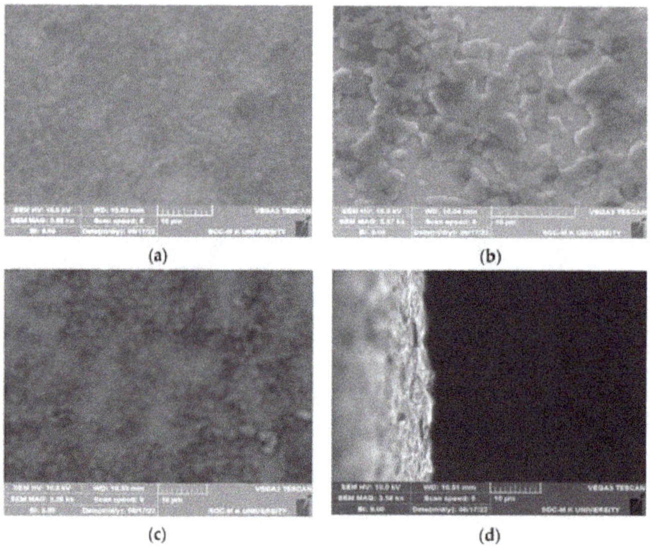

Figure 3. SEM images of Si/PEDOT: PSS fabricated using (a) ethylene glycol, (b) acetonitrile, (c) DMSO, respectively; (d) a cross section image of Si/PEDOT: PSS.

The surface morphologies of the carbon-coated samples were investigated using scanning electron microscopy (SEM). The morphology and thickness of the carbon layer play major roles in the morphology of the resulting surface. Therefore, we attempted to engineer a carbon layer using camphor smoke. **Figure 4** shows the SEM image and cross-section image of the carbon-containing n-type silicon wafer coated with PEDOT: PSS using (a, d) ethylene glycol; (b, e) acetonitrile; and (c, f) DMSO solvent. The cross-section image shows a 1.5 to 2 μm range.

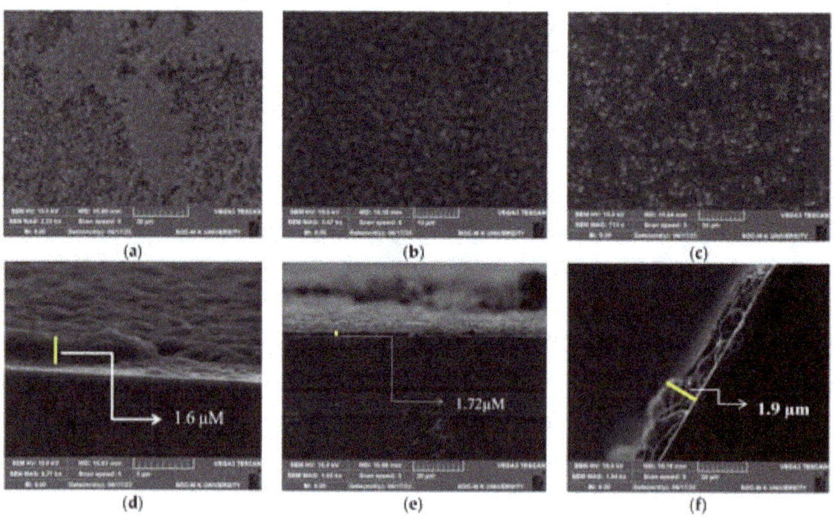

Figure 4. SEM images and cross-section image of the carbon-containing n-type silicon wafer coated with PEDOT: PSS using (a,d) ethylene glycol; (b,e) acetonitrile; and (c,f) DMSO solvent.

3.3. Structural Properties

The structural characterization of Si-PEDOT: PSS was performed by X-ray diffraction (XRD), which gives an idea of the crystallite size, structure, and chemical composition, as shown in **Figure** 5a,b. All three PEDOT: PSS-coated Si wafers were subjected to XRD analysis. The characteristic peak of Si was obtained at 69.5°, and the sharp peak indicated the crystalline nature of Si. Because the Si peaks of the above four data points appear to be the same, the peaks are not identical. There were slight deviations in the Si peak (0.1°) due to the solvent effect. Owing to the presence of different solvents, the forbidden Si (200) peaks were not identical to those of the bare silicon wafer [25]. The peak appearing at 25°-30° corresponds to the SiO_2 layer, indicating the presence of SiO_2 in the wafers. The peak obtained at 32° corresponds to the forbidden Si interface peak obtained owing to the presence of PEDOT: PSS, as shown in **Figure** 5a. **Table 2** presents the d-spacing and the two θ values for the different Si wafers.

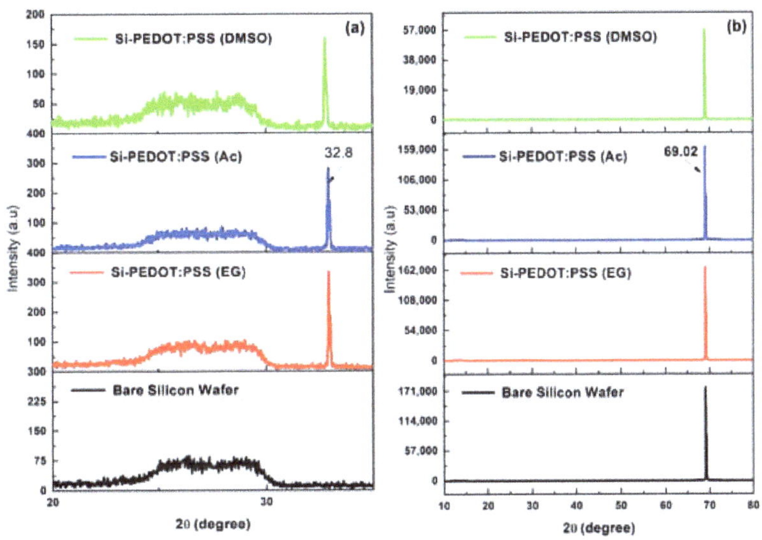

Figure 5. Powder X-ray diffraction of (a) SiO_2 and Si/PEDOT: PSS; (b) Si peak for all the samples.

Table 2. d-spacing and 2θ value calculated from XRD data.

Wafer	d-Spacing (Å)	2θ (degree)
Si-PEDOT: PSS (EG)	1.35828	69.09817
Si-PEDOT: PSS (Ac)	1.35828	69.08817
Si-PEDOT: PSS (DMSO)	1.36027	68.98316

3.4. I–V Analysis

Diode characteristics were measured using a solar simulator with a light equivalent to an air mass of 1.5 (AM 1.5, 100 mW/cm^2). The Voc, Isc, and diode characteristics were calculated using the projected area of the solar cell. **Figure 6**a shows the characteristic I–V curve for the diode formation, although the current obtained (EG) is very low of I_{sc} = 0.2 μA and the V_{oc} is 0.2 V. The characteristic I–V curve shows the formation of a heterojunction between PEDOT: PSS and silicon wafer. **Figure 6**b shows a characteristic I–V curve, which indicates the formation of a heterojunction between PEDOT: PSS and the carbon-coated silicon wafer [**26**]. A comparison of the performance of this study with those reported in the literature is presented in **Table 3**.

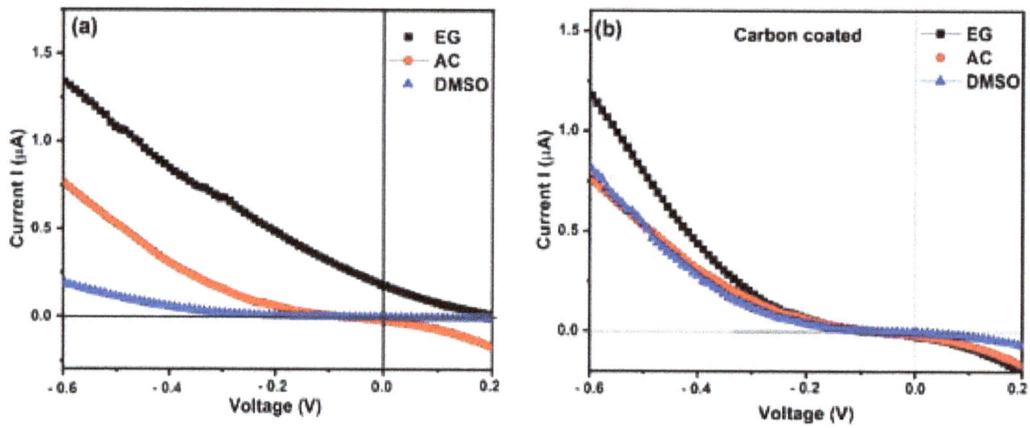

Figure 6. Current-voltage measurement of (a) Si-PEDOT: PSS wafers and (b) carbon coated Si-PEDOT: PSS wafers.

Table 3. Comparison of the performance of this work with others reported in the literature.

S. No.	Findings	PCE	Reference
1	Proper surface termination and light trapping in Si/organic hybrid cells can potentially deliver very high PCE.	10.6%	[8]
2	Effects of H-Si and SiOx-Si surface termination lead to PCE change from 0.04% to 11.3%.	11.3%	[27]
3	Hole-conducting polymer PEDOT: PSS provides a high level of surface passivation on c-Si wafers leading to the efficiencies up to 12.3%.	12.3%	[28]
4	Superior photovoltaic properties of solar cells could be realized the engineering electrode/semiconductor interface.	12.2%	[29]
5	By adding 4% PFI into PEDOT: PSS, the device achieved an FF of 0.70, which improved by 20% compared to the non-treated one.	9.90%	[30]
6	Planar Silicon with a hydrogen-terminated Si (H-Si) surface exhibits high power conversion efficiency (PCE) of 8.27~12.70%.	12.70%	[9]

7	Cells with longer Si NWs lead to poor performance because of lower shunt resistance and higher recombination rate.	12.70%	[22]
8	The minimal-defect interface between PEDOT: PSS and the planar Si substrate Planar-Si HSCs delivered 13.3% PCE.	13.3%	[18]
9	Study on an effective method to improve the conductivity of PEDOT: PSS films with n-Si using EG and Triton.	14.5%	[24]
10	The mixed PEDOT: PSS solution of DMSO and FS31 achieves higher conductivity and a smaller contact angle.	11.07%	[31]
11	Exposure-oxidation treatment of an H-terminated Si substrate enhances the performance of planar hybrid Si/ PEDOT: PSS solar cells.	13.31%	[32]
12	10% concentration of EG in PEDOT: PSS polymer device efficiency is the maximum.	4.69%	[26]
13	The cell performance of the siloxane layer at both interfaces shows the best result of >17% efficiency.	17.34%	[33]
14	Significant improvement in the performance due to the presence of nanostructures leads to the enhancement in light trapping and photo-carrier collection.	8.15%	[34]
15	An attempt to find the best solvent for making PEDOT: PSS hybrid solar cell with silicon. Ethylene glycol shows better result.	I_{sc} = 0.2 μA V_{oc} = 0.2 V	Current work

4. CONCISE SUMMARY

To summarize, we produced heterojunction solar cells by employing three solvents: ethylene glycol, acetonitrile, and dimethyl sulfoxide. The hybrid solar cell was analyzed using UV-DRS (Ultraviolet-Visible Diffuse Reflectance Spectroscopy), SEM (Scanning Electron Microscopy), XRD (X-ray Diffraction), and IV (Current-Voltage) measurements. The analysis revealed that wafers containing ethylene glycol exhibit reduced reflectance and increased absorbance compared to wafers containing other solvents. Increased light absorption results in a wider range of photovoltaic devices, leading to improved cell efficiency. All the manufactured cells have exhibited the characteristic diode curve in current-voltage measurements. The presence of a heterojunction is evidenced by this distinctive diode curve. The cells containing EG exhibited current output on their I–V properties. Improved characterisation results for the manufactured material employed in this study may be achieved by further optimizing the concentration, annealing temperature, and spin-coating RPM.

At a Glance

For this investigation, we created a flat Si/PEDOT: PSS heterojunction solar cell with three distinct solvents-ethylene glycol, acetonitrile, and dimethyl sulfoxide-with the aim of determining the most optimal solvent. The produced samples underwent characterization using diffuse reflectance spectroscopy, scanning electron microscopy, X-ray diffraction, and current-voltage measurements. Analysis of the diffused reflectance spectrum revealed a decrease in reflectance when compared to the silicon wafers without any coating. The absorbance spectrum demonstrates that the Si-coated PEDOT: PSS had a significant increase of over 50% in absorption in the UV region. Similarly, the EG sample exhibited a 20% increase in absorption across the whole visible spectrum. This suggests that the solvent significantly influences the difference in energy levels between Si and Si/PEDOT: PSS. The surface morphology of Si/PEDOT: PSS was analyzed using a scanning electron microscope (SEM). The results revealed that the surfaces exhibited agglomerated and island-formed structures, as well as a carbon-layered Si-PEDOT: PSS surface. The cross-sectional photographs display the magnitude of the PEDOT: PSS layer's thickness on the surface of the silicon wafer. The X-ray diffraction (XRD) pattern exhibits distinct peaks corresponding to silicon at an angle of 69.5°. Additionally, the Si/PEDOT: PSS sample has a prohibited Si (200) peak at an angle of 32°. All manufactured cells have exhibited the characteristic diode curve in current-voltage measurements. The presence of a heterojunction is evidenced by this distinctive diode curve. The cells containing ethylene glycol exhibited a current output of 0.2 μA and a Voc (open-circuit voltage) of 0.2 V.

REFERENCES

1. Andreani, L.C.; Bozzola, A.; Kowalczewski, P.; Liscidini, M.; Redorici, L. Silicon solar cells: Toward the efficiency limits. *Adv. Phys. X* **2019**, *4*, 1548305. **[Google Scholar] [CrossRef] [Green Version]**

2. Almomani, M.S.; Ahmed, N.M.; Rashid, M.; Ibnaouf, K.H.; Aldaghri, O.A.; Madkhali, N.; Cabrera, H. Performance improvement of graded bandgap solar cell via optimization of energy levels alignment in Si quantum dot, TiO2 nanoparticles, and porous Si. *Photonics* **2022**, *9*, 843. **[Google Scholar] [CrossRef]**

3. Battaglia, C.; Cuevas, A.; de Wolf, S. High-efficiency crystalline silicon solar cells: Status and perspectives. *Energy Environ. Sci.* **2016**, *9*, 1552–1576. **[Google Scholar] [CrossRef] [Green Version]**

4. Rajesh, G.; Muthukumarasamy, N.; Velauthapillai, D.; Mohanta, K.; Ragavendran, V.; Batabyal, S.K. Photoinduced electrical bistability of sputter deposited CdZnTe thin films. *Mater. Res. Express* **2018**, *5*, 026412. **[Google Scholar] [CrossRef]**

5. Selvapriya, R.; Abhijith, T.; Ragavendran, V.; Sasirekha, V.; Reddy, V.S.; Pearce, J.M.; Mayandi, J. Impact of coupled plasmonic effect with multishaped silver nanoparticles on efficiency of dye sensitized solar cells. *J. Alloy. Compd.* **2022**, *894*, 162339. **[Google Scholar] [CrossRef]**

6. Selvapriya, R.; Abhijith, T.; Ragavendran, V.; Sasirekha, V.; Reddy, V.S.; Mayandi, J. Screen printed multifunctional TiO_2 photoanode with plasmonic Ag

nanoparticles for performance enhancement of dye sensitized solar cell. *Mater. Lett.* **2020**, *276*, 128194. **[Google Scholar] [CrossRef]**

7. Sasikala, R.; Kandasamy, M.; Suresh, S.; Ragavendran, V.; Sasirekha, V.; Pearce, J.M.; Murugesan, S.; Mayandi, J. Enhanced dye-sensitized solar cell performance using strontium titanate perovskite integrated photoanodes modified with plasmonic silver nanoparticles. *J. Alloy. Compd.* **2021**, *889*, 161693. **[Google Scholar] [CrossRef]**
8. He, L.; Jiang, C.; Wang, H.; Lai, D.; Rusli. High efficiency planar Si/organic heterojunction hybrid solar cells. *Appl. Phys. Lett.* **2012**, *100*, 073503. **[Google Scholar] [CrossRef]**
9. Zhao, Y.; Xie, D.; Xu, J.; Feng, T.; Zhang, X.; Tianling, R.; Zhu, M.; Zhu, H. PEDOT: PSS/planar-Si hybrid solar cells with 12.70% effciency. In Proceedings of the PIERS, Guangzhou, China, 25–28 April 2014. **[Google Scholar]**
10. Wang, H.; Wang, J.; Prakoso, A.B.; Hong, L.; Tan, Y.H.; Tan, C.S.; Rusli, R. High-efficiency planar thin-film Si/PEDOT: PSS hybrid solar cell. *IEEE J. Photovolt.* **2016**, *6*, 217–222. **[Google Scholar] [CrossRef]**
11. Nagamatsu, K.A.; Avasthi, S.; Jhaveri, J.; Sturm, J.C. Fellow, a 12% efficient silicon/PEDOT: PSS heterojunction solar cell fabricated at <100 °C. *IEEE J. Photovolt.* **2014**, *4*, 260–264. **[Google Scholar]**
12. Avasthi, S.; Lee, S.; Loo, Y.; Sturm, J.C. Role of majority and minority carrier barriers silicon/organic hybrid heterojunction solar cells. *Adv. Mater.* **2011**, *23*, 5762–5766. **[Google Scholar] [CrossRef] [PubMed]**
13. He, L.; Jiang, C.; Wang, H.; Lai, D.; Rusli. Si nanowires organic semiconductor hybrid heterojunction solar cells toward 10% efficiency. *ACS Appl. Mater. Interfaces* **2012**, *4*, 1704–1708. **[Google Scholar] [CrossRef] [PubMed]**
14. Sailor, M.J.; Ginsburg, E.J.; Gorman, C.B.; Kumar, A.; Grubbs, R.H.; Lewis, N.S. Thin films of n-Si/Poly-(CH3)3Si-Cyclooctatetraene: Conducting-polymer solar cells and layered structures. *Science* **1990**, *249*, 1146–1149. **[Google Scholar] [CrossRef] [PubMed]**
15. Wang, W.; Schiff, E.A. Polyaniline on crystalline silicon heterojunction solar cells. *Appl. Phys. Lett.* **2007**, *91*, 133504. **[Google Scholar] [CrossRef] [Green Version]**
16. El-Nahass, M.M.; Zeyada, H.M.; Abd-El-Rahman, K.F.; Darwish, A.A.A. Fabrication and characterization of 4-tricyanovinyl-N,N-diethylaniline/p-silicon hybrid organic–inorganic solar cells. *Sol. Energy Mater. Sol. Cells* **2007**, *91*, 1120–1126. **[Google Scholar] [CrossRef]**

17. McGillivray, D.; Thomas, J.P.; Abd-Ellah, M.; Heinig, N.F.; Leung, K.T. Performance enhancement by secondary doping in PEDOT: PSS/Planar-Si hybrid solar cells. *ACS Appl. Mater. Interfaces* **2016**, *8*, 34303–34308. **[Google Scholar] [CrossRef]**

18. Thomas, J.P.; Leung, K.T. Defect-minimized PEDOT:PSS/Planar-Si solar cell with very high efficiency. *Adv. Funct. Mater.* **2014**, *24*, 4978–4985. **[Google Scholar] [CrossRef]**

19. Friedel, B.; Keivanidis, P.E.; Brenner, T.J.K.; Abrusci, A.; McNeill, C.R.; Friend, R.H.; Greenham, N.C. Effects of layer thickness and annealing of PEDOT:PSS layers in organic photodetectors. *Macromolecules* **2009**, *42*, 6741–6747. **[Google Scholar] [CrossRef]**

20. Shiu, S.-C.; Chao, J.-J.; Hung, S.-C.; Yeh, C.-L.; Lin, C.-F. Morphology dependence of silicon nanowire/Poly (3,4 ethylene dioxythiophene):Poly (styrene sulfonate) heterojunction solar cells. *Chem. Mater.* **2010**, *22*, 3108–3113. **[Google Scholar] [CrossRef]**

21. Pietsch, M.; Jäckle, S.; Christiansen, S. Interface investigation of planar hybrid nSi/PEDOT: PSS solar cells with open circuit voltages up to 645 mV and efficiencies of 12.6 %. *Appl. Phys. A* **2014**, *115*, 1109–1113. **[Google Scholar] [CrossRef]**

22. He, L.; Jiang, C.; Wang, H.; Lai, D.; Tan, Y.H.; Tan, C.S.; Rusli. Effects of nanowire texturing on the performance of Si/organic hybrid solar cells fabricated with a 2.2 μm thin-film Si absorber. *Appl. Phys. Lett.* **2012**, *100*, 103104. **[Google Scholar] [CrossRef]**

23. Wang, H.; Wang, J.; Hong, L.; Tan, Y.H.; Tan, C.S. Thin film silicon nanowire/PEDOT: PSS hybrid solar cells with surface treatment. *Nanoscale Res. Lett.* **2016**, *11*, 311. **[Google Scholar] [CrossRef] [Green Version]**

24. Zhang, C.; Zhang, Y.; Guo, H.; Zhang, Z.; Zhang, C. Hole-transporting layer treatment of planar hybrid n-Si/PEDOT:PSS solar cells with power conversion efficiency up to 14.5%. *Nanostructured Sol. Cells* **2017**, *2017*, 3192197. **[Google Scholar] [CrossRef] [Green Version]**

25. Zaumseil, P. High-resolution characterization of the forbidden Si 200 and Si 222 reflections. *J. Appl. Cryst.* **2015**, *48*, 528–532. **[Google Scholar] [CrossRef] [PubMed] [Green Version]**

26. Singh, P.; Nakra, R.; Sivaiah, B.; Sardana, S.K.; Prathap, P.; Rauthan, C.M.S.; Srivastava, S.K. Effect of ethylene glycol doping on performance of PEDOT:PSS/μT-n-Si heterojunction solar cell. *AIP Conf. Proc.* **2018**, *1961*, 020003. **[Google Scholar] [CrossRef]**

27. LiHe, n.; Jiang, C.; Wang, H.; Lei, H.; Lai, D.; Rusli. 11.3% efficient planar Si-PEDOT: PSS hybrid solar cell with a thin interfacial oxide. In Proceedings of the 2012 38th IEEE Photovoltaic Specialists Conference, Austin, TX, USA, 3–8 June 2012. **[Google Scholar]** **[CrossRef]**

28. Schmidt, J.; Titova, V.; Zielke, D. Organic-silicon heterojunction solar cells: Open-circuit voltage potential and stability. *Appl. Phys. Lett.* **2013**, *103*, 183901. **[Google Scholar]** **[CrossRef]**

29. Zhang, Y.; Zu, F.; Lee, S.-T.; Liao, L.; Zhao, N.; Sun, B. Heterojunction with organic thin layers on silicon for record efficiency hybrid solar cells. *Adv. Funct. Mater.* **2014**, *4*, 1300923. **[Google Scholar]** **[CrossRef]**

30. Zhu, Y.; Song, T.; Zhang, F.; Lee, S.T.; Suna, B. Efficient organic-inorganic hybrid Schottky solar cell: The role of built-in potential. *Appl. Phys. Lett.* **2013**, *102*, 113504. **[Google Scholar]** **[CrossRef]**

31. Jiang, X.; Zhang, P.; Zhang, J.; Wang, J.; Li, G.; Fang, X.; Yang, L.; Chen, X. High performance of PEDOT:PSS/n-Si solar cells based on textured surface with AgNWs electrodes. *Nanoscale Res. Lett.* **2018**, *13*, 53. **[Google Scholar]** **[CrossRef]** **[Green Version]**

32. Zhang, C.; Zhang, Y.; Guo, H.; Jiang, Q.; Dong, P.; Zhang, C. Efficient planar hybrid n-Si/PEDOT:PSS solar cells with power conversion efficiency up to 13.31% achieved by controlling the SiOx interlayer. *Energies* **2018**, *11*, 1397. **[Google Scholar]** **[CrossRef]**

33. Yoon, S.-S.; Khang, D.-Y. High efficiency (>17%) Si-Organic hybrid solar cells by simultaneous structural, electrical, and interfacial engineering via low-temperature processes. *Adv. Energy Mater.* **2017**, *8*, 170265. **[Google Scholar]** **[CrossRef]**

34. van Trinh, P.; Anh, N.N.; Cham, N.T.; Tu, l.; van Hao, N.; Thang, B.H.; van Chuc, N.; Thanh, C.T.; Minh, P.N.; Fukata, N. Enhanced power conversion efficiency of an n-Si/PEDOT:PSS hybrid solar cell using nanostructured silicon and gold nanoparticles. *RSC Adv.* **2022**, *12*, 10514–10521. **[Google Scholar]** **[CrossRef]** **[PubMed]**

6
Chapter

INVESTIGATION OF THE PHOTOLUMINESCENCE PROPERTIES OF INAS QUANTUM DOTS THAT HAVE BEEN OVERGROWN BY A GAAS LAYER AT LOW TEMPERATURE, WHILE VARYING THE ARSENIC PRESSURE

1. INTRODUCTION: AN OVERVIEW

Semiconductor quantum dots (QDs) are known to greatly enhance the efficiency of optoelectronic and nanophotonic devices due to their spatial confinement in all three dimensions, resulting in strong quantum confinement [1,2,3,4]. InAs quantum dot (QD) lasers demonstrate a low threshold current density [5,6], a high internal quantum efficiency [7,8], and a stable threshold current and slope efficiency at high temperatures [9,10]. Nevertheless, QD lasers currently exhibit lower output power and modulation bandwidth compared to quantum well lasers [11]. Moreover, the effectiveness of potential quantum information devices based on quantum dots (QDs) is heavily influenced by the shape and makeup of the QDs [12]. Photoluminescence (PL) spectra are often obtained from QDs that have been grown on top of other layers, and their geometrical characteristics may vary compared to QDs viewed directly on the surface. Hence, it is imperative to investigate the mechanisms involved in QD capping. Prior research utilizing transmission electron microscopy (TEM) and cross-sectional scanning tunneling microscopy (X-STM) techniques has shown that the overgrowth of InAs quantum dots (QDs) involves significant mass transfer and intermixing [11, 13, 14, 15, 16]. The study demonstrated that the addition of a single monolayer (ML) of GaAs atop the QD array resulted in a substantial alteration of their size and density, as well as a distinct bimodal size distribution [17]. A separate investigation revealed a sudden reduction in the average height of InAs quantum dots (QDs) from 10 to 1.5 nm following the addition of a 2-nanometer-thick layer of GaAs [14]. The observed mass transfer in these tests is driven by the need to decrease system energy through strain relaxation. Various strategies were employed to maintain the arrangement of the initial quantum dots (QDs). An instance of maintaining the pyramidal shape of InAs quantum dots (QDs) with sharp heterointerfaces was achieved by allowing an excessive development of InAs QDs with a GaAs layer at a temperature of 300 °C [13]. Nevertheless, employing low-temperature capping can lead to subpar structural and optical characteristics of quantum dots (QDs). Different temperature sequences of overgrowth were also employed to assess their impact on the properties of quantum dots (QDs). According to the findings, in order to prevent intermixing and flushing of the tops of quantum dots (QDs), the overgrowth temperature should be equal to or lower than the QD growth temperature [11,16,18]. The study also demonstrated that capping at low growth rates results in greater intermixing and reduction in QD height compared to high growth rates [15]. These findings suggest that the QDs and capping layers have a tendency to mix together and form a single, uninterrupted layer if certain approaches are not used to prevent this from happening. Similarly, any pause in growth causes the continuous dissolution of quantum dots (QDs) in the matrix material [13]. However, there has been significantly less focus on the relationship between the quantum dot (QD) photoluminescence (PL) spectra and the arsenic beam equivalent pressure during overgrowth. This factor is also anticipated to influence the dissolution of QDs. This chapter examines the impact of arsenic pressure (PAs) on the formation of InAs quantum dots (QDs) using a low-temperature GaAs (LT-GaAs) layer with varying thicknesses (H). The focus is on how these factors affect the photoluminescence (PL) spectra of the structures.

We revealed that a high arsenic pressure makes it possible to achieve a single broad line in the PL spectrum corresponding to unimodal size distribution of QDs, whereas the spectrum has two separate lines (~1080 and ~1150 nm) at a low arsenic pressure for both capping layer thicknesses (5 and 10 nm). An additional longwave line (~1150 nm) is attributed to the contribution of largened QDs. Their production can be impeded by augmenting the arsenic flow, which hampers the movement of particles and diminishes the presence of sizable quantum dots in the photoluminescence spectrum.

2. METHODOLOGY

The specimens were cultivated using a SemiTEq STE 35 solid-source molecular beam epitaxy system on GaAs(001) epi-ready substrates. Group-III component sources were utilized in the form of standard effusion cells. A valved cracker cell with As4 mode was utilized for group-V arsenic. Following the customary process of removing the oxide, a GaAs buffer layer with a thickness of 250 nanometers was produced at a temperature of 580 °C. A solitary layer of InAs quantum dots (QDs) was placed between a GaAs inner layer that was 100 nanometers thick and an

Al0.33Ga0.67As outer layer that was 50 nanometers thick. The growth temperatures for the inner and outer layers were 580 °C and 600 °C, respectively. The quantum dots (QDs) were created using the Stranski-Krastanov growth mode following the deposition of 2.5 monolayers (MLs) of Indium Arsenide (InAs) at a temperature of 500 °C and a deposition rate of 0.05 MLs per second. The presence of InAs three-dimensional islands was confirmed by the transition from streaky to spotty patterns in the reflection high-energy electron diffraction pattern. Following the production of quantum dots (QD), a growth interruption of 30 seconds was implemented using the arsenic flux.

Then, four different low-temperature (500 °C) GaAs capping procedures were carried out for samples 1-4: (1) 5 nm at $P_{As} = 3 \times 10^{-5}$ Pa; (2) 10 nm at $P_{As} = 3 \times 10^{-5}$ Pa; (3) 5 nm at $P_{As} = 1 \times 10^{-5}$ Pa; (4) 10 nm at $P_{As} = 1 \times 10^{-5}$ Pa. Thereafter, the substrate temperature was elevated to 580 °C to grow a high-temperature GaAs layer with a thickness of 95 nm for samples 1 and 3 and 90 nm for samples 2 and 4. Above the $Al_{0.33}Ga_{0.67}As$ cladding layer, a 10-nanometer-thick high-temperature GaAs layer was grown, and the structure was annealed at 610 °C to reduce defects. For reference, an additional sample was grown with an array of uncovered QDs formed straightly on a GaAs buffer layer (sample 0).

A schematic QD PL structure grown in this study is illustrated in **Figure 1**.

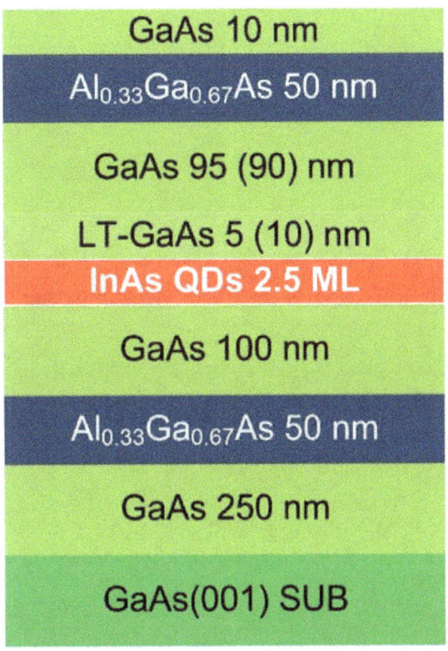

Figure 1. Schematic representation of a typical PL structure in which the LT-GaAs capping layer differs in the thickness and the arsenic pressure provided during the growth.

The reference sample was studied in FEI Nova Nanolab scanning electron microscope (SEM) and NT-MDT Ntegra atomic force microscope (AFM) for the characterization of shape, size and surface density of uncovered QDs.

For PL studies, the samples were placed in a flow Janis ST-500 cryostat to control the sample temperature (T_{PL}) in the range of 77–300 K. The PL was excited by a YLF:Nd^{+3} laser operating in the cw mode (λ = 527 nm). The laser power was varied in the range from 0.084 to 40 mW. PL signal was detected using an SOL instrument's MS 5204i monochromator and a single-channel InGaAs detector using synchronous detection (SRS 830 Stanford Research Systems).

3. RESULTS AND DISCUSSION

The QD growth conditions were chosen so that the surface density of QDs was sufficiently large to provide high PL intensity, but not too large to stimulate QD coalescence. It is commonly known that InAs dots arranged quite close to each other merge due to Ostwald ripening [7,11]. A value of the surface density of uncovered QDs that we observe after 30 s interruption at the growth temperature before quenching is 4.1×10^{10} cm^{-2} (**Figure 2**). Meanwhile, the density of the coalesced QDs is found to be 3.8×10^{8} cm^{-2}. An average diameter of uncovered QDs is 23 nm, and their average height is 5.5 nm. QDs with these parameters are typically supposed to have separate energy levels contributing to distinct peaks in the PL spectrum after the capping procedure [14,19].

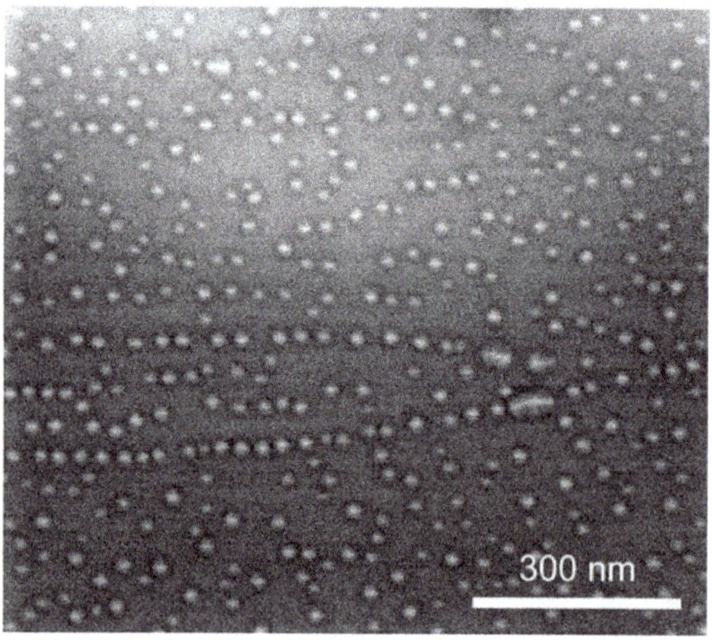

Figure 2. SEM image of an array of uncovered InAs QDs formed on the GaAs surface (sample 0).

PL spectra taken from the overgrown structures demonstrate their strong dependence on the arsenic pressure used during overgrowth of QDs by the LT-GaAs layer (**Figure 3**a). For each room-temperature spectrum, the lines corresponding to volume GaAs and InAs wetting layer are at ~870 and 935 nm, respectively, which is in good agreement with previous studies [20]. However, longer-wavelength lines corresponding to the energy levels of QDs differ in their position and relative intensity (**Table 1**).

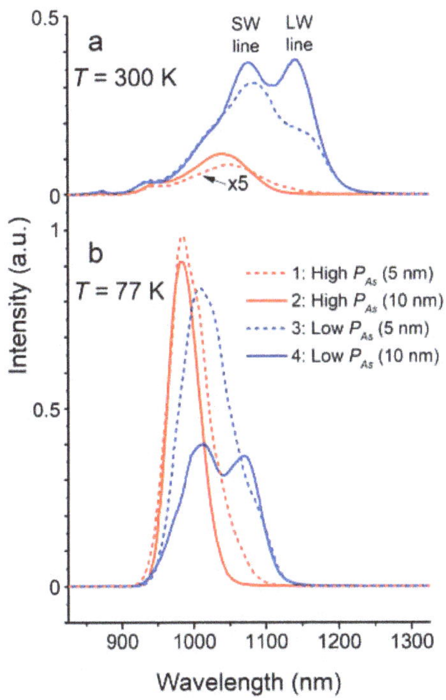

Figure 3. 300 K (a) and 77 K (b) PL spectra of structures with InAs QDs overgrown by the GaAs layer under various growth conditions according to the sample numbers in the legend: red lines: $P_{As} = 3 \times 10^{-5}$ Pa; blue lines: $P_{As} = 1 \times 10^{-5}$ Pa; solid lines: 10 nm; dashed lines: 5 nm.

Table 1. Summary of the PL parameters of InAs QDs covered by the LT-GaAs layer under various growth conditions.

T_{PL} (K)	P_{As} (Pa)	*H* = 5 nm			*H* = 10 nm		
		W	LW Line (nm)	Integrated Intensity (a.u.)	SW Line (nm)	LW Line (nm)	Integrated Intensity (a.u.)
300	3×10^{-5}	1049	-	0.03	1038	-	0.16
	1×10^{-5}	1083	~1160	0.61	1077	1140	0.76
77	3×10^{-5}	983	-	0.88	983	-	0.64
	1×10^{-5}	1009	~1080	1.00	1011	1070	0.62

When QDs are overgrown at $P_{As} = 3 \times 10^{-5}$ Pa, an unambiguous single broad line is observed in the room-temperature PL spectrum at 1049 and 1038 nm (shortwave (SW) line) for the LT-GaAs thicknesses of 5 and 10 nm, respectively (**Figure 3**a), whereas at $P_{As} = 1 \times 10^{-5}$ Pa, two longer wave lines are pronounced. The first line (SW) is approximately the same for both LT-GaAs thicknesses and shifted to ~1080 nm. The longwave (LW) line is at a wavelength of ~1160 and ~1140 nm for 5 and 10 nm LT-GaAs, respectively. Although the appearance of an additional LW line is often attributed to an array of coalesced QDs [11], we consider that its contribution is not sufficient to give a pronounced PL peak, taking into account the two-orders-of-magnitude-lower surface density of coalesced QDs compared with basic ones. Moreover, the coalescence of QDs before capping is equally inherent in each of the PL structures under consideration, but the additional LW lines appear only after the overgrowth at $P_{As} = 1 \times 10^{-5}$ Pa. Therefore, this behavior is not due to the initial size distribution of InAs islands, but to mass transfer processes during the overgrowth.

The red shift of the SW line, as well as the appearance of an additional LW line in the low-pressure LT-GaAs PL spectra, is due to the redistribution of the QD material and formation of new groups of QD sizes. At the high arsenic pressure, the diffusion of group-III atoms is suppressed, and therefore, InAs islands are predominantly conserved. Although the shape, size and composition of InAs QDs can change due to the indium segregation and strain-relaxing mass transfer, their uniformity in size is assumed to be retained. At the same time, an InAs QD covered by an intermediate InGaAs layer with changing In content is formed as a result of QD and LT-GaAs intermixing. As shown by TEM and X-STM studies [11,13], this form of QDs is typical for capping at the QD growth temperature. When the temperature is reduced to ultracold 300 °C, the QD shape remains almost perfectly pyramidal [13]. Meanwhile, an increase in the temperature causes more intensive intermixing, which leads to blurring of the original pyramidal shape of QDs.

Similar behavior is observed when the arsenic pressure during the overgrowth is decreased. The diffusion length of group-III atoms increases, leading to more intensive mass transfer and coalescence of QDs located close to each other. As a result, two types of QDs appear in the volume and contribute to LW lines: 1140 nm for 10 nm LT-GaAs and ~1160 nm for 5 nm LT-GaAs (**Figure 3**a). These lines are attributed to the appearance of QDs of quite a large size which contain an increased number of energy levels in the potential well, providing lower energy electron transitions with emission at longer wavelengths.

Integrated room-temperature PL intensity of QDs obtained at the low arsenic pressure during the overgrowth essentially exceeds the intensity of QDs which were overgrown at the high arsenic pressure (**Table 1**). However, this difference is not so significant at 77 K PL (**Figure 3**b). This indicates that excess arsenic accumulates in the capping layer grown at the high arsenic pressure and acts as additional nonradiative recombination centers for charge carriers migrating faster over the semiconductor volume at room temperature than at 77 K. It also should be noted that the integrated room-temperature PL intensity of QDs overgrown by 5 nm LT-GaAs is weaker than for 10 nm LT-GaAs, especially for the high arsenic pressure. It can be associated with higher defectiveness of thinner capping layers, which has been reported elsewhere. After the low-temperature overgrowth of QDs by a layer of 2–7 nm thickness, shallow depressions are formed above the tops of QDs and preserved on the surface at a stage of substrate heating up to a high temperature used for the overgrowth by a thick GaAs layer [7,11,14,17]. After the high-temperature overgrowth of the layer with these depressions, multiple dislocations are produced in the area above QDs [7,11]. In our study, the defectiveness of the structure increases when the 5 nm LT-GaAs layer is grown at the high arsenic pressure because, in this case, the excess arsenic incorporated into the thin capping layer (5 nm) desorbs, leading to the expansion of the depressions above QDs. Although the concentration of excess arsenic atoms decreases, an increased density of dislocations prevents efficient radiative recombination of carriers in QDs to a greater extent. However, the overgrowth of QDs by the thicker LT-GaAs layer (10 nm) allows inhibition of the depression formation and subsequent generation of dislocations [11].

The intensity maximum of the LW line in the room-temperature spectra for the low arsenic pressure is observed to be higher than the intensity maximum of the SW line in the case of the 10 nm capping layer, but it is lower in the case of the 5-nanometer-thick layer (**Figure 3**a). The integrated PL intensity from QDs overgrown by the 10 nm layer is also higher than from QDs overgrown by the 5 nm layer (**Table 1**). This behavior can be explained by the difference in the surface density of QDs of various sizes formed during the diffusion of In atoms within the overgrown layer mixed with the QD material. In the case of the capping layer of a smaller thickness (5 nm), as noted earlier, a greater number of dislocations are formed in the layer after the high-temperature overgrowth than in the case of the overgrowth by the thicker LT-GaAs layer (10 nm). As a result, the integrated PL intensity at room temperature decreases, as does the intensity of the LW line, which corresponds to larger QDs having a greater number of dislocations in their neighborhood. Therefore, charge carriers in larger QDs are more prone to nonradiative transitions than in smaller QDs. This assumption is confirmed by the fact that the integrated PL intensity from QDs overgrown by the 5 nm layer increases with a decrease in the PL temperature to 77 K (**Figure 3**b).

However, for the 10-nanometer-thick capping layer, the integrated PL intensity remains at the same level, with a slight decrease when moving from 300 to 77 K (**Table 1**). As mentioned earlier, this indicates lower defectiveness of the material compared with the structure with the 5 nm layer. However, it is important to clarify the reason for the reduction in the integrated PL intensity from QDs at 77 K in comparison with the 10 nm layer. This regularity may be due to the continuation of intensive diffusion of In atoms at the stage of heating the 5 nm layer, which leads to an increase in the surface density of smaller QDs responsible for the SW line. Although the density of such QDs overgrown by the 5 nm layer is higher, the integrated PL intensity of QDs at room temperature becomes lower than in the case of the 10 nm layer due to the higher imperfection of the material. When QDs are overgrown by the thicker LT-GaAs layer (10 nm), the QD density remains almost unchanged during the heating, which is confirmed by the retention of two distinct lines in the PL spectrum (SW and LW lines).

In order to estimate the nature of the separate SW and LW lines, the excitation intensity was varied in the range from 0.084 to 40 mW for the two PL temperatures of 300 K and 77 K for the samples with the low-pressure overgrowth by LT-GaAs layers of different thicknesses (**Figure 4**).

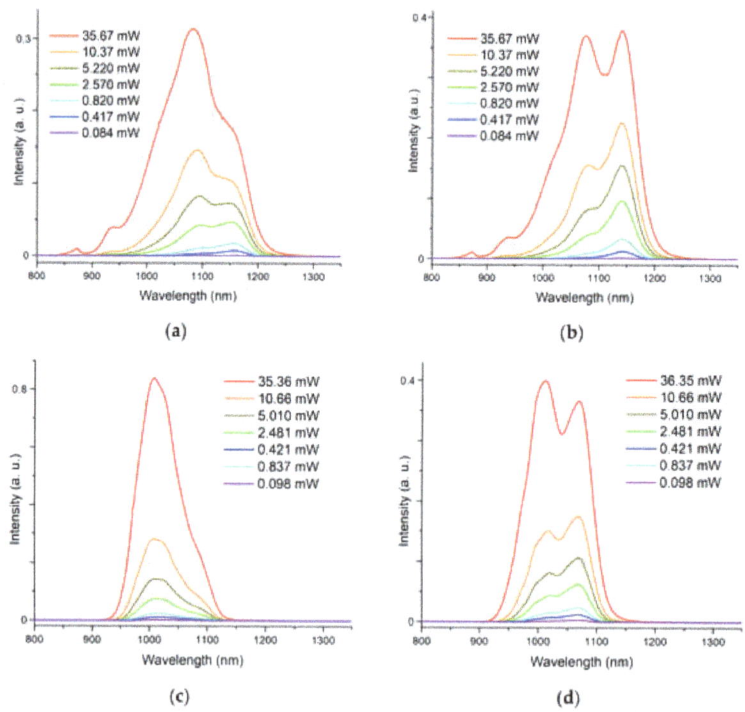

Figure 4. Excitation intensity-dependent PL spectra of QDs overgrown at the low arsenic pressure: (a) *H* = 5 nm (300 K); (b) *H* = 10 nm (300 K); (c) *H* = 5 nm (77 K); (d) *H* = 10 nm (77 K).

For the room-temperature PL spectra of QDs overgrown at the low arsenic pressure, an increase in the ratio of SW to LW lines' maximum intensity is observed, with increasing excitation power for both LT-GaAs layer thicknesses: 5 nm (**Figure 4**a) and 10 nm (**Figure 4**b). This is attributed to the fact that low-energy electron transitions in QDs of larger sizes are more favorable than shortwave transitions in smaller QDs. When the first ones reach saturation at a certain excitation intensity, the probability of high-energy transitions in smaller QDs becomes higher, contributing to the SW line of the PL spectrum.

A similar situation is observed in the cryogenic PL spectra, where the ratio of the SW to LW lines' maximum intensity rises with increasing excitation power for the samples with both LT-GaAs thicknesses, i.e., 5 nm (**Figure 4**c) and 10 nm (**Figure 4**d). However, two LW lines become prominent at wavelengths of ~1020 and ~1080 nm. This fact confirms an assumption about the saturation of low-energy electron transitions, but it also demonstrates that appearance of a new group of QDs with an intermediate size (emitting at ~1020 nm) is possible in the case of the 5 nm LT-GaAs layerThere is no text provided. Therefore, the movement of In atoms inside the thin capping layer is faster when the substrate is heated compared to within the thicker capping layer. Confirmation of the presence of both short-wavelength (SW) and long-wavelength (LW) lines in the cryogenic photoluminescence (PL) spectra, regardless of the excitation power, indicates that the SW line is not a result of transitions from an excited state, but rather arises from transitions within a specific category of small-sized quantum dots (QDs).

4. CONCISE SUMMARY

The influence of the arsenic pressure during low-temperature GaAs overgrowth of InAs QDs on their optical properties was revealed in this study. We demonstrated that overgrowth at a high arsenic pressure provides a single broad line in the QD PL spectrum, whereas two distinct lines are observed in the PL spectrum of QDs overgrown by the LT-GaAs layer at a low arsenic pressure. This behavior is explained by the strain-compensating mass transfer induced by the QD overgrowth. While the diffusion of In atoms is suppressed at the high arsenic pressure, formation of different size groups of QDs occurs during the low-pressure overgrowth. However, the integrated room-temperature PL intensity of QDs is significantly lower for QDs overgrown at the high arsenic pressure because excess arsenic incorporates into the capping layer in this case and leads to the enhancement of nonradiative recombination of charge carriers.

We also reveal a decrease in the integrated intensity of a thinner LT-GaAs capping layer (5 nm) as compared with a thicker layer (10 nm). We attribute it to one more source of nonradiative recombination centers, which has dislocations produced in the area of QD tops. These dislocations are caused by the formation of depressions during the low-temperature overgrowth and expand during the substrate heating of the thin layer (5 nm). Meanwhile, this pit formation can be inhibited using the low-temperature overgrowth by the thicker layer (10 nm). The higher defectiveness of the 5 nm LT-GaAs layer is also found for the overgrowth at the high arsenic pressure, which is associated with the desorption of the excess arsenic, inducing the depression formation.

The presence of the SW line at each excitation power in the cryogenic PL spectra for the samples with the low-pressure overgrowth confirmed the assumption that it is not caused by the electron transitions from an excited energy level, but it is due to the excitation of smaller-sized QDs. The SW line's maximum intensity increases when the low-energy transitions from larger QDs reach saturation, giving rise to the shorter-wavelength transitions from smaller QDs. The analysis of the ratio of relative maximum intensities of the SW and LW lines also indicated a difference in the surface density of QDs obtained after the overgrowth by LT-GaAs layers of different thicknesses. Because the LW line's intensity is relatively lower than the SW line's intensity for the 5 nm LT-GaAs, we can conclude that the concentration of dislocations near larger QDs is higher than that near smaller QDs.

At a Glance

We investigated the impact of arsenic pressure on the optical characteristics of InAs quantum dots during low-temperature GaAs overgrowth. Upon analyzing the photoluminescence spectrum of quantum dots produced under high arsenic pressure, we discovered a solitary wide line that corresponds to a uniform size distribution of quantum dots. Conversely, the spectra of samples containing quantum dots that were produced with low arsenic pressure exhibit two separate peaks at longer wavelengths, specifically around 1080 nm and 1150 nm. We ascribed this behavior to the high-pressure inhibition of atom migration between InAs islands during the overgrowth phase, enabling the retention of the original unimodal size distribution of quantum dots. Under low arsenic pressure, an excessive expansion of quantum dots occurs, resulting in a significant mass transfer. This mass transfer ultimately leads to the creation of arrays of quantum dots that have bigger diameters. The integrated photoluminescence intensity at 300 K is observed to be reduced for quantum dots that are overgrown at higher arsenic pressure. Nevertheless, the disparity in the photoluminescence intensity between the high- and low-pressure overgrowths is not particularly noteworthy at a temperature of 77 K. This suggests that when there is an abundance of arsenic, it becomes part of the capping layer and forms many nonradiative recombination sites, which reduces the intensity of photoluminescence.

REFERENCES

1. Cusack, M.; Briddon, P.; Jaros, M. Electronic Structure of InAs/GaAs Self-Assembled Quantum Dots. *Phys. Rev. B Condens. Matter Mater. Phys.* **1996**, *54*, R2300–R2303. **[Google Scholar] [CrossRef] [PubMed]**

2. Johansen, J.; Stobbe, S.; Nikolaev, I.S.; Lund-Hansen, T.; Kristensen, P.T.; Hvam, J.M.; Vos, W.L.; Lodahl, P. Size Dependence of the Wavefunction of Self-Assembled InAs Quantum Dots from Time-Resolved Optical Measurements. *Phys. Rev. B Condens. Matter Mater. Phys.* **2008**, *77*, 073303. **[Google Scholar] [CrossRef] [Green Version]**

3. Li, S.Y.; He, L. Recent Progresses of Quantum Confinement in Graphene Quantum Dots. *Front. Phys.* **2022**, *17*, 33201. **[Google Scholar] [CrossRef]**

4. Pryor, C. Geometry and Material Parameter Dependence of InAs/GaAs Quantum Dot Electronic Structure. *Phys. Rev. B* **1999**, *60*, 2869–2874. **[Google Scholar] [CrossRef]**

5. Moiseev, E.I.; Kryzhanovskaya, N.V.; Zubov, F.I.; Mikhailovskii, M.S.; Abramov, A.N.; Maximov, M.V.; Kulagina, M.M.; Guseva, Y.A.; Livshits, D.A.; Zhukov, A.E. Record Low Threshold Current Density in Quantum Dot Microdisk Laser. *Semiconductors* **2019**, *53*, 1888–1890. **[Google Scholar] [CrossRef]**

6. Stintz, A.; Liu, G.T.; Li, H.; Lester, L.F.; Malloy, K.J. Low-Threshold Current Density 1.3-Mm InAs Quantum-Dot Lasers with the Dots-in-a-Well (DWELL) Structure. *IEEE Photonics Technol. Lett.* **2000**, *12*, 591–593. **[Google Scholar] [CrossRef]**

7. Liu, W.-S.; Chang, H.; Liu, Y.-S.; Chyi, J.-I. Pinholelike Defects in Multistack 1.3μm InAs Quantum Dot Laser. *J. Appl. Phys.* **2006**, *99*, 114514. **[Google Scholar] [CrossRef]**

8. Ledentsov, N.N.; Kovsh, A.R.; Zhukov, A.E.; Maleev, N.A.; Mikhrin, S.S.; Vasilev, A.P.; Semenova, E.S.; Maximov, M.V.; Shernyakov, Y.M.; Kryzhanovskaya, N.V.; et al. High Performance Quantum Dot Lasers on GaAs Substrates Operating in 1.5 μm Range. *Electron. Lett.* **2003**, *39*, 1126. **[Google Scholar] [CrossRef]**

9. Nishi, K.; Takemasa, K.; Sugawara, M.; Arakawa, Y. Development of Quantum Dot Lasers for Data-Com and Silicon Photonics Applications. *IEEE J. Sel. Top. Quantum Electron.* **2017**, *23*, 1–7. **[Google Scholar] [CrossRef]**

10. Zhukov, A.E.; Kryzhanovskaya, N.V.; Moiseev, E.I.; Nadtochiy, A.M.; Dragunova, A.S.; Maximov, M.V.; Zubov, F.I.; Kadinskaya, S.A.; Berdnikov, Y.; Kulagina, M.M.; et al. Impact of Self-Heating and Elevated Temperature on Performance of Quantum Dot Microdisk Lasers. *IEEE J. Quantum Electron.* **2020**, *56*, 1–8. **[Google Scholar] [CrossRef]**

11. Okumura, S.; Fujisawa, K.; Naruke, T.; Nishi, K.; Onishi, Y.; Takemasa, K.; Sugawara, M.; Sugiyama, M. Impact of Low-Temperature Cover Layer Growth of InAs/GaAs Quantum Dots on Their Optical Properties. *Jpn. J. Appl. Phys.* **2022**, *61*, 085503. **[Google Scholar] [CrossRef]**

12. Arakawa, Y.; Holmes, M.J. Progress in Quantum-Dot Single Photon Sources for Quantum Information Technologies: A Broad Spectrum Overview. *Appl. Phys. Rev.* **2020**, *7*, 021309. **[Google Scholar]** **[CrossRef]**

13. Gong, Q.; Offermans, P.; Nötzel, R.; Koenraad, P.M.; Wolter, J.H. Capping Process of InAs/GaAs Quantum Dots Studied by Cross-Sectional Scanning Tunneling Microscopy. *Appl. Phys. Lett.* **2004**, *85*, 5697–5699. **[Google Scholar]** **[CrossRef]** **[Green Version]**

14. García, J.M.; Medeiros-Ribeiro, G.; Schmidt, K.; Ngo, T.; Feng, J.L.; Lorke, A.; Kotthaus, J.; Petroff, P.M. Intermixing and Shape Changes during the Formation of InAs Self-Assembled Quantum Dots. *Appl. Phys. Lett.* **1997**, *71*, 2014–2016. **[Google Scholar]** **[CrossRef]** **[Green Version]**

15. Utrilla, A.D.; Grossi, D.F.; Reyes, D.F.; Gonzalo, A.; Braza, V.; Ben, T.; González, D.; Guzman, A.; Hierro, A.; Koenraad, P.M.; et al. Size and Shape Tunability of Self-Assembled InAs/GaAs Nanostructures through the Capping Rate. *Appl. Surf. Sci.* **2018**, *444*, 260–266. **[Google Scholar]** **[CrossRef]**

16. Lian, G.D.; Yuan, J.; Brown, L.M.; Kim, G.H.; Ritchie, D.A. Modification of InAs Quantum Dot Structure by the Growth of the Capping Layer. *Appl. Phys. Lett.* **1998**, *73*, 49–51. **[Google Scholar]** **[CrossRef]**

17. Ferdos, F.; Wang, S.; Wei, Y.; Larsson, A.; Sadeghi, M.; Zhao, Q. Influence of a Thin GaAs Cap Layer on Structural and Optical Properties of InAs Quantum Dots. *Appl. Phys. Lett.* **2002**, *81*, 1195–1197. **[Google Scholar]** **[CrossRef]**

18. Songmuang, R.; Kiravittaya, S.; Schmidt, O.G. Shape Evolution of InAs Quantum Dots during Overgrowth. *J. Cryst. Growth* **2003**, *249*, 416–421. **[Google Scholar]** **[CrossRef]**

19. Nishi, K.; Saito, H.; Sugou, S.; Lee, J.-S. A Narrow Photoluminescence Linewidth of 21 MeV at 1.35 Mm from Strain-Reduced InAs Quantum Dots Covered by In0.2Ga0.8As Grown on GaAs Substrates. *Appl. Phys. Lett.* **1999**, *74*, 1111–1113. **[Google Scholar]** **[CrossRef]**

20. Kosogov, A.O.; Werner, P.; Gösele, U.; Ledentsov, N.N.; Bimberg, D.; Ustinov, V.M.; Egorov, A.Y.; Zhukov, A.E.; Kopeev, P.S.; Bert, N.A.; et al. Structural and Optical Properties of InAs/GaAs Quantum Dots Subjected to High Temperature Annealing. *Appl. Phys. Lett.* **1996**, *69*, 3072–3074. **[Google Scholar]** **[CrossRef]**

7 Chapter

IMPROVEMENTS TO THE QUANTUM PROTOCOL CSLOE-2022, MAKING THE CLASSICAL QUANTUM PROTOCOL BB84 MORE SECURE CRYPTOGRAPHICALLY

1. INTRODUCTION: AN OVERVIEW

The chapter focuses only on quantum cryptography. It specifically pertains to quantum cryptography protocols, which experts have been studying for nearly four decades. Quantum cryptography is a technique that has the potential to enhance the distinctive appearance of future telecommunication networks. Nevertheless, it is impossible to accurately anticipate the precise nature of the developed quantum infrastructure and the potential consequences it may bring. Today, quantum cryptography enables us to enhance security against unauthorized access to transmitted information and also to detect such attempts with a higher likelihood. In contemporary society, this element holds great significance as humanity actively pursues the goal of complete digitalization. Simultaneously, there is a rapid global upgrading of technological equipment, computer software, and hardware, which is leading to an increase in the demand on telecommunications networks. The quantities of classified data, sensitive information, and personal user data are on the rise. Simultaneously, the frequency of computer security incidents is inexorably escalating. Adversaries are updating the varieties of risks and methods to execute them, resulting in significant issues with information security and data safeguarding. Ensuring the dependability and functionality of technical equipment in communication networks is getting progressively challenging. The recent hacks in multiple nations have partially disrupted public activity and caused significant disruptions in government operations. Numerous nations see comparable occurrences of computer security on a daily basis, and in certain instances, users are rendered virtually powerless. National governments are deeply alarmed by this predicament, as it poses a significant threat to their vital infrastructure. Inevitably, there will come a time when the progress of modernizing information protection methods will reach a halt, and the widespread adoption of more secure quantum technology will become unavoidable. Consequently, numerous countries worldwide are hopeful about the prospects of quantum cryptography technologies. Regarding this matter, the shift to quantum technologies appears to be unavoidable, and the approach towards scientific research in this field has become significantly more earnest. The authors' research aims to analyze the fundamental principles of quantum cryptography protocols, compare their cryptographic robustness, and explore potential modifications. Two prominent quantum protocols are used as illustrations: the obsolete yet still competitive BB84, and its modified version BB84 Info-Z, which possesses its own distinguishing characteristics. Gaining a comprehensive grasp of the foundations and concepts underlying the operation of quantum protocols is increasingly essential for several information security experts. This text showcases the main areas of focus for the advancement of quantum key distribution systems, as well as the establishment of data transmission protocols that are both secure and based on the principles of quantum physics. This article presents novel hypotheses regarding the potential for universal enhancement of quantum protocols, using the modification of the quantum protocol BB84 as an illustrative example. The authors' approach is applied by utilizing the notion of artificially reconstructing photons and utilizing them to minimize potential hazards and, in certain instances, prevent scenarios associated with weaknesses in quantum protocols. This strategy employs a combination of bogus photon states and the installation of several traps to augment the computational resources utilized by hackers in cyberattacks. Ultimately, it renders these attacks ineffective and futile. An adversary, sometimes referred to as a "eavesdropper," will have significantly limited chances of correctly guessing the intended data transmission qubit and the polarization state of the photon. The simulated replicas, in addition to their primary function, will serve as effective lures for hackers, capable of enticing them into a trap and thereby verifying their presence in the communication channel. In this scenario, it will not be necessary to disconnect the communication channel.

2. PROTOCOL BB84

This quantum protocol is named after the first letters of the surnames of its creators and the year of its publication [1]. The BB84 protocol is designed to transmit secret information encoded in binary. **Figure 1** shows how, in the BB84 protocol, one bit can be encoded in the polarization state of photon. Binary 0 is defined as the polarization of 0° in the rectilinear coordinate basis, or 45° on a diagonal basis. Binary 1 can also be equal to 90° on a rectilinear coordinate basis, or 135° on a diagonal basis. Thereby, one bit can be represented by the polarization of a photon in one of two coordinate bases.

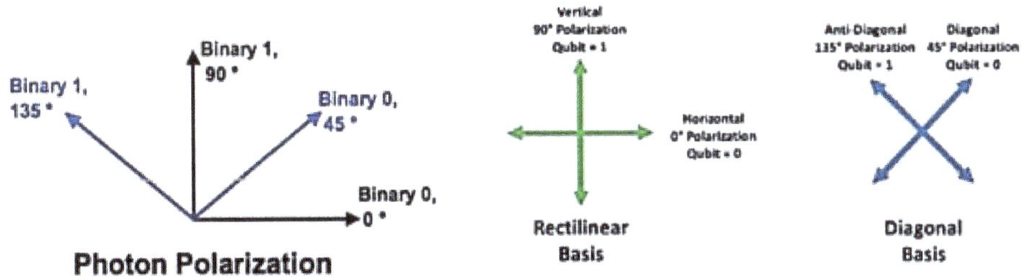

Figure 1. Encoding of the bit in protocol BB84.

Currently, the participants in the transmission of the data process, Alice and Bob, agree to approve some big number n, the threshold of error pa and the linear correction code of error C, with the parity check matrix PC of the order r × n.

The linear key generation function (to enhance confidentiality), represented by P_K matrix of order $m \times n$, is also agreed upon. Both matrices can be known in advance or can be determined during the execution of the protocol, and then can be sent via the classical channel. In turn, the matrix $(r + m) \times n$, the rows of which are the rows P_C and P_K, if taken together, must have the rank $r + m$. Alice randomly selects the sequences of bits: strings i from $2n$ – bit, $b \in F_2^{2n}$, where F_2 defines the field of 2 elements {0; 1}, that is, field of integer numbers modulo 2. Then, the state $\left|i^b\right\rangle = \left|i_1^{b_1}\right\rangle \dots \left|i_{2n}^{b_{2n}}\right\rangle$ is encoded. For each bit, the coordinate basis is randomly selected, rectilinear or diagonal, with which the bit will be encoded. When transmitting the photon from Alice, Bob will inform her about the photon receiving, but will not measure it.

For each photon that Bob receives, he will measure the polarization of the photon on the randomly selected coordinate basis, applying it to his own state. If Bob chose the same coordinate basis of his state for particular photon as Alice did, when he performed the transformation $H^b = H^{b1} \otimes \dots \otimes H^{b2n}$, he switches to the state $|i\rangle = |i_1 \dots i_{2n}\rangle$. Bob should measure the same polarization in the line i^B, and thus output the bit that Alice sent correctly, in the case of absence of noise and signs of eavesdropping on the communication line from the side of an attacker.

To detect eavesdropping, Alice will randomly select n bits that will be used to detect the presence of attackers. By selecting a $2n$–bit string containing n units exactly, Alice must ensure that equality $|s| = n$ was fulfilled. Alice sends to Bob s bits publicly, such, that $sj=0$, which are applicated for testing, and the rest of the bits are employed for generating of the final key. Let us denote the corresponding substrings, appropriate for testing process values i and b, while the substrings appropriate for the generation of a key will be defined as i_s and b_s. For every $j \in [1\dots 2n]$, such, that the value $s_j = 0$, the participants of the data transfer process Alice and Bob publish the value of the bit with jth number.

If Bob chose the incorrect basis of coordinates, then his outcome and, consequently, the bit that he received will be random. It is worth noting that if the mismatch of the bits determined in the published values of the *j*th bit by comparison exceeds the np_a, then they interrupt the protocol execution. The preliminarily fixed parameter p_a of the protocol, in fact, is the relationship of permitted bit flips, intended for the testing process.

Alice and Bob save the values of the residue *n* bits in the strict confidence. The string of Alice is designated as $x=is$ and is called the *information string*. The appropriate string of bits on the side Bob is determined as x^B.

At the second stage, Bob must inform Alice, via any unprotected channel of communication, what coordinate basis he has used to measure every photon. Then, Alice informs Bob if he chose the proper coordinate basis for each photon. At this stage, Alice and Bob throw out the bits appropriating the photons that Bob has measured with another coordinate basis. On the condition that no errors occurred, or no one managed the photons, Alice and Bob now should possess the same string of bits, which is named the *sifted key*.

The example demonstrated below (**Figure 2**) illustrates the bits that Alice selected; the coordinate bases in which she encoded her bits; and the coordinate bases that Bob applicated for his dimension. In addition, **Figure 2** shows the resulting *sifted key*, after the stage when Bob and Alice rejected their bits, as indicated above. However, Alice and Bob, before they finish, have to approve a subset of random bits for the comparison of the provision of the consistency of their actions. For verification, Alice sends Bob an *r*–bit string of error correction. Bob will use ξ to compare or rectify his string x^B as necessary. String $\xi=xPTC$ is named the *syndrome* of *string x* (relative to P_c). If bits correspond, then they are rejected, and other bits create the common secret *m*–bit key.

Alice's bit	0	1	1	0	1	0	0	1
Alice's basis	+	+	X	+	X	X	X	+
Alice's polarization	↑	→	↖	↑	↖	↗	↗	→
Bob's basis	+	X	X	X	+	X	+	+
Bob's measurement	↑	↗	↖	↗	→	↗	→	→
Public discussion								
Shared Secret key	0		1			0		1

Figure 2. Sifted key.

If there is no noise and there are no other dimension errors, a mismatch in any of the compared bits would specify the availability of an *interceptor* presenting in the quantum channel. The reason for this is that if attacker Eve was trying to define the key value, then she would not have any other choice but to measure the photons dispatched by Alice before transmitting these photons to Bob.

Therefore, Eve should attach the separate quantum probe (at quantum sensing [2,3,4,5,6]), which, as she assumed, is in a pure state, and apply the unitary transformation U_j to the composite system. This is true because the theorem on prohibition of cloning (anti-cloning theorem) [7] guarantees that it is impracticable to reproduce a particle of an unknown state.

In view of the fact that Eve will not know which coordinate bases Alice have applicated to encode the bit while she discussed her measurements with Bob, she will be forced to guess. If she measures the states of photons in the wrong coordinate bases, then, as the uncertainty principle of Heisenberg guarantees, the information that was encoded in another coordinate basis will be lost.

Therefore, when a photon reaches Bob, his dimension will be random, and he will receive incorrect information in 50% of cases. Taking into account that the interceptor Eve will choose the basis of measurement incorrectly in 50% cases, on average, then 25% the bits measured by Bob will discern from the bits of Alice. If interceptor Eve captures all possible bits, then, after comparisons of n-bit by Alice and Bob, they would decrease the possibility that attacker Eve can remain unnoticed by a value of $34n$.

Therefore, the probability that the interceptor-attacker Eve has investigated the secret is insignificant if relatively long bit sequences are compared and identified.

3. PROTOCOL BB84-INFO-Z

Modification of BB84 and the considered quantum protocol BB84-Info-Z, is analogous to BB84 [8], except that it applies the following differences:

- Generalized numbers of bits n, n_z and n_x (n-the numbers of informational bits, where Z and X are the test bits, accordingly);
- Section $P = (s, z, b)$ to divide n-bit string i into three non-overlapping sets (I, T_Z и T_X);
- Two special thresholds, which are separate ($p_{A,z}$ and $p_{A,x}$) in place of one threshold (p_A).

Before starting the quantum protocol functioning, Alice and Bob must select some general parameters or public parameters:

- numbers denoted as n, n_z and n_X (specified as the relationship $N = n + n_z + n_x$);

- thresholds of errors denoted as $P_{A,Z}$ and $P_{A,X}$, $r \times n$ (which correspond to the linear code of error correction C);
- matrices of privacy enhancement $m \times n$ (representing the linear key of generation function).

It is important that all rows $R + M$ of the previously considered P_C and P_K matrices are assembled as linearly independent.

Alice randomly selects the section $\mathcal{P}=(s,z,b)$ of the N-bit strings, randomly choosing N-bit strings denoted as, $z, b \in F^N{}_2$, which satisfy the conditions:

$$|\; s \;| = n;\; |\; z \;| = n_Z;\; |\; B \;| = n_X; \text{ and } |\; s + z + b \;| = N.$$

Thereby, the section $\mathcal{PP}$ splits the set of indexes denoted as {1, 2, ..., N} into three non-overlapping and disjointed sets:

- I (information bits, where $s_j = 1$) size n;
- T_Z (test bits Z, where $z_j = 1$) with size n_z;
- T_X (test bits X, where $b_j = 1$) with size n_x.

Alice selects N-bit strings randomly, where $i \in F^N{}_2$, and performs the dispatch of N qubits $\left|i_2^{b_1}\right\rangle, \left|i_2^{b_2}\right\rangle, \ldots, \left|i_N^{b_N}\right\rangle$, one by one, through the quantum information channel. At the same time, Alice uses coordinate basis Z to send the information of test Z-bits, as well as X coordinate basis to send X-bits. Initially, Bob stores every qubit he has received in the quantum memory, without measuring this qubits.

Next, Alice dispatches the string of bits $b = b_1 \ldots b_N$ via the classical channel to Bob. Bob gauges and measures every qubit that he has received and saved. When measuring ith qubit, Bob measures it in the Z-coordinate basis if $b_i = 0$, and measures it in the X-coordinate basis if $b_i = 1$. This string of bits that Bob measured is designated as i^B. If noise and eavesdropping are absent, then the bit string is equal to $i^B = i$.

After that, Alice sends Bob the string of bits, designated as s. The information bits (which will be applied for the final key generating) are n bits $s_j = 1$, while test Z and X bits (which will be applied for testing) are $n_Z + n_X$ with $s_j = 0$. Substrings are denoted by i and b, and correspond to information bits i_s and b_s, accordingly.

Next, Alice and Bob publish together the values of bits, which they obtain for all test bits Z and X, and then the bit values are compared. If, for Alice and Bob, more than $n_Z \cdot p_{a'Z}$ test Z-bits are, or more than $n_X \cdot p_{a'X}$ test X-bits between them are different, then they interrupt this protocol, where $p_{a'Z}$ and $p_{a'X}$ are preliminarily coordinated thresholds of errors. Alice and Bob keep values of residual n bits (information bits where $s_j = 1$) a strict secret. The bit chain of Alice is designated as $x = i_s$, and the bit string of Bob is denoted as x^B.

Then, the syndrome x should be dispatched from Alice to Bob (regarding code C correction of errors and its check on parity of the P_c matrix), which includes r bits and is determined as $\xi = xPTC$. Using the value ξ, Bob rectifies the errors in his string of bits x^B (it is similarly x). The final key is m–bit sequence and is determined as $k = xPTK$. Alice and Bob, together, calculate it. It is obvious that the protocols are very similar.

Let us consider their security against collective attacks **[9,10]**, which are one of the most powerful theoretical cyberattacks.

4. DESCRIPTION OF CYBERATTACK OF EVE AND THE PROPERTIES OF IT

To every j qubit $\left|i_j^{b_j}\right\rangle_{T_j}$ shipped by Alice ($1 \leq j \leq N$), Eve attaches the separate quantum probe (at quantum sensing), which, as it assumed, is in pure state, and applies the unitary transformation U_j to the composite system. Then, she stores her quantum probes in the quantum memory for the subsequent measurements, and dispatches to Bob his part of the system [11].

Therefore, for every qubit there is a certain trial Hilbert space and a certain unitary transformation U_j; they are determined by Eve in advance and, thus, are corrected and fixed for all feasible variants and options of i, b and s.

4.1. Cyberattack of Eve on the Separate Qubit

Because the cyberattack is bitwise, it is possible to focus the analysis on some selected fixed qubit, temporarily discarding subindex j and expressing the general impact of the actions of Eve on the concrete qubit relative to the coordinate basis $|0^b\rangle, |1^b\rangle$

$$U\left|0^E\right\rangle\left|0^b\right\rangle = \left|E_{00}^b\right\rangle\left|0^b\right\rangle + \left|E_{01}^b\right\rangle\left|1^b\right\rangle = \left|\phi_0^b\right\rangle. \tag{1}$$

$$U\left|0^E\right\rangle\left|1^b\right\rangle = \left|E_{10}^b\right\rangle\left|0^b\right\rangle + \left|E_{11}^b\right\rangle\left|1^b\right\rangle = \left|\phi_1^b\right\rangle. \tag{2}$$

where $\left|E_{00}^b\right\rangle, \left|E_{01}^b\right\rangle, \left|E_{10}^b\right\rangle$ and $\left|E_{11}^b\right\rangle$ represent the vectors, or non-normalized states, in the trial Hilbert space of Eve, respective to this concrete qubit. Because the transformation U is unitary, then $|\phi^b{}_0\rangle$ and $|\phi^b{}_1\rangle$ have norm 1 and are orthogonal. This means that

$$\left\langle E_{00}^b \middle| E_{00}^b \right\rangle + \left\langle E_{01}^b \middle| E_{01}^b \right\rangle = 1. \tag{3}$$

$$\left\langle E_{10}^b \middle| E_{10}^b \right\rangle + \left\langle E_{11}^b \middle| E_{11}^b \right\rangle = 1. \tag{4}$$

$$\left\langle E_{00}^b \middle| E_{10}^b \right\rangle + \left\langle E_{01}^b \middle| E_{11}^b \right\rangle = 0 \quad \left\langle E_{10}^b \middle| E_{00}^b \right\rangle + \left\langle E_{11}^b \middle| E_{01}^b \right\rangle = 0. \tag{5}$$

4.2. Spreading the Cyberattack to the Several Qubits–Collective Cyberattack

For every qubit ∈ [1 ... 2*n*] Eve uses the transformation *Uj* in the space $\mathcal{H}^E{}_j \otimes \mathcal{H}_2$, where $\mathcal{H}^E{}_j$ is Eve's trial space, $\mathcal{H}_2$– is the space of qubits. The coordinate basis bj, expressed relative to Eve's point of view, is obtained by tracking Bob from the $|\phi^{bj}{}_0\rangle_j$ and $|\phi^{bj}{}_1\rangle_j$, resulting in the following density matrices:

$$\left(\rho_0^{b_j}\right)_j = \left|E_{00}^{b_j}\right\rangle_j\left\langle E_{00}^{b_j}\right| + \left|E_{01}^{b_j}\right\rangle_j\left\langle E_{01}^{b_j}\right| \tag{6}$$

$$\left(\rho_1^{b_j}\right)_j = \left|E_{10}^{b_j}\right\rangle_j\left\langle E_{10}^{b_j}\right| + \left|E_{11}^{b_j}\right\rangle_j\left\langle E_{11}^{b_j}\right|. \tag{7}$$

If Alice dispatches the string *i* applying the coordinate basis *b*, then the global state of Eve is the tensor product of these states $(\rho^{bj}{}_{ij})_j$. After revealing the test bits [12], Eve requires only those values $(\rho^{bj}{}_{ij})_j$, or which $s_j = 1$. The set $\{j \mid s_j=1\}$ has *n* elements and is global. The global corresponding values *s, b, x* can now be saved as:

$$\rho_x^{b_s} = \left(\rho_{i_{j1}}^{b_{j1}}\right)_{j_1} \otimes \ldots \otimes \left(\rho_{i_{jn}}^{b_{jn}}\right)_{j_n} = \otimes_{l=1}^{n}\left(\rho_{i_{jl}}^{b_l}\right)_{j_l} \tag{8}$$

4.3. Probability of Errors

Supposing that the qubit is under cyberattack by transformation *U*, as determined in (1) and (2), then an error emerges if Alice dispatches 0 and Bob measures 1, or if Alice dispatches 1 and Bob measures 0. Let *k* be the value Bob has measured, *i* the value dispatched by Alice for the concrete qubit, and *b* the coordinate basis applied by Alice for encoding *i*.

Then, the possibility of Bob measuring the error is expressed as:

$$p_\epsilon^b \triangleq \frac{1}{2}\left[\left\langle E_{01}^b | E_{01}^b\right\rangle + \left\langle E_{10}^b | E_{10}^b\right\rangle\right]. \tag{9}$$

4.4. Probability of Errors Occurrence in the Conjugate Basis

Now we study the expression $p_e^{\overline{b}}$, where $\overline{b} = 1 - b$ (when $\overline{0} = 1, \overline{1} = 0$), which responds to the basis, conjugating to the basis that was given and specified as *b*.

Cyberattack *U* is described usually by formulas (1) and (2), in the coordinate basis *b*.

To compute the probability of error when Alice encodes i_j as $\left|i_j^{\overline{b}}\right\rangle$ instead $\left|i_j^b\right\rangle$ it is necessary to represent *U* in the coordinate basis $\overline{b}$. From the Equation (10), it is known that in this circumstance, the probability of errors is determined by the expression:

$$p_\epsilon^{\overline{b}} = \frac{1}{2}\left[\left\langle E_{01}^{\overline{b}} \middle| E_{01}^{\overline{b}}\right\rangle + \left\langle E_{10}^{\overline{b}} \middle| E_{10}^{\overline{b}}\right\rangle\right] \tag{10}$$

Using the fact that

$$|0\rangle^{\bar{b}} = \frac{1}{\sqrt{2}}\left[\left|0^b\right\rangle + \left|1^b\right\rangle\right],\ |1\rangle^{\bar{b}} = \frac{1}{\sqrt{2}}\left[\left|0^b\right\rangle - \left|1^b\right\rangle\right]$$

and using linearity U, expressions are derived from (1) and (2):

$$U\left|0^E\right\rangle\left|0^{\bar{b}}\right\rangle = \frac{1}{\sqrt{2}}\left(\left|E^b_{00}\right\rangle + \left|E^b_{10}\right\rangle\right)\left|0^b\right\rangle + \frac{1}{\sqrt{2}}\left(\left|E^b_{01}\right\rangle + \left|E^b_{11}\right\rangle\right)\left|1^b\right\rangle \tag{11}$$

$$U\left|0^E\right\rangle\left|1^{\bar{b}}\right\rangle = \frac{1}{\sqrt{2}}\left(\left|E^b_{00}\right\rangle - \left|E^b_{10}\right\rangle\right)\left|0^b\right\rangle + \frac{1}{\sqrt{2}}\left(\left|E^b_{01}\right\rangle - \left|E^b_{11}\right\rangle\right)\left|1^b\right\rangle \tag{12}$$

Replacing $|0^b\rangle$ and $|1^b\rangle$ in the right parts of expressions on their magnitudes from the positions $\left|0^{\bar{b}}\right\rangle$ and $\left|1^{\bar{b}}\right\rangle$ I, i.e., $\left|0^b\right\rangle = \frac{1}{\sqrt{2}}\left[\left|0^{\bar{b}}\right\rangle + \left|1^{\bar{b}}\right\rangle\right]$ and $\left|1^b\right\rangle = \frac{1}{\sqrt{2}}\left[\left|0^{\bar{b}}\right\rangle + \left|1^{\bar{b}}\right\rangle\right]$ I, we receive

$$\begin{aligned} U\left|0^E\right\rangle\left|0^{\bar{b}}\right\rangle &= \frac{1}{2}\left[\left|E^b_{00}\right\rangle + \left|E^b_{10}\right\rangle + \left|E^b_{01}\right\rangle + \left|E^b_{11}\right\rangle\right]\left|0^{\bar{b}}\right\rangle \\ &+ \frac{1}{2}\left[\left(\left|E^b_{00}\right\rangle - \left|E^b_{11}\right\rangle\right) + \left(\left|E^b_{10}\right\rangle - \left|E^b_{01}\right\rangle\right)\right]\left|1^{\bar{b}}\right\rangle \end{aligned} \tag{13}$$

$$\begin{aligned} U\left|0^E\right\rangle\left|1^{\bar{b}}\right\rangle &= \frac{1}{2}\left[\left|E^b_{00}\right\rangle - \left|E^b_{11}\right\rangle - \left|E^b_{10}\right\rangle - \left|E^b_{01}\right\rangle\right]\left|0^{\bar{b}}\right\rangle \\ &+ \frac{1}{2}\left[\left(\left|E^b_{00}\right\rangle - \left|E^b_{11}\right\rangle\right) + \left(\left|E^b_{10}\right\rangle - \left|E^b_{01}\right\rangle\right)\right]\left|1^{\bar{b}}\right\rangle \end{aligned} \tag{14}$$

where components for $\left|E^{\bar{b}}_{01}\right\rangle$ and $\left|E^{\bar{b}}_{10}\right\rangle$ are enclosed in parentheses, we can easily see that:

$$p^{\bar{b}}_e = \frac{1}{2}\left[\left\langle E^{\bar{b}}_{01} \mid E^{\bar{b}}_{01}\right\rangle + \left\langle E^{\bar{b}}_{10} \mid E^{\bar{b}}_{10}\right\rangle\right] = \frac{1}{4}\left[\left(\langle E^b_{00}| - \langle E^b_{11}|\right)\left(\left|E^b_{00}\right\rangle - \left|E^b_{11}\right\rangle\right) + \left(\langle E^b_{10}| - \langle E^b_{01}|\right)\left(\left|E^b_{10}\right\rangle - \left|E^b_{01}\right\rangle\right)\right]$$

Let us distribute this result applying Equalities $\langle\varphi|\psi\rangle = \overline{\langle\psi|\varphi\rangle}$ and $\mathcal{Z} + \overline{\mathcal{Z}} = 2Re(\mathcal{Z})$ for $\mathcal{Z} \in C$ (where the complex conjugation is denoted above the line [13]). Using the Equalities (3) and (4), we obtain the following expressions:

$$\begin{aligned} p^{\bar{b}}_e &= \frac{1}{4}\left[2 - \left\langle E^b_{00} \mid E^b_{11}\right\rangle - \left\langle E^b_{11} \mid E^b_{00}\right\rangle - \left\langle E^b_{01} \mid E^b_{10}\right\rangle - \left\langle E^b_{10} \mid E^b_{01}\right\rangle\right] \\ p^{\bar{b}}_e &= \frac{1}{2}\left[1 - \mathrm{Re}\left(\left\langle E^b_{00} \mid E^b_{11}\right\rangle + \left\langle E^b_{01} \mid E^b_{10}\right\rangle\right)\right] \end{aligned} \tag{15}$$

This expression will be applied to relate the perturbation caused by Eve while Alice is encoding ijij bits in the coordinate basis $\bar{b}_j$, so s_j=1, with the information which Eve obtains, while Alice will encode it in this coordinate basis.

According to the principle of «Information against perturbation» [14], the more information Eve receives while the encoding is carried out in the coordinate basis b, the more interference it causes while the bits are encoding and checking in conjugate coordinate basis. Therefore, it is possible to limit Eve's knowledge about the key by limiting the allowable rate (frequency) of errors in this quantum protocol.

5. SECURITY CONFIRMATION OF CLASSICAL PROTOCOL BB84 AGAINST THE COLLECTIVE CYBERATTACKS

Proof of Security

Let us choose such code $\frac{d_{r,m}}{2n} > p_a + \epsilon$, or some small value e; then the expression$2m\sqrt{P\left[\left(\frac{|C_I|}{n} \geq \frac{d_{r,m}}{2n}\right) \wedge \left(\frac{|C_T|}{n} \leq p_a\right)\right]}$willbelessthanvalue $P\ [(\frac{|C_I|}{n} > p_a + \epsilon) \wedge \left(\frac{|C_T|}{n} \leq p_a\right)$; which by itself is exponentially small in n.

We can apply Heffding's selection from [8] (the theorem) for every specific row $c_1 \ldots c_{2n}$, appropriate to all qubit measurements in some valid coordinate basis *b*. Let $\overline{X} = \frac{|C_I|}{n}$ be the average value of selection respective to incorrect bits of information; $\mu = \frac{|C_I|+|C_T|}{2n}$ –is the mathematical expectation $\overline{X}$, which is equal to the expression $2\mu - \overline{X} \leq p_a$, or is identical to inequality $\overline{X} - \mu \geq \mu - p_a$.

To the rows $(\frac{|C_I|}{n} > p_a + \epsilon)$ and $\left(\frac{|C_T|}{n} \leq p_a\right)$ we will rewrite the conditions as:

$$(\overline{X} - \mu > \epsilon + p_a - \mu) \wedge (\overline{X} - \mu \geq \mu - p_a) \tag{16}$$

whence it follows that using Heffding's theorem [15], the relation is obtained:

$$P\left[\left(\frac{|C_I|}{n} > p_a + \epsilon\right) \wedge \left(\frac{|C_T|}{n} \leq p_a\right)\right] \leq P\left[\overline{X} - \mu > \frac{\epsilon}{2}\right] \leq e^{-\frac{1}{2}n\epsilon^2} \tag{17}$$

It is necessary to make sure that the rate of errors in the bits of information is less than the maximum speed at which the error correction code can process.

This condition is necessary for the key to be reliable.

6. SECURITY PROOF FOR THE PROTOCOL BB84–INFO–Z AGAINST THE COLLECTIVE CYBERATTACKS

6.1. General Collective Cyberattack of Eve

Let us assume that, before executing the quantum key distribution (QKD) protocol, Eve chooses to carry out a collective cyberattack [8]. Let the *j*th qubit be given, sent by Alice to Bob. Eve attaches the quantum probe state (at quantum sensing) and applies some unitary operator U_j to the composite system. Then, Eve holds in her quantum memory subsystem E_j, which is the state of her quantum probe; next, she dispatches to Bob the subsystem T_j. This subsystem is the qubit dispatched from Alice to Bob (it may be modified by Eve's cyberattack U_j).

The biggest common collective cyberattack U_j of Eve is directed on the *j*th qubit, presented by orthonormal coordinate basis. The cyberattack described as:

$$U_j\left|0^E\right\rangle E_j\left|0^{b_j}\right\rangle T_j = \left|E_{00}^{b_j}\right\rangle E_j\left|0^{b_j}\right\rangle T_j + \left|E_{01}^{b_j}\right\rangle E_j\left|1^{b_j}\right\rangle T_j \tag{18}$$

$$U_j\left|0^E\right\rangle E_j\left|0^{b_j}\right\rangle T_j = \left|E_{10}^{b_j}\right\rangle E_j\left|0^{b_j}\right\rangle T_j + \left|E_{11}^{b_j}\right\rangle E_j\left|1^{b_j}\right\rangle T_j \tag{19}$$

where $\left|E_{00}^{b_j}\right\rangle E_j, \left|E_{01}^{b_j}\right\rangle E_j, \left|E_{10}^{b_j}\right\rangle E_j$, and $\left|E_{11}^{b_j}\right\rangle E_j$ are non-normalized states in Eve's quantum probe of system E_j, that was fastened and fixed to *j*th qubit.

So, it can be noticed that the quantum probe state can change the initial state of the composite system product, $|0^E\rangle E_j\left|i_j^{b_j}\right\rangle T_j$, to an entangled state. This implies that Eve's cyberattack can cause the entangling of her quantum probe with Bob's probe (at quantum sensing).

Firstly, this can clarify the situation, and bring some kind of information about the state of Bob; secondly, it is the reason for the disturbance, and can be discovered by him. The information received the day before and the disturbance that Eve caused, by their nature, are interrelated with their relationship—this is the main reason why the QKD protocols are secure and safe.

6.2. Proof of Security

As was mentioned earlier [8], the random variable $\widetilde{C}_i$ matches the errors in the string of bits in information bits if they were encoded in the coordinate basis X. Bits of TEST–X are also encoded in basis X. The C_{TX} random variable matches the string of bits of the errors, on these bits. Hence, it is possible to consider the choice of *n*–bits indexes of information (Info) and n_x–bits TEST–X as a random selection (after n, n_z and n_x numbers; and bit indexes TEST–Z, which was selected already) and to apply the theorem of Heffding [9].

Hence, for every string of bits $c_1 \dots c_{n+nx}$, which consists of errors in the bits $n + n_x$ Info and TEST–X, if Info bits were coded in the coordinate basis X, then we can use Heffding's theorem: let us take the sample with size n without changing from combination $c_1, \dots, c_{n+nx}$. In the above discussion [8], the following theorem is actually proven:

Theorem.

Let the values δ > 0 and R > 0; for infinite number of values n, the vectors family is given, which is linearly independent $\{v_1^n, \dots, v_{r_n+m_n}^n\}$*, such, that* $\delta < \frac{d_{r_n,m_n}}{n}, \frac{m_n}{n} \leq R$*. Hence, for anyone Pa,Z, Pa,X>0, such, that* $p_{a,x} + \epsilon_{sec} \leq \frac{\delta}{2}$*, and for any n, nz, nx>0 and two final keys k, k' that are the* m_n*-bit keys, the distance between the states of Eve, appropriate to k and k', meets the following requirements and boundaries:*

$$\langle \Delta_{Eve}^{(p_{a,z},p_{a,x})}(k,k')\rangle \leq 2Rne^{-\left(\frac{n_x}{n+n_x}\right)^2 n\epsilon_{sec}^2}. \qquad (20)$$

6.3. Reliability

By itself, the security is insufficient; it is also necessary that the key be reliable. More specifically, it must be identical for Alice and Bob [16]. It assumes it a need to ensure that quantity of errors in the Info bits was less than the maximum error number, which can be rectified by the error correction code. To do this, it is necessary that the code for error correction can really correct the errors. Consequently, the final key reliability, having exponentially low failure probability, is provided by the inequality below:

$$P\left[\left(\frac{|C_I|}{n} > p_{a,z} + \epsilon_{rel}\right) \wedge \left(\frac{|C_{T_z}|}{n_z} \leq p_{a,z}\right)\right] \leq e^{-2(\frac{n_z}{n+n_z})^2 n \epsilon_{rel}^2} \tag{21}$$

The choice of the Info bits indexes, and the bits of TEST–Z, is random separation of $n + n_z$ bits into two subsets, with sizes n and n_z (provided that the bits indexes of TEST–X was already selected). Thus, it matches Heffding's sample.

7. PROTOCOL CSLOE–2022 (BB84–CSLOE–2022)

It is recommended to become acquainted with the new modification proposed by the authors and named CSLOE–2022, for the old but still effective quantum protocol using the quantum distribution of the key–BB84. For the protocol BB84, it is possible to modify significantly the cryptographic strength and degree of entanglement of the listener, which in their perspective will complicate the possibility of intercepting information from such an interceptor (or cryptanalyst) targeting confidential messages.

It is known that after coordination of the coordinate bases in the classic protocol BB84, an interceptor can receive accurate information about the transmitted state.

The final purpose of modification is to complicate eavesdropping process to the point of uselessness, in terms of spending time and resources, as well as to confirm the guesses about the real possibility of using such a method. The idea consists in the following: it is known that the process of replicating the quantum state, recorded as $\psi \rightarrow \psi \otimes \psi$, (cloning) can be performed perfectly, with probability of 1, then and only then if the coordinate basis to which ψ refers is understood and known [**17**]. Otherwise, the ideal cloning is not possible, since the copies are not perfect. These are contents and consequences of the theorem prohibiting quantum information cloning. This circumstance will be useful.

If it is impossible to reproduce an exact clone of a photon, in order to obtain information from it, it is necessary to measure the characteristics of the original. The only way to measure the characteristics of the photon is to use a detector of single photons. However, as soon the photon reaches the detector, it transmits energy and disappears. That is, the measurement destroys the photon itself [9]. It is worth considering that each photon is unique [18]. However, it is possible to create some kind of photon (kind of similarity). It is known, thanks to quantum teleportation, that it is possible to obtain an exact copy of a photon [17], which, in turn, can be used also to construct such a similarity. For simplicity, we call it a «pseudo-photon». The interceptor will perceive such a pseudo-photon either as a real photon with its own specific set of characteristics, or as some distortion in the channel.

To detect and recognize such a clone in the communication channel, it is necessary to make considerable efforts. In the theory, at the first stage, as in the classic protocol BB84, Alice will communicate with Bob via the quantum communication channel. In turn, Alice will transmit to Bob the modified sequence consisting of cloned pseudo-photons and forming a dictionary (glossary) for each bit with corresponding polarization.

As mentioned earlier, the pseudo-photon will be a kind of photon created artificially [4]. In each concrete case, it is possible to form new sequences that generate the dynamic dictionary, thereby reducing the repetition during encryption.

Knowing that each photon is unique [18], one pseudo-photon cannot be used for each bit, but the whole group with a certain range of values that will be corrected, rectified, adjusted, shifted or expanded can [19].

Each bit, or their sequence, even if they are repeated, will have random pseudo-photons from a certain range of values that are attributed to the concrete bit or bit sequence. When sending the dictionary for decoding, it is possible to send it in parts, to maintain secrecy, and to avoid declassification.

If listening is detected at this first step, the dictionary can be expanded and the intercepted part discarded, as was described earlier [20], or data transmitting can be stopped and a new dictionary created. After successful transmission of the dictionary, it is possible to start sending encoded messages via the communication channel in which real photons, as well as their created copies, will alternate and have absolutely random positions in the sequence. It is worth noting that the protocol can be further complicated. For example, four quantum states can be used to encode bits in two coordinate bases, which corresponds to the quantum protocol BB84 (4 + 2) [21]. Further, such a protocol works according to the classical scenario, but using the dictionary-glossary.

Every time Bob receives a qubit, he reports it to Alice, but does not measure it. Subsequently, for each photon and pseudo-photon which Bob obtains, he will measure the value of polarization on the randomly selected coordinate basis, applying it to his state. If Bob has chosen the same coordinate basis of state for the particular photon, then when he performs transformation H^b, as Alice does, then he goes into the same state. Then, Bob must measure the same polarization in the string i^B. Hence, he can output properly the bit that Alice was going to dispatch in that case, if there is no noise orsigns of eavesdropping in the communication channel.

On the second step, Bob must tell Alice, using any unsecured communication channel, which coordinate basis he applied to quantify and to measure every photon. Then, Alice informs Bob which photons were real by sending the encrypted ranges, and informs him whether he has chosen the right coordinate basis for every original photon.

On this step, Alice and Bob reject the bits corresponding to pseudo-photons and the bits that Bob have quantified and measured in other coordinate bases. If there were no errors and nobody manipulated the photons, Bob and Alice must acquire the same bits string that is named as the sifted key (**Table 1**). The example below (**Figure 3**) demonstrates the bits which Alice selected [22]; the coordinate bases in which she encoded it; and bases that Bob used to measure them. In addition, the sifted key, obtained in the result, is demonstrated after Bob and Alice rejected their bits.

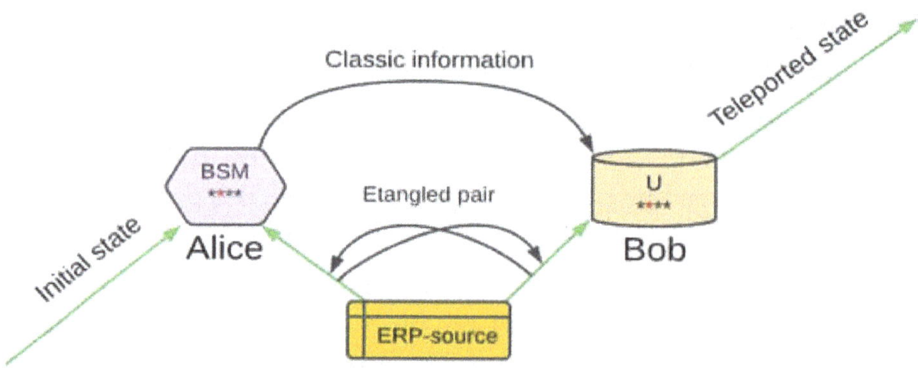

Figure 3. Results of the protocol CSLOE–2022 work.

Table 1. Sifted key.

Bit of Alice	0	1	1	0		1	0	0	1
Basis of Alice	×	+	+	×		+	×	+	×
Polarization of Alice	↗	↑	→	↖		→	↖	↑	↗
Basis of Bob	×	×	×	+		+	×	+	+
Measurement of Bob	↗	↖	↗	→		→	↖	↑	→
Public discussion	×	+	+	*		+	×	+	*
Shared secret key	0					1	0	0	

↑, →, ↖, ↗—is the state of polarization of photon 1, 0 at angles 0°, 45°, 90°, 135°.

The principle of imperfect replication is not new, and it is used often in telecommunications practice. In fact, information transmitted in optical fibers is encoded in the light state; thus, this process is quantum coding [**23**]. The information is strengthened several times on the way from source to receiver; hence, its quality must deteriorate. However, the telecommunication signal includes a large amount of photons being prepared in an identical quantum state. Strengthening in telecommunications boils down to creating several new copies of ψ from $\psi \otimes N$.

That is, the theorem of prohibition of cloning is applied to the amplification of the telecommunication signals, due to the fact that spontaneous radiation is usually available in the amplifiers. Nevertheless, the copy is practically ideal, as the stimulated radiation is the dominant effect. In addition, the sensitivity of modern equipment is quite high, and at this stage is such that quantum limitation must be achieved in the nearest future. Thus, information encoding following the theorem of the prohibition of cloning can also be useful for quantum key distribution protocols [24].

The impossibility of exact and accurate copying of quantum information cannot negate the whole conception of the quantum information. On the contrary, it serves as a demonstration and illustration of power. It is impossible to copy the state of a certain quantum system for intelligent information coding completely using a set of states which are non-orthogonal.

Therefore, if a similar system reaches the receiver undisturbed, that proves that no opponent has copied it. This means that, due to the theorem of the prohibition of cloning, quantum information supports the tools to complete such tasks, which are impossible to solve applying only familiar information. For example, the detection of any eavesdropping device on the communication channel is possible only with the application of the ideas of quantum cryptography.

7.1. Cloning Methods

Let us consider the possible methods of non-ideal cloning of discrete quantum systems, which are performed in the protocol BB84.

There are several variants of such machines:

1. Optimal symmetric universal quantum copying machine (UQCM) proposed by Vladimir Bužek (Buzhek, in various sources is spelled differently)-Mark Hillery (BH) in 1996;
2. Symmetric universal quantum copying machine (UQCM) proposed by Nickolas Gisin-Serge Massar and their scientific group in 1997;
3. Asymmetric universal quantum copying machine (copier)-UQCM.

The symmetric UQCM qubits cloning for $1 \to 2$, developed by Buzhek and Hillery (BH) accepts the cloning qubit as input, and uses the separate qubit as an auxiliary qubit [25]. The action of such a universal quantum-copying machine in the computational base of initial original qubit is described by the expression:

$$\begin{aligned} |0\rangle|\mathrm{R}\rangle|\mathrm{M}\rangle &\to \sqrt{\frac{2}{3}}|0\rangle|0\rangle|1\rangle - \sqrt{\frac{1}{3}}|\psi^{+}\rangle|0\rangle \\ (-|1\rangle)|E\rangle|M\rangle &\to \sqrt{\frac{2}{3}}|1\rangle|1\rangle|0\rangle - \sqrt{\frac{1}{6}}|\psi^{+}\rangle|1\rangle \end{aligned} \tag{22}$$

where $\psi^{+}\rangle = \frac{1}{\sqrt{2}}[|1\rangle|0\rangle + |0\rangle|1\rangle]$. By its linearity, these ratios cause the next action on the most common and general input state: $|\psi\rangle = \alpha|0\rangle + \beta|1\rangle$:

$$|\psi\rangle|R\rangle|M\rangle \rightarrow \sqrt{\frac{2}{3}}|\psi\rangle|\psi\rangle\left|\psi^{\perp}\right\rangle - \sqrt{\frac{1}{6}}\left[|\psi\rangle\left|\psi^{\perp}\right\rangle + \left|\psi^{\perp}\right\rangle|\psi\rangle\right]|\psi\rangle \tag{23}$$

where $\left|\psi^{\perp}\right\rangle = \alpha * |1\rangle - \beta * |0\rangle$.

As can be seen from the Equation (23), α and β can swap their places. In addition, for all input states, this transformation has the similar shape $|\psi\rangle$.

Therefore, the quantum copying machine (QCM) is universal and symmetric. For the original and the copy, its partial states are expressed as:

$$\rho_A = \rho_B = \frac{5}{6}|\psi\rangle\langle\psi| + \frac{1}{6}\left|\psi^{\perp}\right\rangle\left\langle\psi^{\perp}\right| = \frac{1}{2}\left(1 + \frac{2}{3}\vec{m} + \vec{\sigma}\right) \tag{24}$$

The symmetric universal quantum copying machine for NM qubits was developed by Nickolas Gisin and Serge Massar in 1997. It generalizes the Buzhek-Hillery UQCM, and its correctness is determined by the following expression:

$$F_{N\rightarrow M} = \frac{MN + M + N}{M(N+2)}\ (d = 2) \tag{25}$$

which reproduces $F_{1\rightarrow 2} = \frac{5}{6}$ for N=1 and M=2.

Gisin and Massar provided numerical proofs of the optimality of their universal quantum copying machine. Later, Dagmar Bruss, Artur Eckert and Chiara Machiavelli gave analytical proof for this optimality in 1998. They suggested that the output state appertains to the *M* qubits symmetric subspace. Further, Reinhard Werner generalized this outcome for the quantum systems of whatever dimension **[26]**.

The universal and asymmetric cloning relate to such a state of affairs where the resulting clones may have varying accuracy in reproduction.

We should concentrate on universal cloning 1 → 1 + 1. Several researchers studied cases that are more general in 2005. Let us look at some of these ideas with their experimental implementation. In the detailed research of cloning 1 → 1 + 1, Chi-Sheng Niu and Robert Griffiths, in 1998, obtained, in particular, the optimal asymmetric universal quantum copying machine 1 → 1 + 1. Nickolas Cerf in 1998 and 2000 independently obtained the same result. He used the algebraic method, along with Vladimir Bužek and Mark Hillery in 1998 **[27]**, who developed the method of quantum schemes, which was improved compared to the previous construction used for symmetric cloning.

Optimality is illustrated by confirming and proving the accuracy of reproduction of two clones, F_A and F_B, that confirms the inequality of non-cloning:

$$\sqrt{(1-F_A)(1-F_B)} \geq \frac{1}{2} - (1-F_A) - (1-F_B) \tag{26}$$

The authors managed to extend the development of their schemes far beyond the individual case of qubits. Based on the above, to improve the cryptographic strength of protocol BB84, the universal quantum copying machine (UQCM) of Bužek-Hillery is best suited, as it is the simplest quantum copier to implement all the concepts discussed above.

7.2. Comparison of Protocols

Since the protocols BB84, BB84-Info-Z and the CSLOE-2022 modification proposed by the authors have a lot in common, several parameters can be distinguished for their comparison. Let us start by comparing the threshold of errors.

This parameter is necessary to determine whether there was eavesdropping. In the practical implementation of the protocol of quantum keys distribution (QKD), the disadvantages of individual separate components will always show and manifest themselves, and some qubits will be unsuitable for the forming of a secret key.

In addition, listening to the quantum channel makes changes to the transmitted qubits, which also prevents them from being used when forming the secret key.

In the case of the classic protocol BB84, the threshold value of the error rate is 11% [**28**]. For an ideal model, the number of bits received as the result, the final secret key (R) for the bit of the sifted key, is expressed by the relationship:

$$R = 1 - 2\,H(QBER), \tag{27}$$

where the value H is the binary entropy of Shannon and QBER is number of errors measured by Bob. The dependence of R on QBER is demonstrated in the **Figure 4**:

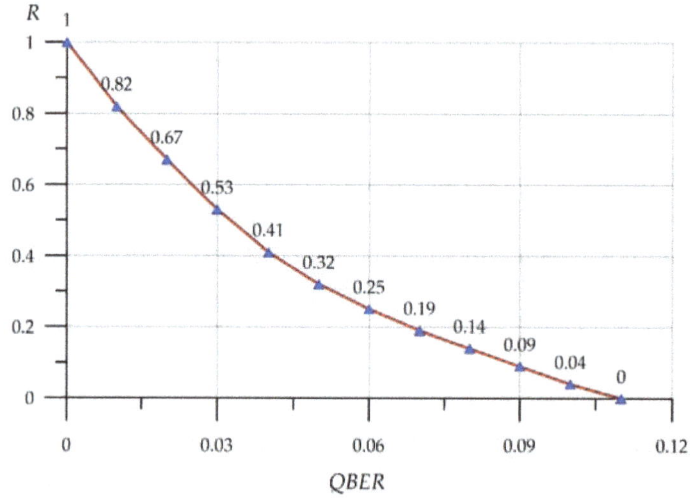

Figure 4. Error threshold for protocol BB84.

This may not always be the case. The quality of execution of the equipment implementing the protocol may allow reducing the threshold.

In the protocol BB84-Info-Z, in addition to the information qubits responsible for key generation, the test qubits X and Z are used [7].

They are necessary for the eavesdropping check. The error threshold for this modification is slightly smaller, at 7.56% [9] (**Figure 5**).

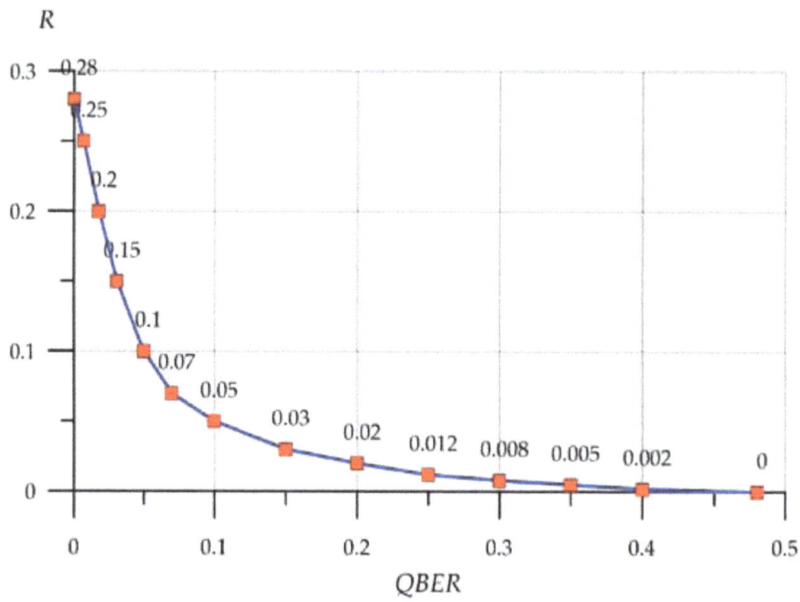

Figure 5. Safe zone of asymptotic rates of errors for BB84-Info-Z protocol.

For the modification proposed by the authors of CSLOE-2022, the threshold of errors is theoretically similar to the original one, since during key formation, imperfect copies of photons used to achieve the entanglement of a listener (eavesdropping interceptor) are discarded, and then the originals are checked, as in the classical protocol. On the other hand, when trying to listen, it is unlikely that the photons will be distorted, since their quantity is significantly lower than the number of copies.

From publication [17] it is known that the code of Steane (CSS), otherwise known as Calderbank–Shor–Steane, is the tool of quantum error correction, introduced by Andrew Steane in 1996. Steane's code uses the classical binary Hamming's code to correct the errors of the qubit flip (X-errors) and the double Hamming code to correct the errors of the phase flip (Z-errors), in practice allowing:

$$1 - 2 * H(\delta)]n \tag{28}$$

qubits, where **δ**-number of measured errors, and *n*-length of qubits sequence.

If expression (28) would be written as:$f(x)$=1-2∗$H(x)$, then the chart (graph) will intersect the axis X at the point 0.11, which gave us the threshold of errors of 11% in the classical quantum protocol BB84 [29].

For the protocol CSLOE-2022, the dependence of the number of bits of the final resulting secret key *R* from the quantity of recorded and fixed errors QBER is preserved, as is shown in **Figure 6**, since all photons, both the original and their copies in the form of pseudo-photons, will be distorted during transmission. Thus, the comparison of protocols by the threshold of errors can be observed in **Table 2**:

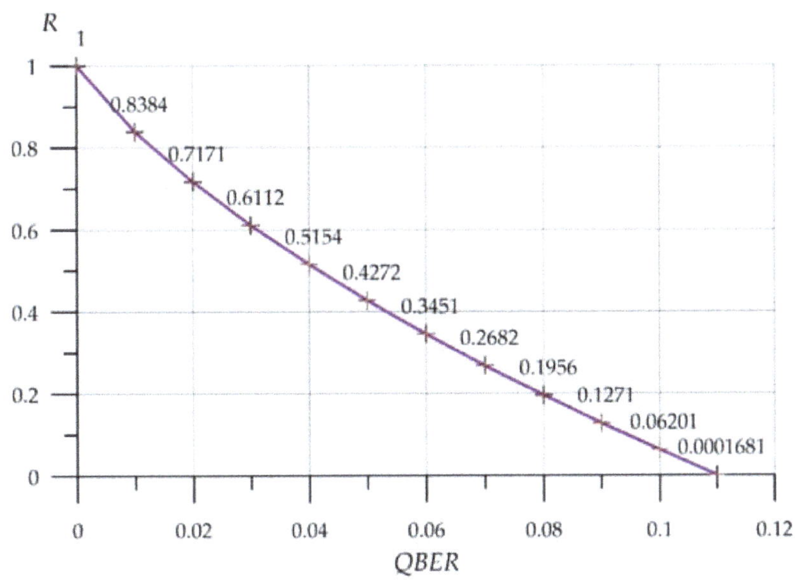

Figure 6. The threshold of errors for the modification CSLOE-2022 proposed by the authors.

Table 2. Comparative characteristics of protocols by threshold of errors.

Protocol	Error Threshold
BB84	11%
BB84–Info–Z	7.56%
CSLOE–2022	11%

The protocol BB84–Info–Z differs for the better from all the others, since a much smaller threshold of errors will allow the eavesdropping intruder to receive significantly less information about the secret key.

Another important parameter is the working distance. The protocols of quantum keys distribution (QKD) operate with single photons, which can be distorted during transmission. Therefore, the working distance is relatively small. For example, for protocol BB84, it is about 70 km [30]. When implementing the protocols DPS and COW, it became possible to reach distances of 250 and 307 km, respectively, but their safety has not yet been proven.

Based on the findings of the research, it was suggested that in the case of using pseudo-photons, it is impossible to declare the numbers of photon losses correctly. If the losses occur from the number of pseudo-photons, and they are used as the trap and entanglement, they may have completely different behavior and completely different interactions, which can lead to both a decrease in losses, or an increase. To declare this confidently without conducting a series of experiments is incorrect and unpredictable. In the case of the protocols BB84 and BB84–Info–Z, the loss of photons is particularly significant. However, the feature of the CSLOE–2022 protocol is its modularity and application of pseudo-photons, which fundamentally distinguishes it from its predecessors.

Therefore, it is possible to make another hypothesis to rely on it, and refer to the fact that the method is based on the cloning of pseudo-photons. At large distances, the transmission of information with photons is carried out with large losses of ($\eta \simeq 0$), and this limitation can be exceeded only by using quantum repeaters. In turn, they will have to be used without limitation for the transmission of recreated pseudo-photons. In addition, it is necessary to be able to combine them with ordinary photons which can be transmitted without repeaters at all.

It is known that quantum communications guarantee the reliable transmission of quantum information and efficient distribution of entanglement, together with the generation of completely secure keys. However, it should be borne in mind that, at long distances, photon transmissions incur significant losses. Quantum repeaters can surpass this limitation. The article [31] discusses the theoretical aspects of the possible limit for ensuring the transmission of information without repeaters.

Relying on this article is possible if the objective is to increase the distance between repeaters. However, the studies themselves given in [**31**] are difficult to use for the clear practical application without conducting of series of experiments, and especially in combination of multi-qubit states, which is what some scientists are trying to practice now.

However, in the case considered by the authors, pseudo-photons are used (that is, recreated from real photons with clear differences), which, in turn, may display completely different, unpredictable behavior. This can lead to both a decrease in the losses or an increase in them. In the case of real photons, for the channel with losses, the article [31] proves that Q_2=K=-$\log_2(1-\eta)$, where η is the coefficient of transmission. In particular, the transmission capacity of the secret key of the channel with losses is the maximum speed achievable with any optical implementation of the quantum key distribution (QKD). At large distances, that is, with large losses, $\eta \simeq 0$, finding the optimal scaling of losses and the speed of $K \simeq 1.44\,\eta$ secret bits per channel use, the fundamental limitation is obtained, which, at this point in time, can be overcome, practically, only with repeaters.

A team of Chinese physicists succeeded in transmitting a secret quantum key at a distance of 511 km in real-world conditions [20]. They were thus able to implement the quantum line of data transmission outside laboratory conditions.

Physicists continue to search for different ways to increase the distance to hundreds and thousands of kilometers, developing repeaters for existing protocols, as well as new ones. For example, the protocol of the twin fields TF (Twin Field). Unlike the standard protocol, for example, BB84, in which Alice directly sends photons to Bob, the protocol TF [32] includes an additional Charlie node, which is located between Alice and Bob. In that case, Alice and Bob carry out the transfer and transmit their information on the weak coherent pulse to Charlie, who compares them and announces whether the received bits coincide or not.

However, Charlie has no information about the bits he has received; he can only compare them and declare whether they have coincided at the moment or not, so Charlie is considered to be an untrusted node.

This approach allows exceeding the known limit of the key generation rate without repeaters. It uses two photon sources, the phases of which, in addition to hundreds of kilometers of optical fiber between them, are not easy to stabilize.

Thus, we can summarize the comparison of protocols BB84, BB84–Info–Z and CSLOE–2022 by working distance and record the results in **Table 3**:

Table 3. Working distance of protocols.

Protocol	Working Distance
BB84	70 km
BB84-Info-Z	70 km
CSLOE-2022	511 km

8. CONCISE SUMMARY

The article shows that the quantum protocol BB84-Info-Z is protected completely from collective attacks—one of the most powerful cyberattacks.

It is found that the results of the quantum protocol BB84 have much in common with the BB84-Info-Z quantum protocol, with two significant exceptions:

The rate of errors should be checked separately so that it remains below the threshold values *pa, z* and *pa, x* for bits TEST-Z and TEST-X; accordingly, when in the quantum protocol BB84, the threshold value of error rate p_a is applied to all bits of TEST jointly [7].

The indexes and information indicators of the interceptor Eve (in security terms) and probability of the error-correcting code failure (in reliability terms) differ from the indexes and indicators in the case of classical quantum protocol BB84 [8].

Consequently, modifying the quantum protocol BB84 to restrict the bits Info to only the coordinate basis Z does not compromise its reliability and security, particularly against collective cyberattacks. It will not have any impact on the threshold of the asymptotic error rate [8]. The BB84-Info-Z protocol can be securely used for distributing a secret key. It has a flawless implementation and is resistant to collective cyberattacks. The authors propose a modification of the classical quantum protocol BB84 (CSLOE-2022) that can be utilized for quantum key distribution. This modification involves the use of imperfect copying, which does not violate the rules of physics but enhances the cryptographic strength of the protocol. Currently, this alteration is merely a concept and requires validation through a sequence of empirical tests with specialized apparatus. Nevertheless, the concept of imperfect replication has been successfully tested in communication channels, as referenced in [33,34,35,36,37,38,39,40,41,42,43,44,45,46,47]. This experimental application enables the transmission of information across far greater distances. However, quantum key distribution necessitates additional criteria for the performance quality of individual components. Consequently, these novel concepts are being proposed, and engineers are assigned precise assignments to validate the accuracy of the new hypothesis. In addition, the authors intend to investigate the optical quantum memory of photons and pseudo-photons in order to explore the potential for storing information within them. From the acquired data, it is imperative to identify the theoretical constraints and devise strategies to mitigate and counteract them, for both photons and pseudo-photons derived from them.

At a Glance

Quantum cryptography techniques enable the secure transmission of data in a communication channel, preventing unwanted access by attackers and allowing the detection of any attempted interception. This scientific field is presently pertinent as it is connected to the issue of security and data safeguarding in contemporary information and communication networks. The article focuses on quantum cryptography and discusses the advancement of quantum protocols in the form of quantum key distribution systems. The development of safe data transfer protocols is demonstrated based on the principles of quantum physics. The authors regarded BB84 as the most well-known protocol for quantum key distribution. A more contemporary variant of this protocol is BB84 Info-Z. A comparative analysis of these factors has also been conducted. The superiority of the BB84-Info-Z quantum protocol over BB84 has been confirmed due to its higher efficiency resulting from a lower error threshold. This reduced error threshold significantly limits the amount of information that can be obtained by an interceptor regarding the secret key. The authors propose a novel concept to enhance the BB84 protocol, which has become outmoded during the past four decades, in response to the growing demands of contemporary quantum cryptography protocols. The update is referred to as CSLOE-2022. It enhances the cryptographic strength and entanglement level of the interceptor (cryptanalyst), making it extremely difficult to intercept information. The primary objective of the CSLOE-2022 modification is to render the eavesdropping process exceedingly intricate, hence rendering it entirely futile for an attacker in terms of time and resource expenditure. The change enables surpassing the established speed limit for key generation without the need for repeaters. This is achieved by utilizing two sources, whose phases, along with the extensive distance of fiber between them, pose significant challenges in terms of stabilization. The comparison of the protocols based on working distance revealed that the BB84 protocol has a maximum distance of 70 km. Similarly, the BB84-Info-Z protocol also has a maximum distance of 70 km. However, the authors propose a modification to the CSLOE-2022 protocol that theoretically allows for a significant increase in the working distance of the quantum protocol to 511 km, which is 7.3 times greater.

REFERENCES

1. Bennett, C.H.; Brassard, G. Quantum Cryptography: Public Key Distribution and Coin Tossing. In Proceedings of the IEEE International Conference on Computers, Systems and Signal Processing, Bangalore, India, 10–12 December 1984. **[Google Scholar]**

2. Aiello, C.D.; Hirose, M.; Cappellaro, P. Composite-pulse magnetometry with a solid-state quantum sensor. *Nat. Commun.* **2013**, *4*, 1419. **[Google Scholar]** **[CrossRef]** **[PubMed]** **[Green Version]**

3. Kröll, S.; Gallo, K.; Hennrich, M. *Research in Quantum Sensing*; Wallenberg Centre for Quantum Technology, Science Advances; Chalmers University of Technology: Gothenburg, Sweden, 2022. **[Google Scholar]** **[CrossRef]** **[Green Version]**

4. Riexinger, F.; Kutas, M.; Haase, B.; Bortz, M. *General Simulation Method for Quantum-Sensing Systems*; Institute for Industrial Mathematics ITWM, Department of Physics and Research Center OPTIMAS: Kaiserslautern, Germany, 2022; Available online: **https://arxiv.org/pdf/2112.07243v1.pdf** (accessed on 28 November 2022).

5. Schowengerdt, R. *Remote Sensing. Models and Methods for Image Processing*; Elsevier: Amsterdam, The Netherlands, 2007. **[Google Scholar]**

6. Molotkov, S.N. On the Resistance of Quantum Cryptography Systems with Phase-Time Coding to Active Probing Attacks. *J. Exp. Theor. Phys.* **2008**, *158*, 1011–1031. **[Google Scholar]**

7. Park, J.L. The concept of transition in quantum mechanics. *Found. Phys.* **1970**, *1*, 23–33. **[Google Scholar]** **[CrossRef]**

8. Boyer, M.; Liss, R.; Mor, T. Composable Security against Collective Attacks of Modified BB84 QKD Protocol with Information Only in One Basis. *Theor. Comput. Sci.* **2020**, *801*, 96–109. **[Google Scholar]** **[CrossRef]**

9. Biham, E.; Mor, T. Security of Quantum Cryptography against Collective Attacks. *Phys. Rev. Lett.* **1997**, *78*, 2256–2259. **[Google Scholar]** **[CrossRef]** **[Green Version]**

10. Boyer, M.; Gelles, R.; Mor, T. Security of the Bennett–Brassard Quantum Key Distribution Protocol against Collective Attacks. *Algorithms* **2009**, *2*, 790–807. **[Google Scholar]** **[CrossRef]** **[Green Version]**

11. Vercruysse, D.; Sapra, N.; Yang, K.; Vukovi, J. Inverse–Designed Photonic Crystal Devices for Optical Beam Steering. *arXiv* **2021**, arXiv:2102.00681. **[Google Scholar]** **[CrossRef]**

12. Buckley, S.; Radulaski, M.; Zhang, J.; Petykiewicz, J.; Biermann, K.; Vučković, J. Nonlinear Frequency Conversion Using High Quality Modes in GaAs Nanobeam Cavities. *Opt. Lett.* **2014**, *39*, 5673–5676. **[Google Scholar]** **[CrossRef]** **[Green Version]**

13. Cerf, N.; Ipe, A.; Rottenberg, X. Cloning of Continuous Quantum Variables. Ecole Polytechnique, CP 165. *Phys. Rev. Lett.* **2000**, *85*, 1754–1757. **[Google Scholar]** **[CrossRef]**

14. Fuchs, C.; Peres, A. Quantum-State Disturbance versus Information Gain: Uncertainty Relations for Quantum Information. *Phys. Rev.* **1996**, *53*, 2038–2045. **[Google Scholar]** **[CrossRef]**

15. Skori'c, B.; Wolfs, Z. Diagrammatic Security Proof for 8-State Encoding. *arXiv* **2021**, arXiv:2103.01936v1. **[Google Scholar]**

16. Morimae, T. *Quantum Randomized Encoding, Verifification of Quantum Computing, No-Cloning, and Blind Quantum Computing*; Yukawa Institute for Theoretical Physics, Kyoto University: Kyoto, Japan, 2020. **[Google Scholar]**

17. Schimpf, C.; Reindl, M.; Huber, D.; Lehner, B.; Silva, S.; Manna, S.; Vyvlecka, M.; Walther, P. Quantum Cryptography with Highly Entangled Photons from Semiconductor Quantum Dots. *arXiv* **2020**, arXiv:2007.12726v1. **[Google Scholar]** **[CrossRef]** **[PubMed]**

18. Tan, X. *Introduction to Quantum Cryptography*; IntechOpen: London, UK, 2013. **[CrossRef]** **[Green Version]**

19. Shor, P.; Preskill, J. *Simple Proof of Security of the BB84 Quantum Key Distribution Protocol*; AT&T Labs Research: Florham Park, NJ, USA, 2000. **[Google Scholar]**

20. Huttner, B.; Imoto, N.; Gisin, N.; Mor, T. Quantum cryptography with coherent states. *Phys. Rev. A* **1995**, *51*, 1863–1869. **[Google Scholar]** **[CrossRef]** **[PubMed]** **[Green Version]**

21. Djordjevic, I. *Quantum Information Processing, Quantum Computing, and Quantum Error Correction*, 2nd ed.; Academic Press: Cambridge, MA, USA, 2021. **[Google Scholar]** **[CrossRef]**

22. Bouwmeester, D.; Pan, J.-W.; Mattle, K.; Eibl, M.; Weinfurter, H.; Zeilinger, A. Experimental quantum teleportation. *Nature* **1997**, *390*, 575–579. **[Google Scholar]** **[CrossRef]** **[Green Version]**

23. Houça, R.; Belouad, A.; El Choubabi, B.; Kamal, A.; El Bouziani, M. Quantum teleportation via a two-qubit Heisenberg XXX chain with x-component of Dzyaloshinskii-Moriya interaction. *J. Magn. Magn. Mater.* **2022**, *563*, 169816. **[Google Scholar]** **[CrossRef]**

24. Yang, L.; Liu, Y.C.; Li, Y.S. Quantum Teleportation of Particles in an Environment. *Chin. Phys. B* **2020**, *29*, 060301. **[Google Scholar] [CrossRef]**
25. Chen, J.; Zhang, C.; Liu, Y.; Li, Y.; Liu, H.; Jiang, H.; Chen, T.; Zhang, Q.; Pan, J. Twin-Field Quantum Key Distribution over 511 km Optical Fiber Linking two Distant Metropolitans areas. *Nat. Photon.* **2021**, *15*, 570–575. **[Google Scholar] [CrossRef]**
26. Nang Paing, S.; Setiawan, J.W.; Tariq, S.; Talha Rahim, M.; Lee, K.; Shin, H. Counterfactual Anonymous Quantum Teleportation in the Presence of Adversarial Attacks and Channel Noise. *Sensors* **2022**, *22*, 7587. **[Google Scholar] [CrossRef] [PubMed]**
27. Gisin, N. *Quantum Randomness. Non-Locality, Teleportation and Other Quantum Wonders*; Alpina non-fiction: Moscow, Russia, 2018; 208p. **[Google Scholar]**
28. Safaryan, O.A.; Lemeshko, K.S.; Beskopylny, A.N.; Cherckesova, L.V.; Korochentsev, D.A. Mathematical Analysis of Parametric Characteristics of the Consensus Algorithms Operation with the Choice of the Most Priority One for Implementation in the Financial Sphere. *Electronics* **2021**, *10*, 2659. **[Google Scholar] [CrossRef]**
29. Wang, Y.; Hu, M.-L. Quantum Teleportation and Dense Coding in Multiple Bosonic Reservoirs. *Entropy* **2022**, *24*, 1114. **[Google Scholar] [CrossRef] [PubMed]**
30. Wen, X.; Chen, Y.; Zhang, W.; Jiang, Z.L.; Fang, J. Blockchain Consensus Mechanism Based on Quantum Teleportation. *Mathematics* **2022**, *10*, 2385. **[Google Scholar] [CrossRef]**
31. Pirandola, S.; Laurenza, R.; Ottaviani, C.; Banchi, L. Fundamental limits of repeaterless quantum communications. *Nat. Commun.* **2017**, *8*, 15043. **[Google Scholar] [CrossRef] [PubMed] [Green Version]**
32. Lucamarini, M.; Yuan, Z.; Dynes, J.; Shields, A. Overcoming the Rate-Distance Limit of Quantum Key Distribution without Quantum Repeaters. *Nature* **2018**, *557*, 400–403. **[Google Scholar] [CrossRef] [PubMed] [Green Version]**
33. Yan, F.; Wang, D. Probabilistic and controlled teleportation of unknown quantum states. *Phys. Lett. A* **2003**, *316*, 297–303. **[Google Scholar] [CrossRef]**
34. Khawasik, M.; El-Sayed, W.G.; Rashad, M.Z.; Younes, A. A Secured Half-Duplex Bidirectional Quantum Key Distribution Protocol against Collective Attacks. *Symmetry* **2022**, *14*, 2481. **[Google Scholar] [CrossRef]**
35. Cardoso-Isidoro, C.; Delgado, F. Shared Quantum Key Distribution Based on Asymmetric Double Quantum Teleportation. *Symmetry* **2022**, *14*, 713. **[Google Scholar] [CrossRef]**

36. Blunt, N.S.; Camps, J.; Crawford, O.; Izs'ak, R.; Leontica, S.; Mirani, A.; Moylett, A.E.; Scivier, S.A.; S¨underhauf, C.; Schopf, P.; et al. A Perspective on the Current State-of-the-art of Quantum Computing for Drug. *arXiv* **2022**, arXiv:2206.00551. **[Google Scholar]**

37. Chamberland, C.; Noh, K.; Arrangoiz-Arriola, P.; Campbell, E.T.; Hann, C.T.; Iverson, J.; Putterman, H.; Bohdanowicz, T.C.; Flammia, S.T.; Keller, A.; et al. Building a Fault-Tolerant Quantum Computer Using Concatenated Cat Codes, PRX Quantum 3. *arXiv* **2022**, arXiv:2012.04108. **[Google Scholar]**

38. Chamberland, C.; Campbell, E.T. Universal Quantum Computing with Twist-Free and Temporally Encoded Lattice Surgery. *PRX Quantum* **2022**, *3*, 010331. **[Google Scholar] [CrossRef]**

39. Kivlichan, I.D.; Gidney, C.; Berry, D.W.; Wiebe, N.; McClean, J.; Sun, W.; Jiang, Z.; Rubin, N.; Fowler, A.; Aspuru-Guzik, A.; et al. Improved Fault-Tolerant Quantum Simulation of Condensed-Phase Correlated Electrons via Trotterization, Quantum 4. *arXiv* **2020**, arXiv:1902.10673. **[Google Scholar] [CrossRef]**

40. Lu, D.; Li, Z.; Yu, J.; Han, Z. A Verifiable Arbitrated Quantum Signature Scheme Based on Controlled Quantum Teleportation. *Entropy* **2022**, *24*, 111. **[Google Scholar] [CrossRef] [PubMed]**

41. Hermans, S.L.; Pompili, M.; Beukers, H.K.C.; Baier, S.; Borregaard, J.; Hanson, R. Qubit Teleportation between Non-nbeighbouring Nodes in a Quantum Network. *Nature* **2022**, *605*, 663–668. **[Google Scholar] [CrossRef]**

42. Sun, Q.-C.; Mao, Y.-L.; Chen, S.-J.; Zhang, W.; Jiang, Y.-F.; Zhang, Y.-B.; Miki, S.; Yamashita, T.; Terai, H.; Jiang, X.; et al. Quantum teleportation with independent sources and prior entanglement distribution over a network. *Nat. Photon.* **2016**, *10*, 671–675. **[Google Scholar] [CrossRef]**

43. Xu, J.; Chen, X.; Xiao, H.; Wang, P.; Ma, M. A Performance-Consumption Balanced Scheme of Multi-Hop Quantum Networks for Teleportation. *Appl. Sci.* **2021**, *11*, 10869. **[Google Scholar] [CrossRef]**

44. Wu, H.; Liu, X.; Zhang, H.; Ruan, X.; Guo, Y. Performance Analysis of Continuous Variable Quantum Teleportation with Noiseless Linear Amplifier in Seawater Channel. *Symmetry* **2022**, *14*, 997. **[Google Scholar] [CrossRef]**

45. Benatti, F.; Floreanini, R.; Marzolino, U. Entanglement and Non-Locality in Quantum Protocols with Identical Particles. *Entropy* **2021**, *23*, 479. **[Google Scholar] [CrossRef]**

46. Raj, R.; Banerjee, S.; Panigrahi, P.K. Remote State Design for Efficient Quantum Metrology with Separable and Non-Teleporting States. *Quantum Rep.* **2021**, *3*, 228–241. **[Google Scholar] [CrossRef]**

47. Liss, R.; Mor, T. Quantum Communication—Celebrating the Silver Jubilee of Teleportation. *Entropy* **2020**, *22*, 628. **[Google Scholar]** **[CrossRef]** **[PubMed]**

8 Chapter

INVESTIGATION OF CHARGE TRANSPORT PROPERTIES IN VTP: PC71BM ORGANIC SCHOTTKY DIODE

1. INTRODUCTION: AN OVERVIEW

Organic electronic devices have progressively entered the commercial sector as substitutes for traditional inorganic electronic gadgets. Many researchers have been motivated to hunt for new materials in order to enhance the performance of organic semiconductor semiconductors. Metal phthalocyanines (MPcs) have garnered significant attention due to their attractive advantages, such as cost-effectiveness, environmentally benign organic composition, and robust thermal and chemical durability. In optoelectronic applications, organic materials have been extensively studied in the past few decades to address various limitations of their inorganic counterparts. The distinctive characteristics of organic materials, such as heightened sensitivity, cost efficiency, eco-friendliness, and the ability to be easily fabricated using methods like spin coating, drop casting, dip coating, and spray coating from liquid solutions, make them advantageous for creating organic photovoltaics (OPVs) [1,2]. In recent years, there has been significant research conducted on bulk heterojunction (BHJ) organic photovoltaics (OPV) and organic photodiodes (OPDs). These devices include blending an acceptor and a donor in the photoactive medium to create a donor/acceptor (D/A) interface at the molecular level [3,4,5]. MPcs, which are comprised of heterocyclic macrocyclic organic compounds, have undergone comprehensive evaluation for numerous applications [6,7]. Multiple metal phthalocyanines, such as VOPcPhO [8,9,10,11,12,13], were employed for the assessment of organic photovoltaics (OPVs) and organic photodetectors (OPDs) [8,9,10,11]. Organic field effect transistors (OFET) and organic photovoltaics (OPVs) have the potential to achieve significant progress by utilizing a basic device known as a metal/organic/metal-based Schottky diode [14]. One electrode in a Schottky diode functions as an Ohmic injector, while the other electrode acts as a blocking element [14]. The J-V characteristic's slope and intercept in the linear and saturation zones can be utilized to compute the ideality factor and carrier mobility, correspondingly [14]. A Schottky diode is a type of diode that consists of a metal-semiconductor junction and exhibits rectification capabilities, similar to a p-n junction diode [15]. A Schottky diode is classified as a majority carrier device due to its rapid switching speed [15]. Both p-type and n-type semiconductors can be utilized to create Schottky diodes using various semiconducting materials, including organic semiconductors [15]. The metals Pd, Pt, Au, Ti, Al, W, Cr, molybdenum, and certain metal silicides are utilized to create a Schottky contact with semiconductors [15]. Prior research has examined, produced, and evaluated different arrangements of Schottky diodes [16,17]. The present study introduces a vanadium metal complex, specifically the MPc compound, which is commonly referred to as VTP. Aside from its distinct extended ligands, the aromatic structure of VTP closely resembles that of VOPcPhO. The ligands in VTP have the ability to alter the molecule's sensitivity to absorbing light [18,19]. Currently, VTP has not been employed as the primary layer in the production of OPVs. This study utilized a mixture of two organic compounds with different absorption capabilities. The inclusion of a network structure is expected to enhance the separation of charges and the transportation of photogenerated charge carriers [20]. VTP is presented as a donor material with unique absorption characteristics, particularly in the lower visible spectrum [21]. The VTP and PC71BM exhibit remarkable hole-transport and electron-transport capabilities, respectively [22], which makes them very desirable as donor and acceptor components. The absorption of both substances is crucial for catching particular light wavelengths within the visible spectrum. The VTP and PC71BM matrix are projected to have a synergistic absorption profile, resulting in an increase in light absorbance in the lower wavelength area of the visible spectrum. Previously, we documented the utilization of a composite mixture consisting of VTP:PC71BM as an organic photodetector (OPD). The increased sensing performances of the OPD can be attributable to the careful selection of donor and acceptor components, namely VTP and PC71BM, respectively [19]. Hence, in this study, the VTP: PC71BM composites were employed to examine the charge transport characteristics and explore the photovoltaic impact of the composite blend.

2. EXPERIMENTAL SECTION

VTP and $PC_{71}BM$ were purchased from Sigma Aldrich (St. Louis, MO, USA) and used without modification. Each material was separately dissolved in chloroform to produce a 10 mg/mL solution, which was then continuously stirred for hours. VTP and $PC_{71}BM$ were mixed in three different volume ratios (1.0:1.0, 1.0:1.2, and 1.0:1.4), then stirred in a nitrogen-filled glove box for an hour. The experimental approach was briefly discussed in our earlier publication [19].

3. RESULTS AND DISCUSSION

Figure 1 depicts (a) the molecular structure of VTP and $PC_{71}BM$, (b) the schematic diagram of OPVs and OPDs, (c) a cross-section Field Effect Scanning Electron Microscopy (FESEM) image of the prepared samples, (d) the energy level diagram reported previously and (e) schematic diagram of energy levels for Schottky diode [**19,23**]. According to **Figure 1**c, the thicknesses of the PEDOT: PSS, active layer (VTP: $PC_{71}BM$), and aluminum (Al) were 40, 65, and 75 nm, respectively.

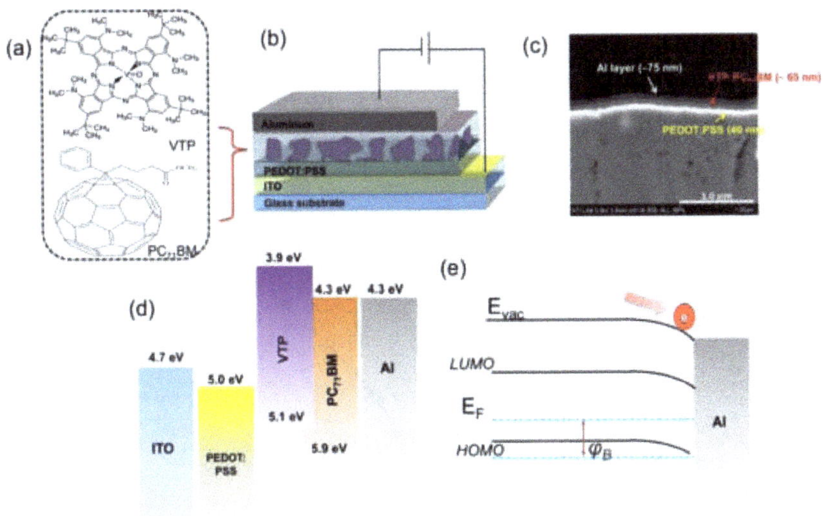

Figure 1. (a) The molecular structure of VTP (Roslan et al., 2018) [19] and $PC_{71}BM$ (b) device structure [19], (c) cross-section image of the prepared sample [19], (d) Energy level diagram [19], and (e) schematic diagram of energy levels of Schottky diode.

The measurement of current-voltage (I-V) characteristics in a dark setting is extremely useful for evaluating junction properties. This analysis produced significant parameters such as the ideality factor, rectification ratio (RR), reverse saturation current, barrier height, shunt and series resistance. The illumination setting was omitted due to minor fluctuations in the light intensity that could cause significant noise to the device, making production difficult. In the case of dark I-V measurements, charge carriers are injected into the circuit using current rather than light-generated carriers. **Figure 2** depicts the semi-logarithmic plot of I-V curves of the ITO/PEDOT: PSS/VTP/Al device, which was analyzed in dark conditions at room temperature. According to **Figure 2**, the device displayed a rectifying behavior that resembled a non-linear and asymmetric diode at room temperature (300 K). The RR is the forward current to reverse current (I_F/I_R) ratio at a given applied voltage. Its value is determined by the bias, which reflects the increased charge injection into the active layer. The RR value of the tested device was 1.6 at ± 0.5± 0.5 V, which is attributable to the formation of the space charge layer at the interface [24]. The recorded RR value is comparable with the MPc group, as reported in past studies [20,25,26]. Predominantly, in relation to the Schottky barrier diode, the thermionic emission theory predicts that the *I-V* characteristics at the forward bias are given as follows:

$$I = I_o \exp\left(\frac{qV}{nkT}\right)\left[1 - \exp\left(\frac{-qV}{nkT}\right)\right] \tag{1}$$

where *Io* is given as follows:

$$I_o = AA^{*}T^{2}\exp\left(\frac{-q\Phi_b}{k_B T}\right) \tag{2}$$

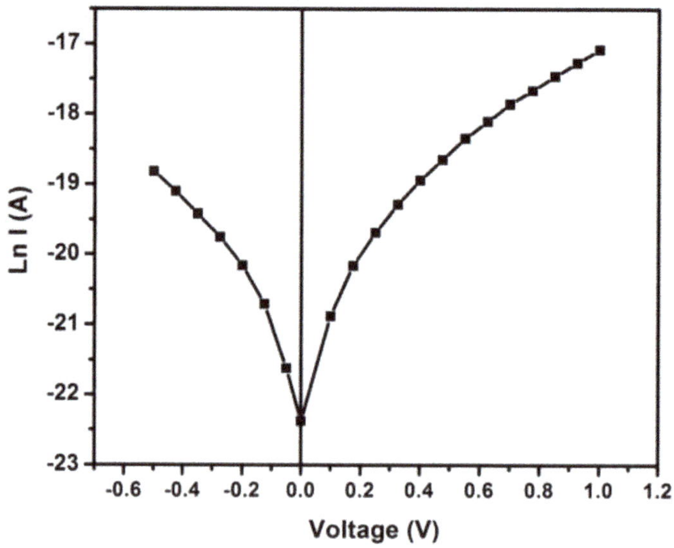

Figure 2. Current-voltage characteristics of VTP-based solar cell in semi-logarithmic scale.

Io denotes the saturation current, *V* refers to the forward bias voltage, *Φb* reflects the zero-bias barrier height, *kB* exemplifies the Boltzmann constant, *T* represents the temperature in Kelvin, *A* depicts the active area, and *n* is the ideality factor of the diode device. *A*∗ signifies the effective Richardson constant, which is attainable from the Richardson–Dushman correlation ($A=4\pi emk^2/h^3$), where *m* is the mass of the electron, *e* is the elementary charge, and *h* is Plank's constant. Scott and Malliaras [27] asserted that most organic semiconductor materials are ideally found at 10^{-2} A/cm^2 K^2 [27]. The reverse saturation current is obtained from the y-intercept of the semi-log *I-V* curve, which resulted in 0.2 nA upon weighing in its natural log.

$$\Phi_b = \frac{k_B T}{q} \ln\left(\frac{AA^*T^2}{I_o}\right) \quad (3)$$

The effective barrier height of the junction Φ_b was obtained from Equation (3), which refers to the contact that is present between the metal and semiconductor interface. This was calculated using Equation (2), with the value of the reverse saturation current at 0.673 eV. The conformity of the diode device to pure thermionic emission was determined by incorporating the ideality factor, *n*, by using the same semi-log *I-V* characteristics. For an ideal diode, the *n* value reflects unity, despite it deviating from its ideal value at times, wherein the observed value is greater than unity. The value of the ideality factor in this study was obtained from the slope of the linear region in the forward bias of semi-log *I-V* curves using the following equation:

$$n = \frac{q}{kT} \frac{dV}{d(\ln I)} \quad (4)$$

The ideality factor recorded in this study is 2.93. The *n* value exceeded '2' due to the following reasons: the non-homogeneous barrier, the non-homogeneous thickness of the organic film, temperature [28], and the prevalent current in the single junction photovoltaic device because of recombination [29]. Prior studies reported that the ideality factor of Pc derivatives such as CuPc, MgPc, and AlPc is massive, primarily due to recombination issues [30].

The shunt and series resistance (R_{sh} and R_s, respectively) are essential factors that dictate the performance of a device. These resistance values are obtained from the plot of the junction resistance, R_J, versus the voltage, *V*; $R_J = \frac{dV}{dJ}$. Massive shunt resistance is required to decrease the leakage of current via cells (e.g., pinholes), apart from suppressing the recombination of charge carriers at the device interface [31,32,33]. This scenario, however, contradicts the low series resistance that reflects internal resistance, which allows and enhances high current to flow through the device, whereby a low R_s denotes a low resistivity of the organic material [34,35]. Figure 3 illustrates that the values of the shunt resistance, R_{sh}, and series resistance, R_s, can be retrieved from the graph of the junction resistance (*R*) versus voltage (*V*). The resistance decreased exponentially with an increment in the forward bias. The R_{sh} corresponded to the resistance within the vicinity of zero bias, thus resulting in 745 Ω/cm². The *R* curve was saturated upon a continuous supply of forward bias. Therefore, the applied external electric field compensated for the rectifying potential barrier between the cathode and anode, whereby the current that flowed through the device was restricted by R_s. Hence, the R_s value was deduced by extrapolating the saturated part of the *R* curve towards the interception with the resistance axis, in which the result was recorded at 32 Ω/cm² [36].

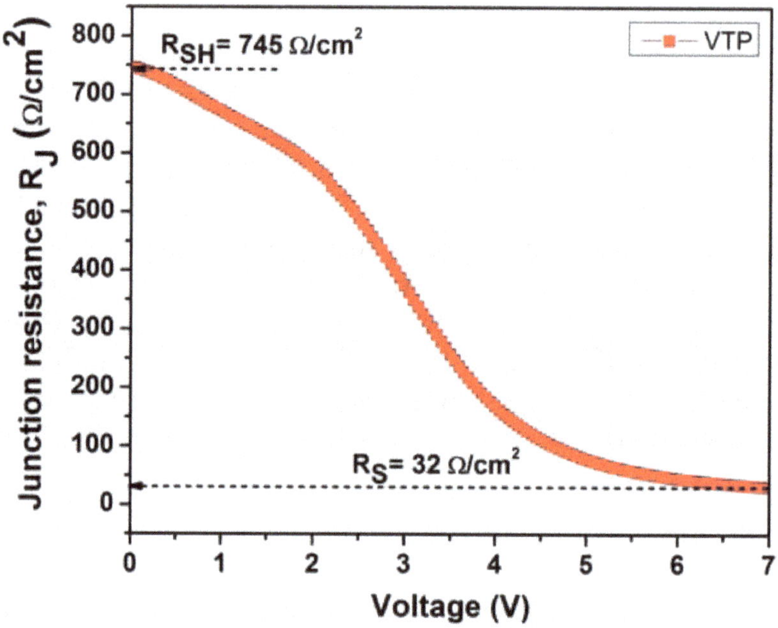

Figure 3. Junction resistance versus biased voltage for single junction VTP device.

A similar analysis was executed for the blended film to determine its R_{sh} and R_s, as exhibited in Figure 4. Meanwhile, Table 1 represents the estimated values of both the shunt and series resistance values of all the diode devices. As a result, the value of R_s decreased from 19.40 to 10.47 and 7.42 Ω/cm^2 by increasing the concentrations of $PC_{71}BM$, respectively. This reduction reflects the increasing surface roughness morphology of the highest doping $PC_{71}BM$, mainly because the R_s value is closely related to the active layer morphology, intrinsic resistance, and thickness of the active layer [**32**]. The processing additives improved the device performance by dissolving the aggregates of $PC_{71}BM$ in the VTP: $PC_{71}BM$ matrix. This facilitates the integration of the $PC_{71}BM$ molecule into the VTP, thus resulting in a greater donor–acceptor interface. The enhanced surface interface between D/A in the active layer contributes to an improved migration of excitons between the interface and to a favorable charge separation. It is worth highlighting here that R_{sh} is correlated with pin holes and traps that are present in the thin film morphology, hence causing charge carrier recombination and current leakage. According to Table 1, the R_{sh} value increased from 1636.12 to 1724.40 and, finally, to 3838.74 Ω/cm^2, with an increment of $PC_{71}BM$. The increment in R_{sh} value reflects that the incorporation of $PC_{71}BM$ in the active layer generated additional percolating pathways that helped the charge carriers to diffuse at a longer distance.

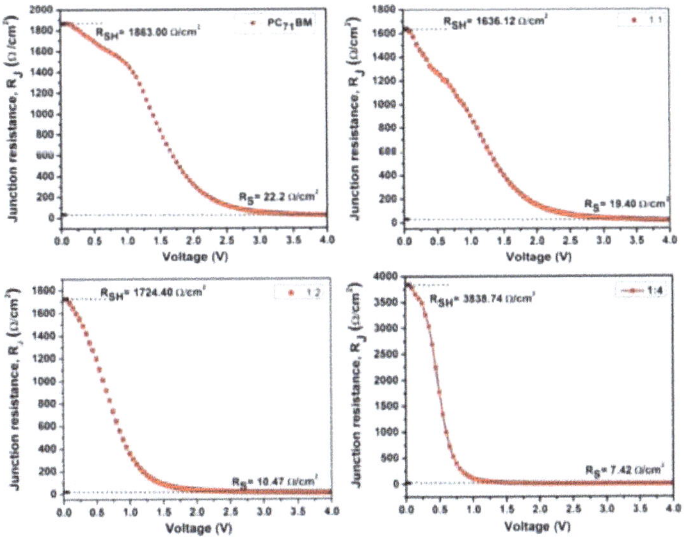

Figure 4. Junction resistance versus biased voltage for $PC_{71}BM$ and blended thin films of diode devices.

Table 1. Extracted R_s and R_{sh} values from junction resistance versus biased voltage.

Resistance (Ω/cm²)	VTP	$PC_{71}BM$	1:1	1:2	1:4
$R_{series,}$ R_s	32.00	22.20	19.40	10.47	7.42
$R_{shunt,}$ R_{sh}	745.00	1863.00	1636.12	1724.40	3838.74

Studies pertaining to charge transport in organic thin film devices can be classified into two types: charge carrier injection at the interface and charge carrier transport in the bulk system. The characterization of the forward bias *I-V* curves in dark conditions yielded significant characteristics in light of the transport mechanism, which was responsible for the conduction process that took place at the electrode interface. The effective carrier mobility for all devices was determined by adopting the space-charge-limited conduction (SCLC) approach with a positive voltage of up to 10 *V* in dark conditions. **Figure 5** displays the current density versus voltage curves excluding illumination, which was re-plotted in a double logarithmic scale to identify the mechanism that dominated the transport charge in the VTP diode. It is common for the double log forward bias *J-V* to reveal the power law behavior of $J \sim\sim V^m$, where 'm' denotes the slope of each region. The *m* value varies with the injection level and is linked with the distribution of trapping centers. The region $m = 1$ reflects the ohmic region, while $m = 2$ is the SCLC region and $m > 2$ represents the trapped-charge limited-current region [37].

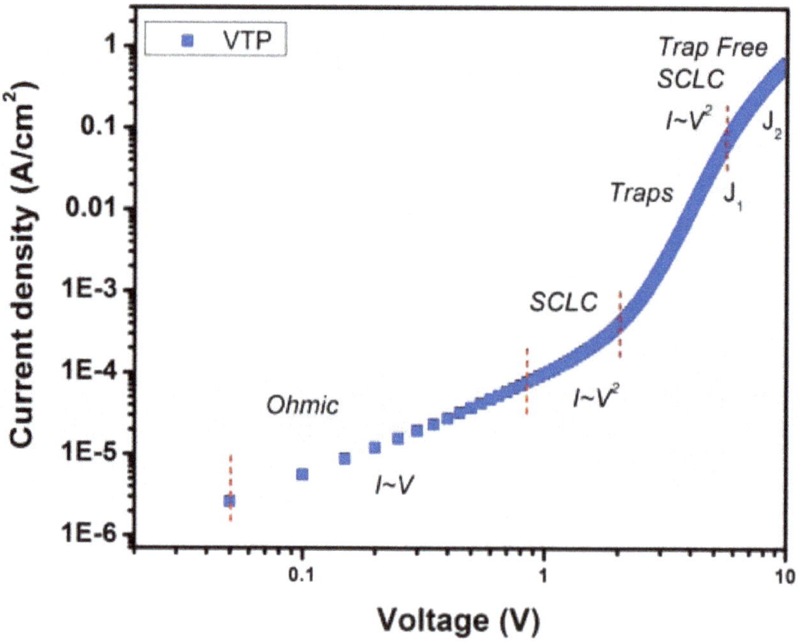

Figure 5. The double log scale of *J-V* characteristics of the VTP-based diode device measured in dark conditions.

Figure 5 illustrates the characteristics of the VTP-based diode device in order to elucidate the conduction mechanism. The double logarithmic graph is distributed in four regions (marked with dotted red lines), with various regions having varied slopes: ohmic, SCLC, traps, and trap-free SCLC regions (TF-SCLC). At the low voltage region, the slope was ~1.2, signifying the ohmic conduction mechanism. The current increased slowly with the applied voltage, as the injected effective-charge carrier density was lower than the background thermal carrier density [38]. In region 2 (intermediate voltage range), the conduction mechanism of the device was dominated by SCLC, wherein the slope was ~2.08. In this scenario, the voltage was more significant than that of region 1. The density of the injected free charge in this region was higher than the thermally generated free-charge carrier density; hence, there was an increment in the current. Typically, the formation of SCLC is attributed to the low mobility of charge carriers in an organic thin film, thus localizing the injected carriers by the trap states to limit the carrier conduction of the device [**39**]. The presence of traps was due to structural defects that derived from the non-uniformity and sub-atomic structure of the VTP layer. Under such conditions, the mobility of the charge carriers for the VTP diode devices was determined using the following equation:

$$J = \frac{9}{8}\varepsilon\varepsilon_O\mu\left(\frac{V^2}{d^3}\right) \tag{5}$$

where J reflects the current density, V is the voltage, d denotes the active cell thickness, ε represents the permittivity of free space, ε_0 signifies the relative dielectric constant (ε_0=8.854×10^{-12} Fm^{-1}), and μ stands for the charge carriers' mobility. **Table 2** shows the estimated charge carrier mobility. Next, the exponential increase of current in region 3 at higher voltages gave the slope ~5.6. This condition was dominated by the trap-filled limit (TFL) mechanism, whereby all deep traps were filled by the injected electrons, hence making the existing trap sites be fully occupied. As for region 4, where a high voltage was applied, the slope of the plot decreased sharply (~3.5) as the diode device approached the TFL at a high injection level. At the high injection level, the conduction mechanism was similar to that in TF-SCLC. Here, the injected carriers occupied most of the traps, and the accumulation of the space charge near the interfacing electrode culminated in the generation of a field that hindered further rejection. The VTP diode between the interfaces' contact functioned as if there was no trap in it, and the current varied with the square of the voltage [40,41,42]. The estimated current density in this region is expressed via Child's law [34,43]:

$$J = \frac{9}{8}\varepsilon\varepsilon_0\mu\theta\left(\frac{V^2}{d^3}\right) \tag{6}$$

where θ denotes the trapping factor defined as (J_1/J_2), in which J_1 and J_2 are the initial and final values of the current density in the TF-SCLC region, as portrayed in **Figure 6** and **Figure 7**, respectively. **Figure 6** portrays similar analyses applied to the remaining thin films, while **Table 2** summarizes the estimated values of the charge mobility for all the diode devices.

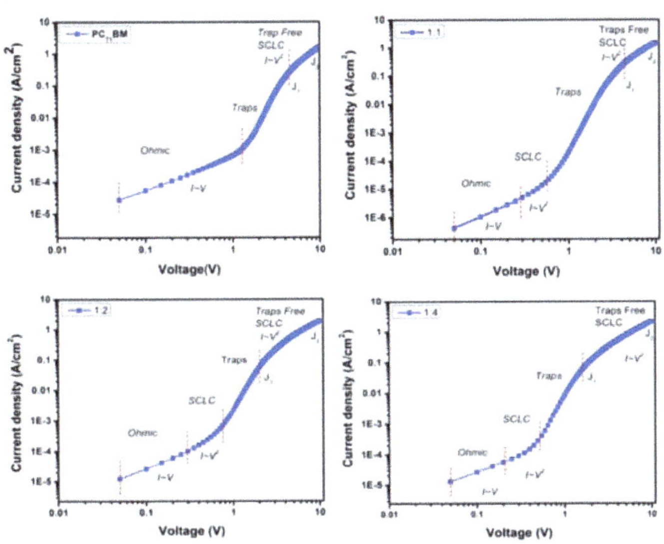

Figure 6. The double log of J-V characteristics of $PC_{71}BM$ and blended film measured in dark conditions.

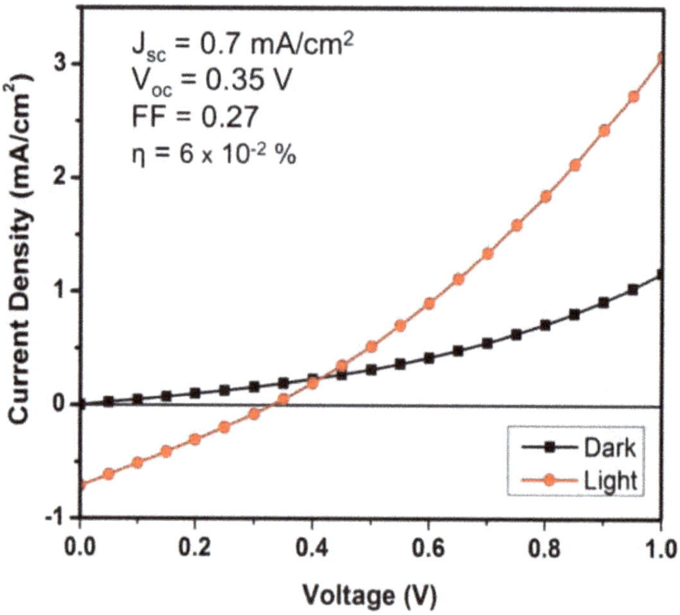

Figure 7. The *J-V* characteristics of ITO/PEDOT: PSS/VTP: $PC_{71}BM$/Al photovoltaic device under 100 $mWcm^{-2}$ illumination.

Table 2. Extracted mobility (cm^2/Vs) values for VTP, $PC_{71}BM$, and their varied ratios from diode measurements.

REGIME	SCLC	TF-SCLC
OPVs	Mobility (cm^2/V·s)	Mobility (cm^2/V·s)
VTP	4.24×10^{-7}	5.00×10^{-5}
$PC_{71}BM$	–	1.20×10^{-4}
1:1 (VTP: $PC_{71}BM$)	2.02×10^{-7}	7.06×10^{-5}
1:2	1.07×10^{-6}	7.55×10^{-5}
1:4	2.38×10^{-6}	7.63×10^{-5}

Figure 7 shows the *I-V* performance in the dark and under light settings for the ITO/PEDOT: PSS/VTP: $PC_{71}BM$/Al photovoltaic device. The photovoltaic characteristics, such as J_{SC}, V_{OC}, *FF*, and PCE (η), are portrayed in **Figure 7.** The device's efficiency is due to its enhanced charge production, which is achieved by a wider absorption spectrum. The dissociation of excitons in the VTP: PC71BM device due to the BHJ system was another contributing element to the improved performance efficiency. This limitation affects both single-layer and bilayer solar cells, mainly due to the fact that the average distance traveled by carriers is less than 20 nm before recombination occurs [44]. The specific blending mechanism achieved an efficiency of around 0.06%. This may be attributed to a limited absorption intensity within the visible-range spectrum, which is inadequate to

facilitate charge separation, hence resulting in a low conversion efficiency of light energy into electrical energy. This is consistent with the device's poor charge carrier mobility. The prospective application of VTP:PC71BM has been employed in the OPD device, as documented in our earlier study [19]. The incorporation of extended ligands into the macrocycle of VTP enhanced the sensitivity of the photodetector, resulting in a high-quality device.

4. CONCISE SUMMARY

To summarize, this study builds upon prior research on the utilization of VTP: PC71BM as a prospective application in optical devices. The current-voltage (I-V) characteristics exhibited by a single vertical tunneling transistor (VTP) exhibit rectifying behavior. The parameters of the Schottky diode, including the ideality factor, barrier height, shunt resistance, and series resistance, were determined. Additionally, the optimal composite blend of VTP: PC71BM was employed for its photovoltaic impact. This illustrates that while the same composite mix exhibits significant promise for organic photodetectors, as previously reported, it does not, however, achieve satisfactory efficiency in organic photovoltaics. This phenomenon is likely caused by the comparatively limited movement of the composite mix, resulting in a decreased absorption intensity in the visible-range spectrum. The results of the current studies are anticipated to facilitate further extensive examinations of composite blends in organic devices.

At a Glance

This study examined the electrical conductivity characteristics of VTP, an organic compound consisting of vanadyl 3,10,17,24-tetra-tert-butyl-1,8,15,22-tetrakis(dimethylamino)-29H,31H phthalocyanine. The I-V profile demonstrated by single VTP shows a rectifying behavior, and Schottky diode parameters such as the ideality factor, barrier height, shunt, and series resistance were calculated. In addition, the charge transport characteristics of a single-layer of VTP and its mixture with phenyl C71 butyric acid methyl ester (PC71BM) were assessed using the I-V traditional method and diode analysis. Furthermore, the enhanced diode characteristics of VTP: PC71BM were selected to assess its photovoltaic efficacy. The current density-voltage (J-V) characteristics were evaluated in both dark and light conditions to determine the key parameters of the photovoltaic effect. The results demonstrate that the optimized VTP:PC71BM composite blend exhibited a comparatively low photovoltaic efficiency. Nevertheless, the inclusion of elongated ligands results in a remarkably high sensitivity when utilized in the organic photodetector device, as documented in our prior research.

REFERENCES

1. Ahmad, Z.; Sayyad, M.H.; Yaseen, M.; Aw, K.C.; M-Tahir, M.; Ali, M. Potential of 5,10,15,20-Tetrakis (3',5'-di-tertbutylphenyl) porphyrinatocopper (II) for a multifunctional sensor. *Sens. Actuators B Chem.* **2011**, *155*, 81–85. **[Google Scholar] [CrossRef]**

2. Fakir, M.S.; Ahmad, Z.; Sulaiman, K. Modification of optical band gap and surface morphology of NiTsPc thin films. *Chin. Phys. Lett.* **2012**, *29*, 126802. **[Google Scholar] [CrossRef]**

3. Ahmad, Z.; Suhail, M.H.; Muhammad, I.I.; Al-Rawi, W.K.; Sulaiman, K.; Zafar, Q.; Hamzah, A.S.; Shaameri, Z. MEH-PPV/Alq3-based bulk heterojunction photodetector. *Chin. Phys. B* **2013**, *22*, 100701. **[Google Scholar] [CrossRef]**

4. Arbab, E.A.; Taleatu, B.A.; Mola, G.T. Ternary molecules blend organic bulk heterojunction solar cell. *Mater. Sci. Semicond. Process.* **2015**, *40*, 158–161. **[Google Scholar] [CrossRef]**

5. Koster, L.J.; Smits, E.C.P.; Mihailetchi, V.D.; Blom, P.W. Device model for the operation of polymer/fullerene bulk heterojunction solar cells. *Phys. Rev. B* **2005**, *72*, 085205. **[Google Scholar] [CrossRef] [Green Version]**

6. Hamui, L.; Sánchez-Vergara, M.E. Innovative Implementation of an Alternative Tetrathiafulvene Derivative for Flexible Indium Phthalocyanine Chloride-Based Solar Cells. *Micromachines* **2021**, *12*, 633. **[Google Scholar] [CrossRef] [PubMed]**

7. Roslan, N.A.; Abdullah, S.M.; Halizan, M.Z.M.; Bawazeer, T.M.; Alsenany, N.; Alsoufi, M.S.; Majid, W.H.A.; Supangat, A. VTP as an Active Layer in a Vertical Organic Field Effect Transistor. *J. Electron. Mater.* **2018**, *47*, 2184–2191. **[Google Scholar] [CrossRef]**

8. Ahmad, Z.; Abdullah, S.M.; Sulaiman, K. Bulk heterojunction photodiode: To detect the whole visible spectrum. *Measurement* **2013**, *46*, 2073–2076. **[Google Scholar] [CrossRef]**

9. Farooq, A.; Karimov, K.S.; Ahmed, N.; Ali, T.; Khalid Alamgir, M.; Usman, M. Copper phthalocyanine and metal free phthalocyanine bulk heterojunction photodetector. *Phys. B Condens. Matter* **2015**, *457*, 17–21. **[Google Scholar] [CrossRef]**

10. Hamdan, K.S.; Abdullah, S.M.; Sulaiman, K.; Zakaria, R. Effects of silver nanoparticles towards the efficiency of organic solar cells. *Appl. Phys. A* **2013**, *115*, 63–68. **[Google Scholar] [CrossRef]**

11. Zafar, Q.; Fatima, N.; Karimov, K.S.; Ahmed, M.M.; Sulaiman, K. Realizing broad-bandwidth visible wavelength photodiode based on solutionprocessed ZnPc/PC71BM dyad. *Opt. Mater.* **2017**, *64*, 131–136. **[Google Scholar] [CrossRef]**

12. Abdullah, S.M.; Ahmad, Z.; Aziz, F.; Sulaiman, K. Investigation of VOPcPhO as an acceptor material for bulk heterojunction solar cells. *Org. Electron.* **2012**, *13*, 2532–2537. **[Google Scholar] [CrossRef]**
13. Zafar, Q.; Ahmad, Z.; Sulaiman, K.; Hamzah, A.S.; Rahman, Z.A. A MEHPPV/VOPcPhO composite based diode as a photodetector. *Sens. Actuators A Phys.* **2014**, *206*, 138–143. **[Google Scholar] [CrossRef]**
14. Rani, V.; Yadav, S.; Ghosh, S. Organic Schottky Diode: Characterization of Traps. *AIP Conf. Proc.* **2015**, *1665*, 120023. **[Google Scholar] [CrossRef]**
15. Srivastava, A.; Chakrabarti, P. An organic Schottky diode (OSD) based on a-silicon/polycarbazole contact. *Synth. Met.* **2015**, *207*, 96–101. **[Google Scholar] [CrossRef]**
16. Mott, N.F. Note on the contact between a metal and an insulator or semi-conductor. *Proc. Camb. Philos. Soc.* **1938**, *34*, 568–572. **[Google Scholar] [CrossRef]**
17. Rideout, V.L. A review of the theory, technology and applications of metal-semiconductor rectifiers. *Thin Solid Film.* **1978**, *48*, 261–399. **[Google Scholar] [CrossRef]**
18. Guo, L.; Ma, G.; Liu, Y.; Mi, J.; Qian, S.; Qiu, L. Optical and non-linear optical properties of vanadium oxide phthalocyanine films. *Appl. Phys. B* **2002**, *74*, 253–257. **[Google Scholar] [CrossRef]**
19. Roslan, N.A.; Abdullah, S.M.; Abd Majid, W.H.; Supangat, A. Investigation of VTP:PC71BM organic composite as highly responsive organic photodetector. *Sens. Actuators A Phys.* **2018**, *279*, 361–366. **[Google Scholar] [CrossRef]**
20. Zafar, Q.; Aziz, F.; Sulaiman, K. Eco-benign visible wavelength photodetector based on phthalocyanine-low bandgap copolymer composite blend. *RSC Adv.* **2016**, *6*, 13101–13109. **[Google Scholar] [CrossRef]**
21. Aziz, F.; Ahmad, Z.; Najeeb, M.A.; Malik, H.A.; Abdullah, S.M.; Touati, F.; Sulaiman, K. Colloidal distribution of the PCPDTBT and VOPcPhO in the organic amalgam thin films and their optical properties. *Appl. Phys. A* **2017**, *123*, 773. **[Google Scholar] [CrossRef]**
22. Ebenhoch, B.; Thomson SA, J.; Genevičius, K.; Juška, G.; Samuel, I.D.W. Charge carrier mobility of the organic photovoltaic materials PTB7 and PC71BM and its influence on device performance. *Org. Electron.* **2015**, *22*, 62–68. **[Google Scholar] [CrossRef] [Green Version]**
23. Wang, H.; Wang, X.; Fan, P.; Yang, X.; Yu, J. Enhanced power conversion efficiency of P3HT:PC71BM bulk heterojunction polymer solar cells by doping a high-mobility small organic molecule. *Int. J. Photoenergy* **2015**, *2015*. **[Google Scholar] [CrossRef] [Green Version]**

24. Ahmad, Z.; Sayyad, M.H.; Wahab, F.; Sulaiman, K.; Shahid, M.; Chaudry, J.A.; Munawar, M.A.; Aziz, F. Enhancement of electronic and charge transport properties of NiPc by potassium-tetrasulpho group. *Phys. B Condens. Matter* **2013**, *413*, 21–23. **[Google Scholar] [CrossRef]**

25. Benhaliliba, M.; Ocak, Y.S.; Benouis, C.E. Effect of metal on characteristics of MPc organic diodes. *J. Nano Electron. Phys.* **2014**, *6*, 04009-1–04009-3. **[Google Scholar]**

26. Aziz, F.; Ahmad, Z.; Abdullah, S.M.; Sulaiman, K.; Sayyad, M.H. Photovoltaic effect in single-junction organic solar cell fabricated using vanadyl phthalocyanine soluble derivative. *Pigment. Resin Technol.* **2015**, *44*, 26–32. **[Google Scholar] [CrossRef]**

27. Scott, J.C.; Malliaras, G.G. Charge injection and recombination at the metal–organic interface. *Chem. Phys. Lett.* **1999**, *299*, 115–119. **[Google Scholar] [CrossRef]**

28. Al-Ta'ii, H.M.J.; Amin, Y.M.; Periasamy, V. Humidity influenced capacitance and resistance of an Al/DNA/Al Schottky diode irradiated by alpha particles. *Sci. Rep.* **2016**, *6*, 25519. **[Google Scholar] [CrossRef] [Green Version]**

29. Yakuphanoglu, F. Photovoltaic properties of hybrid organic/inorganic semiconductor photodiode. *Synth. Met.* **2007**, *157*, 859–862. **[Google Scholar] [CrossRef]**

30. Rajaputra, S.; Vallurupalli, S.; Singh, V.P. Copper phthalocyanine based Schottky diode solar cells. *J. Mater. Sci. Mater. Electron.* **2007**, *18*, 1147–1150. **[Google Scholar] [CrossRef]**

31. Ito, M.; Palanisamy, K.; Kumar, A.; Murugesan, V.S.; Shin, P.-K.; Tsuda, N.; Yamada, J.; Ochiai, S. Characterization of the organic thin film solar cells with active layers of PTB7/PC71BM prepared by using solvent mixtures with different additives. *Int. J. Photoenergy* **2014**, *2014*. **[Google Scholar] [CrossRef]**

32. Murugesan, V.S.; Ono, S.; Tsuda, N.; Yamada, J.; Shin, P.-K.; Ochiai, S. Characterization of organic thin film solar cells of PCDTBT:PC71BM prepared by different mixing ratio and effect of hole transport layer. *Int. J. Photoenergy* **2015**, *2015*. **[Google Scholar] [CrossRef] [Green Version]**

33. Servaites, J.D.; Ratner, M.A.; Marks, T.J. Organic solar cells: A new look at traditional models. *Energy Environ. Sci.* **2011**, *4*, 4410–4422. **[Google Scholar] [CrossRef]**

34. Khan, S.M.; Sayyad, M.H.; Karimov, K.S. Investigation of temperature dependent electrical properties of p-VOPc/n-si heterojunction under dark conditions. *Ionics* **2011**, *17*, 307–313. **[Google Scholar] [CrossRef]**

35. Lim, L.W.; Aziz, F.; Muhammad, F.F.; Supangat, A.; Sulaiman, K. Electrical properties of Al/PTB7-Th/n-Si metal-polymer-semiconductor Schottky barrier diode. *Synth. Met.* **2016**, *221*, 169–175. **[Google Scholar] [CrossRef]**

36. Brus, V.V.; Proctor, C.M.; Ran, N.A.; Nguyen, T.Q. Capacitance spectroscopy for quantifying recombination losses in nonfullerene small molecule bulk heterojunction solar cells. *Adv. Energy Mater.* **2016**, *6*, 1502250. **[Google Scholar] [CrossRef]**

37. Chiguvare, Z.; Parisi, J.; Dyakonov, V. Current limiting mechanisms in indium-tin-oxide/poly3-hexylthiophene/aluminum thin film devices. *J. Appl. Phys.* **2003**, *94*, 2440–2448. **[Google Scholar] [CrossRef]**

38. Missoum, I.; Ocak, Y.; Benhaliliba, M.; Benouis, C.; Chaker, A. Microelectronic properties of organic Schottky diodes based on MgPc for solar cell applications. *Synth. Met.* **2016**, *214*, 76–81. **[Google Scholar] [CrossRef]**

39. Shang, D.; Wang, Q.; Chen, L.; Dong, R.; Li, X.; Zhang, W. Effect of carrier trapping on the hysteretic current-voltage characteristics in Ag/La 0.7 Ca 0.3 MnO_3/Pt heterostructures. *Phys. Rev. B* **2006**, *73*, 245427. **[Google Scholar] [CrossRef]**

40. Güllü, Ö.; Aydoğan, □.; Türüt, A. Fabrication and electrical characteristics of Schottky diode based on organic material. *Microelectron. Eng.* **2008**, *85*, 1647–1651. **[Google Scholar] [CrossRef]**

41. Janardhanam, V.; Jyothi, I.; Lee, J.-H.; Kim, J.-Y.; Reddy, V.R.; Choi, C.-J. Electrical properties and carrier transport mechanism of Au/n-GaN Schottky contact modified using a copper pthalocyanine (CuPc) interlayer. *Mater. Trans.* **2014**, *55*, 758–762. **[Google Scholar] [CrossRef]**

42. Jomaa, T.B.; Beji, L.; Ltaeif, A.; Bouazizi, A. The current–voltage characteristics of heterostructures formed by MEH-PPV spin-coated on n-type GaAs and n-type porous GaAs. *Mater. Sci. Eng. C* **2006**, *26*, 530–533. **[Google Scholar] [CrossRef]**

43. Najeeb, M.A.; Abdullah, S.M.; Aziz, F.; Ahmad, Z.; Shakoor, R.; Mohamed, A.; Khalil, U.; Swelm, W.; Al-Ghamdi, A.A.; Sulaiman, K. A comparative study on the performance of hybrid solar cells containing ZnSTe QDs in hole transporting layer and photoactive layer. *J. Nanoparticle Res.* **2016**, *18*, 384–392. **[Google Scholar] [CrossRef]**

44. Cai, W.; Gong, X.; Cao, Y. Polymer solar cells: Recent development and possible routes for improvement in the performance. *Solar Energy Mater. Sol. Cells* **2010**, *94*, 114–127. **[Google Scholar] [CrossRef]**

9 Chapter

LOGIC OPERATIONS AT A WAVELENGTH OF 1.55 µM USING K-SHAPED SILICON WAVEGUIDES

1. INTRODUCTION: AN OVERVIEW

All-optical gates are crucial components for constructing lightwave broadband communications networks [1]. They allow signal-processing activities to be executed without the need for optoelectronic conversions at the photonic nodes. The optical domain enables the achievement of various signal processing tasks, including packet processing, pseudorandom binary sequence generation, encryption/decryption, error detection and correction, arithmetic operations, construction of optical memory elements, digital comparison, buffering, implementation of any other Boolean function, and construction of combinational logic circuits. These tasks are made possible by utilizing XOR, AND, OR, NOT, NOR, NAND, and XNOR logic operations. Silicon photonics claims to be able to produce cost-efficient platforms that are both effective and have low loss. Silicon-on-silica technique involves the application of a thin layer of crystalline silicon onto an insulating layer made of silica, also known as silicon dioxide. Due to the significant infrared transparency of silicon and refractive index difference between silicon (i.e., core with $n_{silicon} \approx 3.48$ at 1.55 µm) and silica (i.e., cladding with $n_{silica} \approx 1.444$ at 1.55 µm), silicon-on-silica optical waveguides have unique optical features [15]. Various optical waveguides have been recently used for implementing both all-optical logic gates and all-optical networks [16,17,18,19,20,21,22,23]. Hence, this article presents a simulation of seven fundamental logic operations, namely XOR, AND, OR, NOT, NOR, NAND, and XNOR, utilizing K-shaped waveguides that operate at a wavelength of 1.55 µm often used in telecommunications. This waveguide consists of four terminals, with each terminal having one output port and three input ports. The waveguide is made up of silicon that is printed on silica. Silicon is widely recognized for its comparatively low optical loss of 2 dB/cm for wavelengths up to 8 µm. However, the optical loss of silica grows significantly after 3.6 µm, as stated in reference [15]. The constructive and destructive interferences resulting from the phase difference between the input beams play a crucial role in achieving the desired logic operations. To illustrate the execution of logic operations, we obtained finite-difference time-domain (FDTD) solutions using commercially available software. The convolutional optimally matched layer was used as an absorbing boundary condition. The performance of logic operations is evaluated using the contrast ratio (CR) metric. Based on the simulation results, the waveguide used can achieve better capacity ratios (CRs) at a data rate of 120 Gb/s. As a result, it can beat previously published designs [16,17,18,19,20].

2. K-SHAPED WAVEGUIDE

This study involves the construction of a K-shaped waveguide consisting of four terminals composed of three silicon slots arranged on a silica substrate. An electromagnetic wave polarized in the transverse magnetic mode at a wavelength of 1.55 μm excites the three input ports. The incident beams have the same wavelength and strength. Figure 1 illustrates the schematic representation of the K-shaped silicon-on-silica waveguide, including its field intensity distributions.

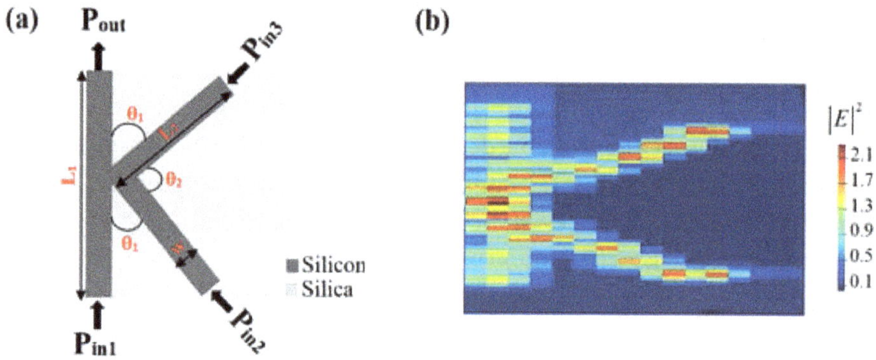

Figure 1. (a) Schematic depiction and (b) field-intensity distributions of K-shaped silicon-on-silica waveguide.

To record the simulation outcomes, the FDTD monitors are employed. Setting the threshold transmission (T_{th}) value to 0.12 is necessary at first. The formula for the output transmission (T) is $T=I_{out}/I_{in}$ [16], where $I_{out}=|E_{out}|^2$ is the intensity at P_{out}, and $I_{in} = I_1 + I_2 + I_3$ is the sum of the intensities at the three input ports. The input beams must satisfy the requirements for phase-matching in order to maximize T. In essence, this implies ensuring sure that the interacting waves are kept in the proper relative phase throughout the direction of propagation. However, before high CR logic gates can be accomplished, the phase-matching condition necessitates a specific selection of the input wavelength and waveguide characteristics. The phase-matching in silicon waveguides is induced by the contributions of the waveguide birefringence, material dispersion, waveguide dispersion, and cross- and phase-self modulations [24]. It is, therefore, feasible to achieve phase-matching by designing the waveguide such that the birefringence and material dispersion terms cancel one another, according to the phase-matching analysis of silicon waveguides, as reported in [24]. When $T > T_{th}$, P_{out} generates a logical output of '1', while in all other cases, it generates a '0'. The CR is an important metric for logic devices and is defined as $CR(dB) = 10\,ln[P^1_{mean}/P^0_{mean}]$ [25], where P^1_{mean} and P^0_{mean} represent the mean peak powers of output logic bits '1' and '0', respectively. Compared to other metrics, such as the extinction ratio, the CR offers a better and more accurate evaluation of the performance of the optical logic operations [26]. For the proposed waveguide, **Table 1** lists the default parameters' values used in the simulation.

Table 1. Simulation Parameters

t	Definition	Value	Unit
L_1	Length of long slot	2.5	µm
L_2	Length of short slot	1.0	µm
W	Width of slot	0.22	µm
D	Thickness of slot	0.3	µm
θ_1	Angle between long and short slots	50	degree
θ_2	Angle between short slots	80	degree
$n_{silicon}$	Silicon refractive index at 1.55 µm	3.48	-
n_{silica}	Silica refractive index at 1.55 µm	1.444	-
λ	Operating wavelength	1.55	µm
T_{th}	Threshold transmission	0.12	-

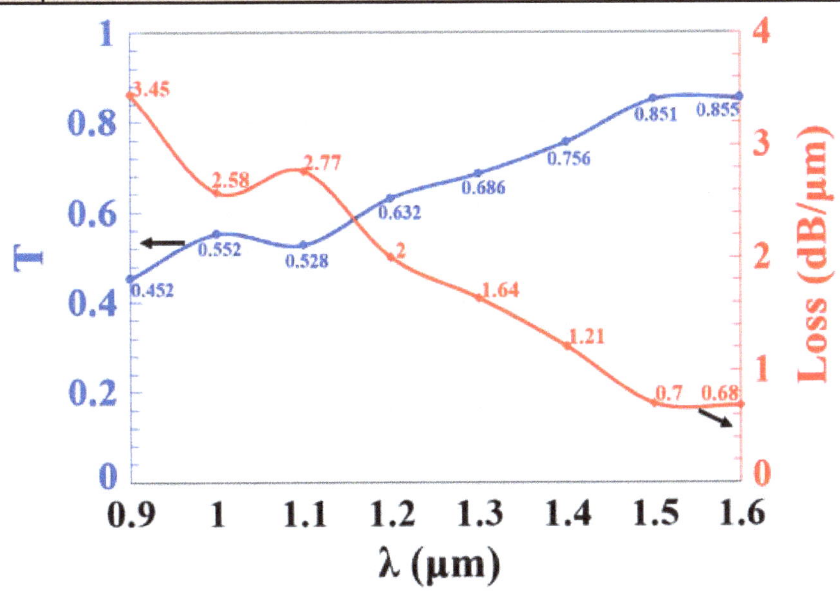

Figure 2. Normalized spectral transmission (T) and loss versus operating wavelength (λ), using K-shaped silicon-on-silica waveguide.

When all incident beams (i.e., two beams and a reference or clock light) are launched at the three input ports with the same phase of 180°, the normalized spectral transmission (T) and loss as a function of the operating wavelength (λ) are shown in **Figure 2**. The employed waveguide achieves a high T = 0.852 and a low loss = 0.69 dB/μm at 1.55 μm. Such small propagation losses are a direct result of the scattering at the inner slots' interfaces and absorption within the materials. **Figure 2** also shows that this waveguide achieves a high T and a low loss at a wide range of telecommunication wavelengths, from 1.3–1.6 μm.

Relaxed tolerances are crucial for both manufacturing and operating conditions. Manufacturing tolerances refer to the management of the geometrical dimensions during processing and their ensuing effect on device performance. Operation tolerances describe how the device responds to variations in wavelength, polarization, temperature, input field distribution, and refractive index [27,28]. Most laser sources have a significant practical wavelength tolerance. For example, a 1550 nm fiber laser may have a wavelength tolerance of ±20 nm, resulting in an actual wavelength of 1550 ± 20 nm [29]. **Figure 3** shows the dependence of the loss on the wavelength tolerance using the proposed waveguide. This result has been taken based on Equations (4)–(7) from Ref. [30].

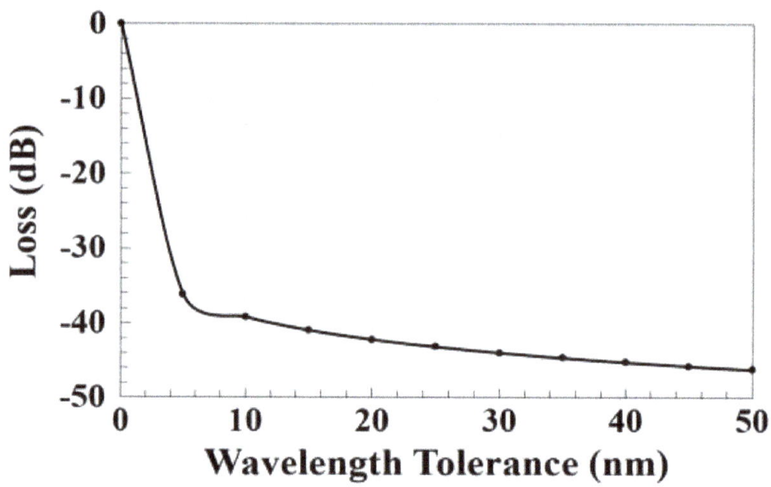

Figure 3. Optical loss versus wavelength tolerance, using K-shaped silicon-on-silica waveguide.

In this waveguide, we have used three acute angles, with a sum that is 180°, to perform the K letter. These angles (i.e., θ_1 and θ_2) play an important role in the K-shaped design in order to implement the considered logic gates with high CRs. Thus, the effect of the angle between the long and short slots (θ_1) on a normalized spectral transmission (T) at an operating wavelength of 1550 nm is simulated, as shown in **Figure 4**. It is clear from **Figure 4** that the highest T occurred at θ_1 = 50° (i.e., θ_2 = 80°), as optimized in this simulation.

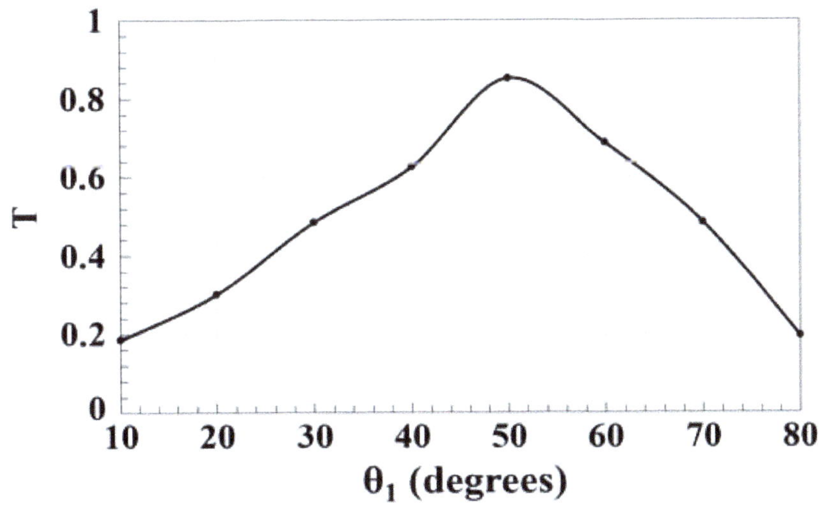

Figure 4. Normalized spectral transmission (T) versus angle between long and short slots (θ_1), using K-shaped silicon-on-silica waveguide.

3. LOGIC OPERATIONS

3.1. XOR

To carry out the XOR, AND, and OR logic operations, a reference beam (REF) must be injected into P_{in2} of **Figure 1**, while the other two beams are launched into P_{in1} and P_{in3}. The REF (all '1's) is used to introduce a reference phase difference between the input signals, resulting in either constructive or destructive interference. Constructive interference happens when all input beams are injected at the same phase (resulting in an output of '1'); destructive interference happens when they are launched at different phases (resulting in an output of '0'). As a result, for an XOR operation, P_{out} produces a '1' (meaning $T > T_{th}$) because of the constructive interference that occurs between the input beams when the combination of these input beams (01, 10) is injected along with the REF at the same phase (i.e., $\Phi_1 = \Phi_3 = \Phi_{REF} = 180°$). The destructive interference between the incident beams causes a '0' output to be produced at P_{out} (meaning $T < T_{th}$) when the combination (11), with the REF at different phases (i.e., $\Phi_1 = 0°$, $\Phi_3 = 90°$, and $\Phi_{REF} = 180°$), is launched. This results in the XOR logic function. We notice the presence of light at ports having '0' input, which is a natural result because the inner interfaces of the three input ports of the K-shaped waveguide are all opposite, and, therefore, the light is deflected inside them in an outward direction. The XOR field intensity distributions are displayed in **Figure 5**, using a K-shaped silicon-on-silica waveguide at 1.55 μm.

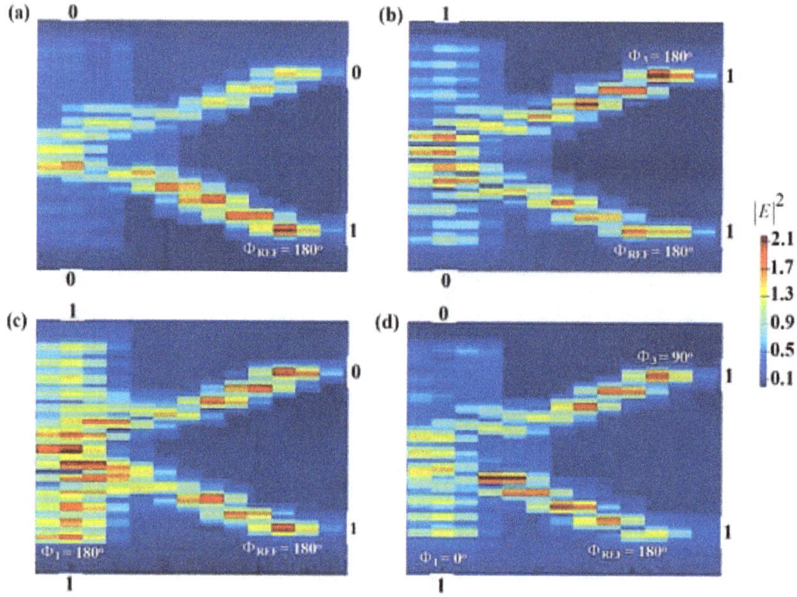

Figure 5. XOR field-intensity distributions, using K-shaped silicon-on-silica waveguide at 1.55 μm: (a) '00' input, (b) '01' input, (c) '10' input, and (d) '11' input.

Due to the large difference between the mean peak powers of '1' and '0', the suggested waveguide achieves a high CR = 34 dB. The XOR simulation outcomes, employing the K-shaped silicon-on-silica waveguide at 1.55 μm, are shown in **Table 2.**

Table 2. XOR simulation outcomes (T_{th} = 0.12).

P_{in1}	P_{in3}	P_{in2} (REF)	T	P_{out}	CR (dB)
0	0	1	0.021	0	34
0	1	1	0.464	1	
1	0	1	0.852	1	
1	1	1	0.023	0	

3.2. AND

Similar to the XOR operation, the AND operation involves injecting two beams into P_{in1} and P_{in3} as well as launching the REF (all '1's) from P_{in2}. P_{out} creates a '1' output, due to constructive interference, when all incident beams are released into the proposed waveguide at the same phase (i.e., $\Phi_1 = \Phi_3 = \Phi_{REF}$ = 180°). In contrast, when these incident beams are injected at a different phase, P_{out} outputs a '0' because of destructive interference. This results in the AND operation. In **Figure 6**, the AND field intensity distributions are shown, using a K-shaped silicon-on-silica waveguide at 1.55 μm.

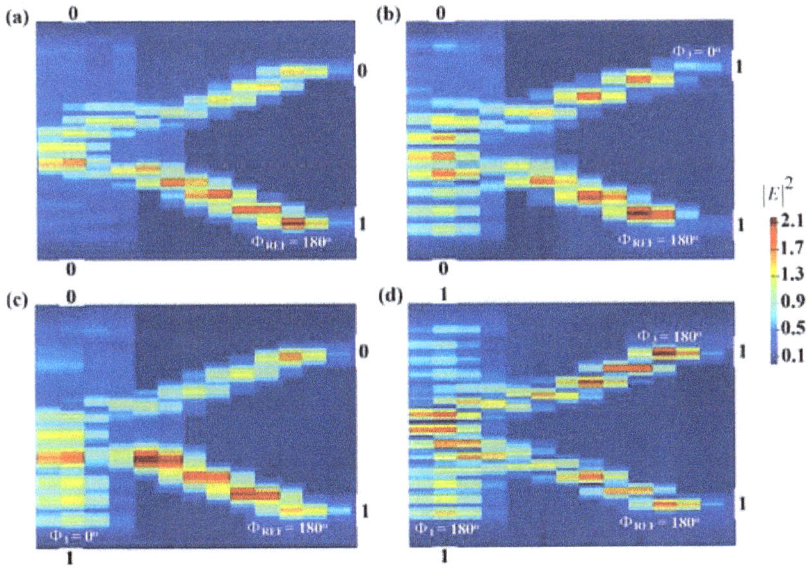

Figure 6. AND field intensity distributions, using K-shaped silicon-on-silica waveguide at 1.55 μm: (a) '00' input, (b) '01' input, (c) '10' input, and (d) '11' input.

For the logic AND operation, the proposed waveguide achieves CR = 31 dB at 1.55 μm. The further results of the AND simulation are listed in **Table 3.**

Table 3. AND simulation outcomes (T_{th} = 0.12).

P_{in1}	P_{in3}	P_{in2} (REF)	T	P_{out}	CR (dB)
0	0	1	0.021	0	31
0	1	1	0.022	0	
1	0	1	0.023	0	
1	1	1	0.521	1	

3.3. OR

When the combination of input beams (01, 10, or 11) is inserted with REF at the same phase of 180°, the result of the P_{out} becomes a '1'. Thus, the OR logic function between the two input beams is realized. **Figure 7** depicts the OR field intensity distributions, using the proposed waveguide at 1.55 μm.

The suggested waveguide obtains a high CR = 33.73 dB due to the significant difference between the mean peak powers of '1' and '0'. **Table 4** provides an overview of the outcomes of the OR simulation at 1.55 μm, in terms of T and CR.

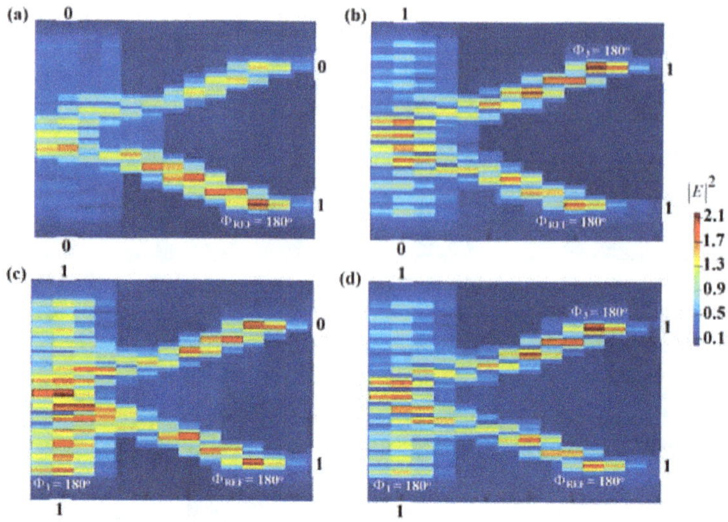

Figure 7. OR field intensity distributions, using K-shaped silicon-on-silica waveguide at 1.55 μm: (a) '00' input, (b) '01' input, (c) '10' input, and (d) '11' input.

Table 4. OR simulation outcomes (T_{th} = 0.12).

P_{in1}	P_{in3}	P_{in2} (REF)	T	P_{out}	CR (dB)
0	0	1	0.021	0	33.73
0	1	1	0.464	1	
1	0	1	0.852	1	
1	1	1	0.521	1	

The REF is essential for realizing the XOR, AND, and OR operations. Therefore, using the suggested waveguide at 1.55 μm, we have compared the performance of these three operations in terms of CR in the presence of the REF beam (i.e., REF = '1') and the absence of it, meaning there is no input beam injected into P_{in2}. **Table 5** indicates the necessity of using the REF to obtain higher CRs.

Table 5. CR with and without REF.

Operation	CR (dB) with REF	CR (dB) without REF
XOR	34	7.1
AND	31	6.4
OR	33.73	7

3.4. NOT

To carry out all inverted logic operations, including NOT, NOR, NAND, and XNOR, a clock light (Clk) with an angle of 0° must be sent into the proposed waveguide from P_{in1} of **Figure 1**. The Clk introduces an additional phase shift on the traveling beams, which changes the waveguide balance and results in an output. One beam is injected into P_{in3} at an angle of 180° to perform the NOT operation. Due to the destructive interference that occurs as a result of the input beams' various phase conditions, when P_{in3} is set to '1', P_{out} produces a logical '0' (i.e., $T < T_{th}$). When P_{in3} is 'OFF', the Clk (all '1's) outputs a logical '1' (i.e., $T > T_{th}$) at P_{out}, instead of going through a differencing phase. In this manner, the NOT gate is performed. Using a K-shaped silicon-on-silica waveguide, **Figure 8** illustrates the NOT field intensity distributions at 1.55 μm.

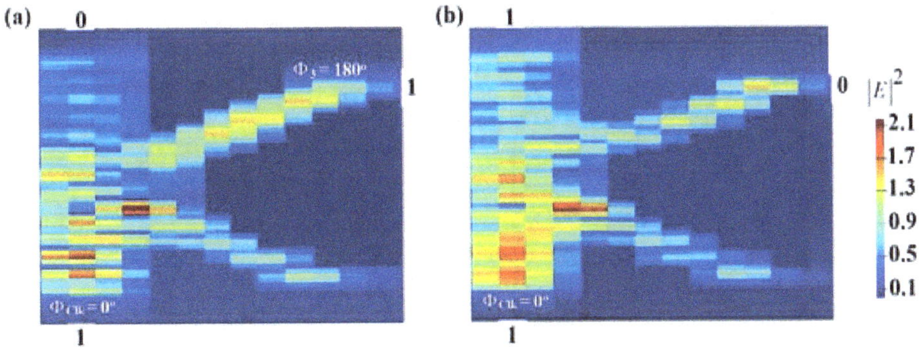

Figure 8. NOT field intensity distributions, using K-shaped silicon-on-silica waveguide at 1.55 μm: (a) '1' input and (b) '0' input.

The suggested waveguide results in a high CR = 30.5 dB for NOT operation. **Table 6** provides a summary of the outcomes of the NOT simulation, using the proposed waveguide at 1.55 μm.

Table 6. NOT simulation outcomes (T_{th} = 0.12).

P_{in1} (Clk)	P_{in3}	T	P_{out}	CR (dB)
1	1	0.032	0	30.5
1	0	0.675	1	

3.5. NOR

Two beams are launched into P_{in2} and P_{in3} to perform the NOR (NOT-OR) operation, and P_{in1} is launched with Clk (all '1's), as shown in **Figure 1**. When the input beams (01, 10, or 11) are combined and injected at different angles, destructive interference results in a logical '0' at P_{out}. If the launched beams' combination is (00), the Clk beam with Φ_{Clk} = 0° will cancel the phase balance of the three inputs, resulting in a logical '1' at P_{out}. Thus, the NOR logic operation is realized, as shown in **Figure 9**.

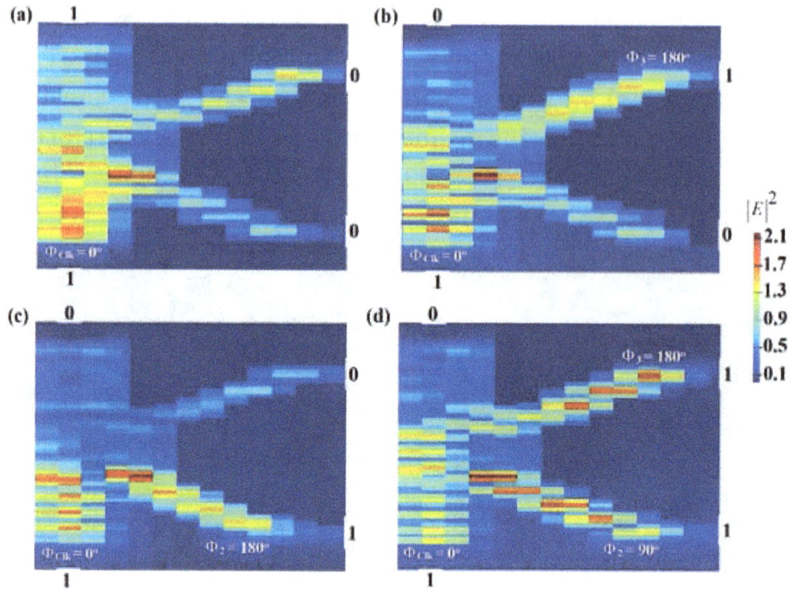

Figure 9. NOR field intensity distributions, using K-shaped silicon-on-silica waveguide at 1.55 μm: (a) '00' input, (b) '01' input, (c) '10' input, and (d) '11' input.

The suggested waveguide achieves a high CR = 33 dB for the NOR operation, as a result of the large disparity between P^1_{mean} and P^0_{mean}. A summary of the simulation outcomes for this logic operation is given in **Table 7**.

Table 7. NOR simulation outcomes (T_{th} = 0.12).

P_{in1} (Clk)	P_{in2}	P_{in3}	T	P_{out}	CR (dB)
1	0	0	0.675	1	33
1	0	1	0.032	0	
1	1	0	0.022	0	
1	1	1	0.022	0	

3.6. NAND

The NAND (NOT-AND) can be produced by injecting the Clk into P_{in1} and the other two beams into P_{in2} and P_{in3}, respectively. When both P_{in2} and P_{in3} are 'OFF' (i.e., 00), the Clk with a Φ_{Clk} = 0° cancels the phase balance of the three inputs, causing P_{out} to become '1'. Constructive interference simply occurs when Clk and (01, 10) are launched at the same angle of 0°, yielding an output of '1'. A '0' output is produced when (11) is launched with Clk at various phases, such as Φ_2 = 90°, Φ_3 = 180°, and Φ_{Clk} = 0°, as illustrated in **Figure 10**.

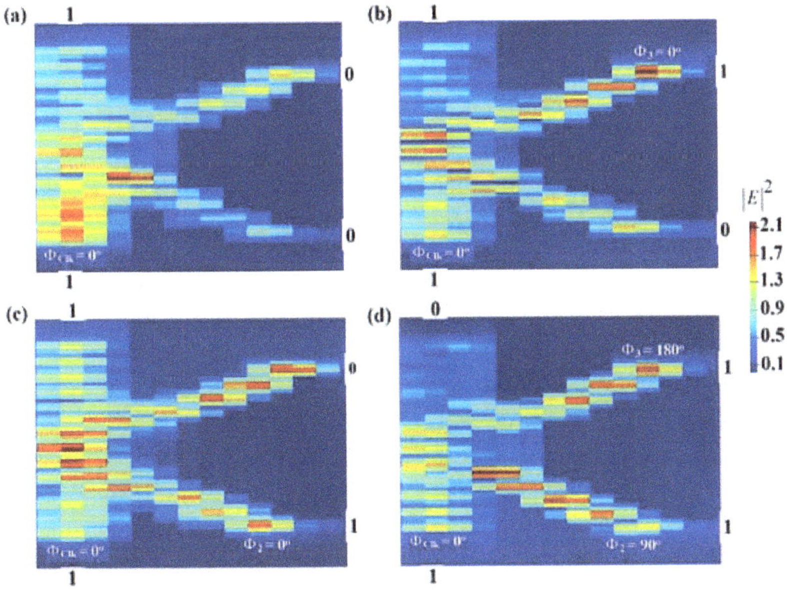

Figure 10. NAND field intensity distributions, using K-shaped silicon-on-silica waveguide at 1.55 μm: (a) '00' input, (b) '01' input, (c) '10' input, and (d) '11' input.

A summary of the NAND simulation outcomes utilizing the suggested waveguide, which achieves a high CR = 34 dB at 1.55 μm, is shown in **Table 8**.

Table 8. NAND simulation outcomes (T_{th} = 0.12).

P_{in1} (Clk)	P_{in2}	P_{in3}	T	P_{out}	CR (dB)
1	0	0	0.675	1	34
1	0	1	0.464	1	
1	1	0	0.852	1	
1	1	1	0.022	0	

3.7. XNOR

Similar to the NOR and NAND operations, the Clk enters P_{in1} to create the XNOR (exclusive-XOR) logic function, while the other two beams are injected from P_{in2} and P_{in3}, respectively. Constructive interference causes P_{out} to emit a '1' when the combination of the input beams (11) is introduced with the Clk at the same phase of 0°. In contrast, P_{out} produces a '0' when the input beams' combinations, (01) or (10), are inserted with a different phase, as depicted in Figure 11.

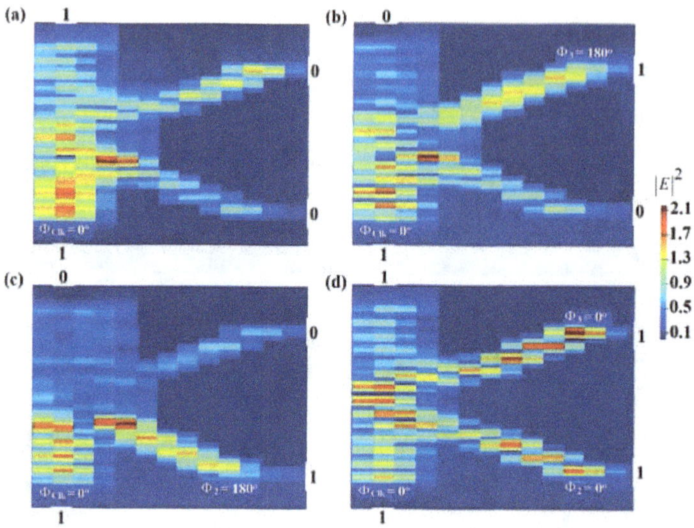

Figure 11. XNOR field intensity distributions, using K-shaped silicon-on-silica waveguide at 1.55 μm: (a) '00' input, (b) '01' input, (c) '10' input, and (d) '11' input.

Table 9 summarizes the XNOR simulation outcomes with a high CR = 31 dB, using the suggested waveguide.

Table 9. XNOR simulation outcomes (T_{th} = 0.12).

P_{in1} (Clk)	P_{in2}	P_{in3}	T	P_{out}	CR (dB)
1	0	0	0.675	1	31
1	0	1	0.032	0	
1	1	0	0.022	0	
1	1	1	0.521	1	

The Nyquist formula gives the speed of a transmission system as 2B $\log_2[M]$ [16], where M is the total number of signal levels, and B is the optical bandwidth, which is defined as $B=(c/\lambda 2)\Delta\lambda$, where c is the speed of light in vacuum, λ = 1.55 μm is the optical carrier wavelength, and Δλ is the signal's spectral width. Note $\log_2[M]$ is in a binary form, i.e., $\log_2[M] = \log[M]/\log[2]$. This means that in our case, where B = 30 GHz and for four signal levels (00, 01, 10, and 11), the predicted speed is 120 Gb/s.

Silicon and silica components are readily available, making it easier and more affordable to build the suggested waveguide. As a result, assuming that the necessary technology is available and that the major outcomes of this simulation are valid, the experimental verification of the suggested waveguide may be completed. This is a technology problem that can be resolved in practice, so it is not a crucial obstacle. Several logic operations, on the other hand, have been experimentally implemented based on various optical waveguides and components [31,32,33,34,35,36,37,38].

Table 10 compares the functionality of the considered waveguide, for realizing the intended logic operations at various wavelengths, to that of several waveguides reported on the same topic. This table suggests that the proposed waveguide can achieve faster logic operations with higher CRs than the other listed schemes.

Table 10. At various wavelengths, a comparison of our design and other waveguide-based logic function designs.

Operations	Design	Wave length (nm)	CR (dB)	Ref.
XOR, AND, OR, NOR, NAND, XNOR	Dielectric-loaded waveguides	471	24.41–33.39	[16]
OR, NOT, AND, XOR	Metallic waveguide arrays	632.8	9.3–20	[17]
NOT, XOR, AND, OR, NOR, NAND, XNOR	Nanoring insulator–metal–insulator waveguides	1550	-1.1–18.75	[18]
NOT, XOR, AND, OR, NOR, NAND, XNOR	Dielectric–metal–dielectric design	900 and 1330	5.37–22	[19]
AND, OR, NAND, NOR, XOR, Fan-Out, Half adder, Full adder	Photonic crystal circiuts	1550	5.54–11.56	[20]
XOR, AND, OR, NOT, NOR, XNOR, NAND	K-shaped silicon waveguides	1550	30.5–34	This work

4. CONCISE SUMMARY

Simulations were conducted using K-shaped silicon-on-silica waveguides to demonstrate seven fundamental logic operations: XOR, AND, OR, NOT, NOR, NAND, and XNOR. The simulations were performed at the 1.55 μm telecommunications wavelength. The operations were simulated using FDTD solutions acquired from commercially accessible software. The successful implementation of these logical processes depends on the constructive and destructive interferences resulting from the appropriate phase difference of the optical input beams. The suggested K-shaped waveguide outperforms existing reported waveguides in terms of achieving logic operations with a greater contrast ratio and operating speed.

At a Glance

Silicon possesses unique features that make it the optimal choice as a semiconductor material for a diverse range of electronic devices. This study showcases the simulation of various fundamental optical logic operations, such as XOR, AND, OR, NOT, NOR, XNOR, and NAND, utilizing K-shaped compact silicon waveguides. These waveguides are operated at the 1.55 μm telecommunications wavelength. This waveguide consists of three waveguide strips, all composed of silicon and printed on silica. The desired logic processes can be achieved by manipulating the phase of the incident beams. The contrast ratio (CR) is used as a quantitative measure to assess the performance of the activities under consideration. The suggested K-shaped waveguide achieves greater capacity ratios (CRs) and a speed of around 120 gigabits per second (Gb/s) compared to previous waveguides that have been described.

REFERENCES

1. Houbavlis, C.; Zoiros, K.E.; Kalyvas, M.; Theophilopoulos, G.; Bintjas, C.; Yiannopoulos, K.; Pleros, N.; Vlachos, K.; Avramopoulos, H.; Schares, L.; et al. All-optical signal processing and applications within the Esprit project DO ALL. *J. Lightwave Technol.* **2005,** *23*, 781–801. **[Google Scholar] [CrossRef]**
2. Clavero, R.; Mart´ınez, J.M.; Ramos, F.; Mart, J. All-optical packet routing scheme for optical label-swapping networks. *Opt. Express* **2004,** *12*, 4326–4332. **[Google Scholar] [CrossRef] [PubMed]**
3. Ji, W.; Zhang, M.; Ye, P. All-optical-packet header and payload separation for unslotted optical-packet-switched networks. *J. Lightwave Technol.* **2007**, *25*, 703–709. **[Google Scholar] [CrossRef]**
4. Ma, S.; Sun, H.; Chen, Z.; Dutta, N.K. High-speed all-optical PRBS generation based on quantum-dot semiconductor optical amplifiers. *Opt. Express* **2009,** *17*, 18469–18477. **[Google Scholar] [CrossRef]**
5. Singh, S.; Lovkesh; Ye, X.; Kaler, R.S. Design of ultrafast encryption and decryption circuits for secured optical networks. *IEEE J. Quantum Electron.* **2012,** *48*, 1547–1553. **[Google Scholar] [CrossRef]**
6. Aikawa, Y.; Shimizu, S.; Uenohara, H. Demonstration of all-optical divider circuit using SOA-MZI-type XOR gate and feedback loop for forward error detection. *J. Lightwave Technol.* **2011,** *29*, 2259–2266. **[Google Scholar] [CrossRef]**
7. Kim, S.H.; Kim, J.H.; Choi, J.W.; Son, C.W.; Byun, Y.T.; Jhon, Y.M.; Lee, S.; Woo, D.H.; Kim, S.H. All-optical half adder using cross gain modulation in semiconductor optical amplifiers. *Opt. Express* **2006,** *14*, 10693–10698. **[Google Scholar] [CrossRef]**
8. Gayen, D.K.; Bhattachryya, A.; Chattopadhyay, T.; Roy, J.N. Ultrafast all-optical half-adder using quantum-dot semiconductor optical amplifier-based Mach–Zehnder interferometer. *J. Lightwave Technol.* **2012,** *30*, 3387–3393. **[Google Scholar] [CrossRef]**
9. Berrettini, G.; Nguyen, A.T.; Lazzeri, E.; Meloni, G.; Scaffardi, M.; Pot, L.; Bogoni, A. All-optical digital circuits exploiting SOA-based loop memories. *IEEE J. Sel. Top. Quantum Electron.* **2012,** *18*, 847–858. **[Google Scholar] [CrossRef]**

10. Wang, Y.; Zhang, X.; Dong, J.; Huang, D. Simultaneous demonstration on all-optical digital encoder and comparator at 40 Gb/s with semiconductor optical amplifiers. *Opt. Express* **2007**, *15*, 15080–15085. **[Google Scholar] [CrossRef]**
11. Scaffardi, M.; Ghelfi, P.; Lazzeri, E.; Pot, L.; Bogoni, A. Photonic processing for digital comparison and full addition based on semiconductor optical amplifiers. *IEEE J. Sel. Top. Quantum Electron.* **2008**, *14*, 826–833. **[Google Scholar] [CrossRef]**
12. Zoiros, K.E.; Houbavlis, T.; Kalyvas, M. Ultra-high speed all-optical shift registers and their applications in OTDM networks. *Opt. Quantum Electron.* **2004**, *36*, 1005–1053. **[Google Scholar] [CrossRef]**
13. Kumar, S.; Willner, A.E. Simultaneous four-wave mixing and cross-gain modulation for implementing an all-optical XNOR logic gate using a single SOA. *Opt. Express* **2006**, *14*, 5092–5097. **[Google Scholar] [CrossRef] [PubMed]**
14. Jung, Y.J.; Son, C.W.; Jhon, Y.M.; Lee, S.; Park, N. One-level simplification method for all-optical combinational logic circuits. *IEEE Photon. Technol. Lett.* **2008**, *20*, 800–802. **[Google Scholar] [CrossRef]**
15. Mashanovich, G.Z.; Milošević, M.M.; Nedeljkovic, M.; Owens, N.; Xiong, B.; Teo, E.J.; Hu, Y. Low loss silicon waveguides for the mid-infrared. *Opt. Express* **2011**, *19*, 7112–7119. **[Google Scholar] [CrossRef] [Green Version]**
16. Yao, C.; Kotb, A.; Wang, B.; Singh, S.; Guo, C. All-optical logic gates using dielectric-loaded waveguides with quasi-rhombus metasurfaces. *Opt. Lett.* **2020**, *45*, 3769–3772. **[Google Scholar] [CrossRef]**
17. Yanga, W.; Shi, X.; Xing, H.; Chen, X. All-optical logic gates based on metallic waveguide arrays. *Res. Phys.* **2018**, *11*, 837–841. **[Google Scholar] [CrossRef]**
18. Abdulnabi, S.H.; Abbas, M.N. All-optical logic gates based on nanoring insulator–metal–insulator plasmonic waveguides at optical communications band. *J. Nanophotonics* **2019**, *13*, 016009. **[Google Scholar] [CrossRef] [Green Version]**
19. Al-Musawi, H.K.; Al-Janabi, A.K.; Al-Abassi, S.A.W.; Abusiba, N.A.A.; Al-Fatlawi, N.A.Q. Plasmonic logic gates based on dielectric-metal-dielectric design with two optical communication bands. *Optik* **2020**, *223*, 165416. **[Google Scholar] [CrossRef]**
20. Caballero, L.P.; Povinelli, M.L.; Ramirez, J.C.; Guimarães, P.S.S.; Neto, O.P.V. Photonic crystal integrated logic gates and circuits. *Opt. Express* **2022**, *30*, 1976. **[Google Scholar] [CrossRef]**
21. Asakawa, K.; Sugimoto, Y.; Nakamura, S. Silicon photonics for telecom and data-com applications. *Opto-Electron. Adv.* **2020**, *3*, 200011. **[Google Scholar] [CrossRef]**

22. Sun, C.; Yu, Y.; Zhang, X. Ultra-compact waveguide crossing for a mode-division multiplexing optical network. *Opt. Lett.* **2017**, *42*, 4913–4916. **[Google Scholar]** **[CrossRef]** **[PubMed]**
23. Green, W.M.J.; Rooks, M.J.; Sekaric, L.; Vlasov, Y.A. Ultra-compact, low RF power, 10 Gb/s silicon Mach-Zehnder modulator. *Opt. Express* **2007**, *15*, 17106–17113. **[Google Scholar]** **[CrossRef]** **[PubMed]**
24. Dimitropoulos, D.; Raghunathan, V.; Claps, R.; Jalali, B. Phase-matching and nonlinear optical processes in silicon waveguides. *Opt. Express* **2004**, *12*, 149–160. **[Google Scholar]** **[CrossRef]** **[Green Version]**
25. Kotb, A.; Guo, C. 100 Gb/s all-optical multifunctional AND, XOR, NOR, OR, XNOR, and NAND logic gates in a single compact scheme based on semiconductor optical amplifiers. *Opt. Laser Technol.* **2021**, *137*, 106828. **[Google Scholar]** **[CrossRef]**
26. Zoiros, K.E.; Papadopoulos; Houbavlis, T.; Kanellos, G.T. Theoretical analysis and performance investigation of ultrafast all-optical Boolean XOR gate with semiconductor optical amplifier-assisted Sagnac interferometer. *Opt. Commun.* **2006**, *258*, 114–134. **[Google Scholar]** **[CrossRef]**
27. Passaro, V.M.N.; Notte, M.L. Optimizing SOI slot waveguide fabrication tolerances and strip-slot coupling for very efficient optical sensing. *Sensors* **2012**, *12*, 2436–2455. **[Google Scholar]** **[CrossRef]** **[Green Version]**
28. Prinzen, A.; Waldow, M.; Kurz, H. Fabrication tolerances of SOI based directional couplers and ring resonators. *Opt. Express* **2013**, *21*, 17212–17220. **[Google Scholar]** **[CrossRef]**
29. Available online: **https://www.edmundoptics.com/p/1550nm-0-250mw-fiber-coupled-laser/12219/** (accessed on 13 November 2022).
30. Al-Hetara, A.M.; Shamsan, Z.A. Optical wavelength and dimensions tolerance criterion for multimode interference couplers. *WSEAS Trans. Commun.* **2014**, *13*, 567–571. **[Google Scholar]**
31. Kita, S.; Nozaki, K.; Takata, K.; Shinya, A.; Notomi, M. Ultrashort low-loss Ψ gates for linear optical logic on Si photonics platform. *Commun. Phys.* **2020**, *3*, 33. **[Google Scholar]** **[CrossRef]** **[Green Version]**
32. Donzella, V.; Sherwali, A.; Flueckiger, J.; Grist, S.M.; Fard, S.T.; Chrostowski, L. Design and fabrication of SOI micro-ring resonators based on sub-wavelength grating waveguides. *Opt. Express* **2015**, *23*, 4791–4803. **[Google Scholar]** **[CrossRef]** **[PubMed]**

33. Pan, D.; Wei, H.; Xu, H. Optical interferometric logic gates based on metal slot waveguide network realizing whole fundamental logic operations. *Opt. Express* **2013**, *21*, 9556. **[Google Scholar] [CrossRef] [PubMed] [Green Version]**

34. Gao, L.; Chen, L.; Wei, H.; Xu, H. Lithographically fabricated gold nanowire waveguides for plasmonic routers and logic gates. *Nanoscale* **2018**, *10*, 14771. **[Google Scholar] [CrossRef] [PubMed] [Green Version]**

35. Fu, Y.; Hu, X.; Lu, C.; Yue, S.; Yang, H.; Gong, Q. All-optical logic gates based on nanoscale plasmonic slot waveguides. *Nano Lett.* **2012**, *12*, 5784–5790. **[Google Scholar] [CrossRef] [PubMed]**

36. Hou, Z.; Sun, Y.; Li, Q.; Xudong, F.; Cheng, R. Smart bio-gel optofluidic Mach–Zehnder interferometers multiphoton-lithographically customized with chemo-mechanical-opto transduction and bio-triggered degradation. *Lab Chip* **2020**, *20*, 3815–3823. **[Google Scholar] [CrossRef]**

37. Liang, D.; Zhang, X.; Li, M.; Lin, Z.; Dai, H.; Wu, Z.; Pu, J. Visually Adjusting Coupling Conditions in Light-Emitting Micro-Components. *IEEE Photon. Technol. Lett.* **2019**, *31*, 1425–1428. **[Google Scholar] [CrossRef]**

38. Kowsari, A.; Saghaei, H. Resonantly enhanced all-optical switching in microfibre Mach–Zehnder interferometers. *Electron. Lett.* **2018**, *54*, 229–231. **[Google Scholar] [CrossRef]**

10 Chapter

AN ANALYSIS IS CONDUCTED TO EXAMINE THE IMPACT OF THE BARRIER LAYER ON THE UNIFORMITY OF SWITCHING AND SYNAPTIC PLASTICITY IN ALN-BASED CONDUCTIVE BRIDGE RANDOM ACCESS MEMORY

1. INTRODUCTION: AN OVERVIEW

The conductive bridge random access memory (CBRAM) has shown significant promise for energy-efficient non-volatile storage with cognitive computing capabilities, thanks to its high scalability, large dynamic memory window, excellent analog conductance switching, and low-off state current in comparison to valence change memory [1,2,3,4]. The CBRAM consists of a straightforward metal-insulator-metal (MIM) structure. It has a top electrode composed of an electrochemically active metal such as Cu or Ag, a bottom electrode made of an inert metal like Pt or W, and an insulating layer in between. This insulating layer enables the creation and dissolution of nanoscale metallic conductive filaments[5,6,7,8,9]. The cation (Ag^{+}, Cu^{2+}) migration process in filament evolution is functionally equivalent to Ca^{2+} dynamics of biological synapse and it was previously reported by demonstrating synaptic behaviors including short-term plasticity (STP), long-term plasticity (LTP), and spike-time dependent plasticity (STDP) characteristics [10,11,12,13,14,15]. Although there have been multiple efforts to imitate the human brain using resistive switching architecture, the goal of achieving device performance that is comparable to biological counterparts remains ongoing. In order to be considered suitable for neuromorphic computing, a memristive device must demonstrate consistent and predictable switching behavior in both space and time, have a long retention period, and be capable of updating its conductance in a linear manner. The imperfections in the insulating layer and the interface between the electrode and insulator cause ambiguity in the conductive filament's behavior. Furthermore, the insufficient ability to control the injection of metal ions and the formation of many filaments are factors that contribute to the diversity in spatial and temporal switching [16,17,18].

As a switching layer in CRAM devices, a variety of chalcogenides (Ag2S, Cu2S), oxides (HfOx, TaOx, and SiOx), and nitride (SiNx and AlNx) based materials have been employed. For these reasons, AlN is a potential contender among these storage media because of its large energy bandgap (6.0 eV), high thermal conductivity (285 W/m.K), and greater thermal stability, all of which result in enhanced switching behavior. It is also much more compatible with nitride electrodes (TiN and TaN), which are common in semiconductor production foundries. Moreover, AlN-based memory could be easily integrated with power electronics devices, including nitride compound semiconductor-based high electron mobility transistors (HEMTs) **[19]** that can drive high-speed memristors. The anion migration-based AlN RRAM has shown energy-efficient ultra-high switching speeds and high-density multilevel storage capabilities [**20,21**]. A recent paper has described the use of the AlN CBRAM technology for non-volatile storage applications, utilizing high-performance cation migration-based resistive switching operation. The primary issue in CBRAM devices is the switching instability caused by the unregulated movement of metal-ion species from the active electrode to the switching layer during switching cycles, leading to drift and diffusion. To enhance the stability of the CBRAM device, many methods were utilized, including the addition of a metal layer [22], the use of a composite electrode [23], the incorporation of a metal nano-island array [24], and the alloying of conduction channels [25]. The insertion of the barrier layer is a method that is both cost-effective and straightforward to produce, especially when compared to other procedures. Transition metals including titanium, tantalum, and tungsten have been thoroughly researched as diffusion barrier materials since they do not mix well with copper [26,27,28]. Metal nitrides serve as more effective barrier layers than their corresponding metals due to the enhanced chemical stability achieved through nitridation [29]. The inhibitory characteristics of WNx in copper connection and microelectronic circuits have already been documented [30,31]. WNx has demonstrated superior barrier qualities for the copper electrode due to the absence of copper-nitrogen reactions and the ability of smaller nitrogen molecules to occupy empty spaces, hence lowering copper diffusion [32]. In this study, we suggest incorporating the WNx barrier layer to improve the uniformity of switching by restricting the diffusion of Cu into the switching layer. The addition of a barrier layer resulted in a substantial improvement in the reliability parameters of the system, including endurance and data retention. In order to showcase the appropriateness of the CBRAM technology for neuromorphic applications, the long-term potentiation and depression synaptic activity are depicted with exceptional linearity.

2. MATERIALS AND METHODS

The schematic device processing commenced by depositing a 500 nm thermally grown SiO2 layer on an RCA cleaned Si wafer as an isolation layer. A DC sputtering system was used to deposit a 20 nm Ti adhesion layer and a 60 nm thick Pt layer for the bottom electrode. Subsequently, a 10 nm thick AlN switching layer was applied at room temperature in a mixed environment of argon (Ar) and nitrogen (N2) using DC sputtering. The process was carried out at a working pressure of 7.6 mTorr. Before creating the pattern on the top electrode, a layer of WNx diffusion barrier, with thicknesses of 7 and 10 nm, was formed using DC sputtering at room temperature. This was done by maintaining the flow rates of Ar and N2 gases at 24 sccm and 12 sccm, respectively. Afterwards, a layer of copper with a thickness of 80 nm was applied as the upper electrode using DC sputtering at room temperature, thereby finalizing the structure of the stacked device. The barrier and the Cu top electrode were both printed using a circular metal shadow mask with a diameter of 120 μm. The electrical direct current (DC) and pulse characteristics were measured using the Agilent B1500/B1530 semiconductor parameter analyzer. In the electrical measuring setup, the Cu top electrode was consistently biased with both DC and pulse signals, while the Pt bottom electrode remained grounded. The device's structural information was obtained using a JOEL JEM-2010 transmission electron microscope (TEM).

3. RESULTS AND DISCUSSION

Figure 1a illustrates the schematic arrangement of the proposed Cu/WNx/AlN/Pt CBRAM device. The transmission electron microscopy (TEM) image of the device depicted in Figure 1b unambiguously reveals the presence of numerous levels in the stacked structure. The device undergoes a transition from a state of high resistance to a state of low resistance, which is referred to as the forming process. The behavior of devices without a barrier layer (WOBL) and with a barrier layer (WBL), as shown in Figure 1c, reveals that WOBL devices have a higher leakage current and a lower forming voltage. This is because copper diffuses into the switching layer during the deposition of the top electrode [33,34]. By using the WNx barrier layer, which effectively inhibits the diffusion of copper, the leakage current is minimized and the forming voltage is elevated. The outcome demonstrates that when a thicker WNx barrier layer (10 nm) is present, there is an undesirable increase in forming voltage and the resistive switching operation becomes volatile. Therefore, a WBL device with a barrier layer thickness of 7 nm was compared to the WOBL device for further analysis. Figure 1d displays the voltage distribution for 10 devices during the forming process. The mean forming voltage of the WOBL device is 1.2 V, while the WBL device has a forming voltage of 2.5 V. The device-to-device fluctuation in Vforming is more pronounced in the WOBL device and is mostly caused by uneven spatial copper diffusion into the AlN switching layer.

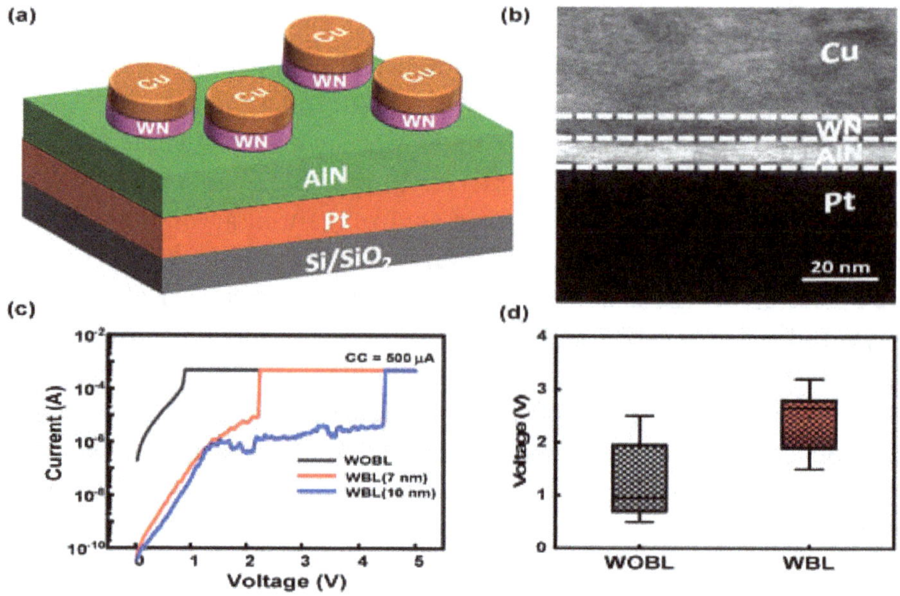

Figure 1. (a) Schematic structure of the Pt/AlN/WN/Cu RRAM device. (b) The TEM cross-sectional image of the device. (c) The forming characteristics of WOBL and WBL devices. (d) Forming voltage distribution for memristor without/with barrier layer for 10 devices.

Figure 2 a, b show the typical DC current-voltage characteristics of the WOBL and WBL devices. Both devices exhibit bipolar switching behavior with a positive bias sweep of 1.2 V at a compliance current (CC) of 1 mA for set operation and a negative sweep of -0.8 V for reset operation, respectively. From the measured 100 DC switching cycles, we observed that the WOBL device shows larger cycle-to-cycle variation in HRS/LRS compared to the WBL device. Without the barrier layer, multiple incomplete conductive paths are formed by the diffusion of copper into the AlN layer during the forming process, which subsequently causes variation in set/reset switching cycles. The device-to-device and cycle-to-cycle on/off resistance distribution is plotted in **Figure** 2c,d with each device tested for 30 DC cycles. The on-state resistance, LRS, of WOBL devices is significantly lower compared to WBL devices. This indicates filament formation through a large amount of Cu ion migration. The WOBL devices show the worst-case coefficient of variances (σ/μ) of 167% and 33% for HRS and LRS, respectively. In the case of the WBL, the variations improved significantly to 38% and 18%, respectively. This result confirms the switching stability with the insertion of the barrier layer. The resistance at LRS increases with the increase of temperature, which confirms the metallic nature of the conductive filament shown in **Figure** 2f. From the resistance-temperature plot, the temperature resistance coefficient is calculated by using Equation (1) [35]:

$$R(T) = R_0[1 + \alpha(T - T_0)] \tag{1}$$

where R_0 is the resistance at temperature T_0 and α is the temperature coefficient of resistance. The estimated α = 0.003 confirms the formation of a Cu filament during the on state.

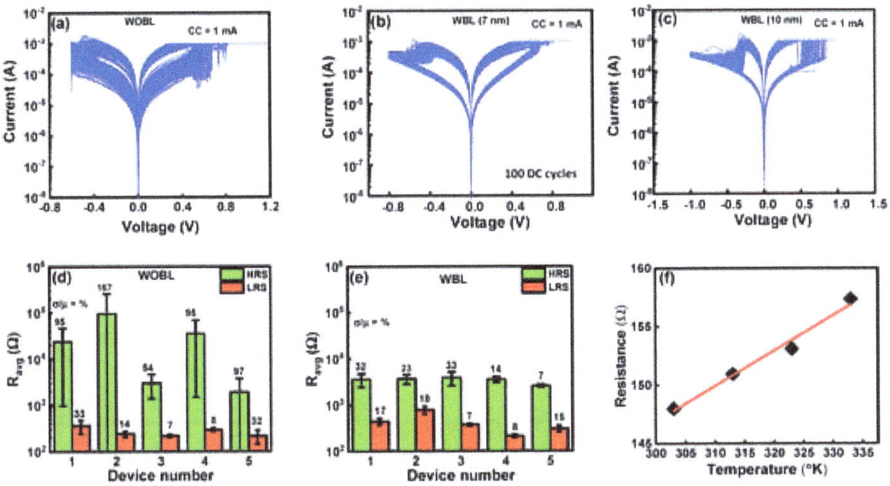

Figure 2. Typical DC I-V characteristics for 100 cycles; (a) without barrier layer; (b) with barrier layer (7 nm); and (c) with barrier layer (10 nm). The statistical distribution of LRS and HRS for memristor; (d) without barrier layer; (e) with barrier layer; and (f) temperature-dependent resistance in LRS.

The conduction mechanism of the Cu/WN_x/AlN/Pt CBRAM cell was investigated by analyzing the ln(I)~ln(V) graph in different regions as shown in **Figure 3**. The I-V curve between 0 and 0.25 V with a slope of 1.2 is well-fitted with the ionic conduction Equation (2) given by [36,37]:

$$\ln(j) \propto \ln(V) - (\ln T + a/T) \quad (2)$$

where *J*, *V*, *a*, and *T* are current density, bias voltage, constant, and temperature. This equation explains that ionic conduction occurs due to the oxidation of electrochemically active metal Cu into Cu ions and then the migration of Cu ions to the Pt bottom electrode with the application of positive bias to the top Cu electrode. The Cu ions are reduced at the bottom Pt electrode and the metallic conductive filaments grow from the bottom electrode to the top electrode, making the connection between the electrodes. As the applied bias voltage increases, the diameter of the filament becomes thicker, resulting in higher slopes of 2.1 and 3.7 between 0.25 to 0.58 V and 0.58 to 0.65 V, respectively. Due to the presence of the WN_x barrier layer, there is a gradual increment in slope. During biasing from 0.55 to 0 V, the current plot fitted well with the ohmic conduction mechanism which can be expressed by Equation (3):

$$\ln(j) \propto \ln(V) - a/T \quad (3)$$

where J, V, a, and T are current density, bias voltage, constant, and temperature. The current conduction from 0 to -0.47 V follows Ohm's law as -0.47 V was not sufficient to rapture the Cu filament. During reset operation with bias, from -0.48 to -0.74 V, there is a slope of -2.2 which indicates the breaking of the Cu filament, while from -0.74 to 0 V there is a positive slope of 1.3, showing the drift of Cu ions from the filament side to the Cu electrode. Similar conduction mechanisms were demonstrated using an Ag-based conductive bridge random access memory system in a prior study [**36**].

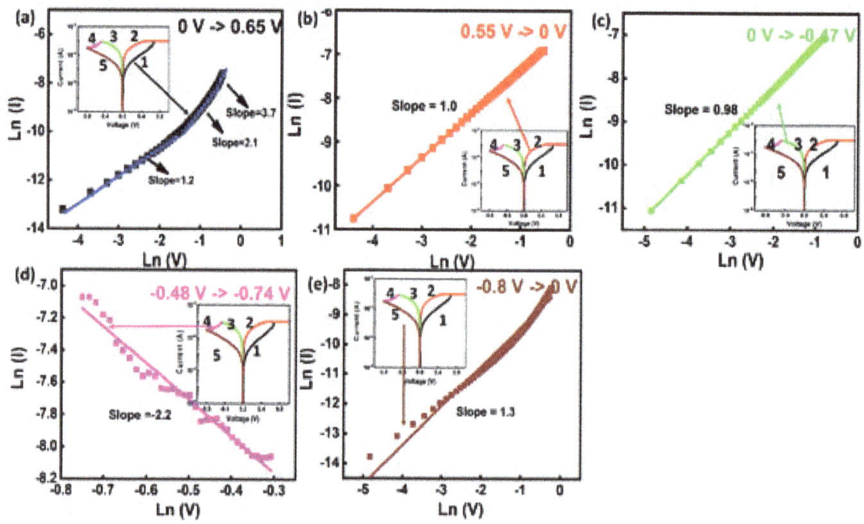

Figure 3. Conduction mechanism of WBL device in; (a) set operation HRS region from 0 to 0.65 V; (b) LRS region from 0.55 to 0 V, (c,e) reset operation from 0 to -0.47 V and from -0.47 to -0.74 V (d) reset operation HRS region from 0.74 to 0 V.

Figure 4 illustrates the schematic of the resistive switching mechanism for WOBL and WBL CBRAM devices, providing an explanation for the stability of the switching process. Due to the absence of a barrier layer between the Cu atoms and the AlN switching layer, a greater number of Cu atoms diffuse into the AlN switching layer for the WOBL device compared to the WBL sample in its original state, as depicted in Figure 4a,b. Under positive bias, the process of Cu ionization initiates at the upper electrode and moves towards the Pt electrode. At the Pt electrode, the Cu ions are then converted back into Cu atoms, resulting in the formation of a conductive bridge within the CBRAM device. Once the conductive route between the electrodes is fully formed, the device transitions to the low resistance state (LRS). The WOBL layer has several partially formed conductive filaments, along with a small number of fully formed conductive filaments. During the reset process, the filament undergoes electrochemical disintegration, causing the Cu ions to migrate back to the top electrode. The variety in the switching behavior of the WOBL device is demonstrated by considering two competing filamentary routes. Assuming CF1 is an initially established conductive pathway that undergoes ongoing structural modifications during multiple set-reset switching cycles. The gap formed in the filament during the reset event is g1. The selected conductive path will be CF1 only if g1 < g2. The CF2 conductive path undergoes minimal structural change as very little current flows. After several set-reset cycles, the CF2 becomes the selected conductive path with g2 < g1, which leads to a new LRS/HRS value. The primary cause of switching instability is the alteration in the conductive route, which becomes more prominent when there are numerous incomplete conductive filaments present. The existence of the WNx barrier layer restricts the diffusion of Cu atoms into the switching layer, hence preventing the formation of numerous conductive pathways, as illustrated in Figure 4c. Reduced filamentary pathways lead to enhanced switching stability, a crucial criterion for synaptic devices.

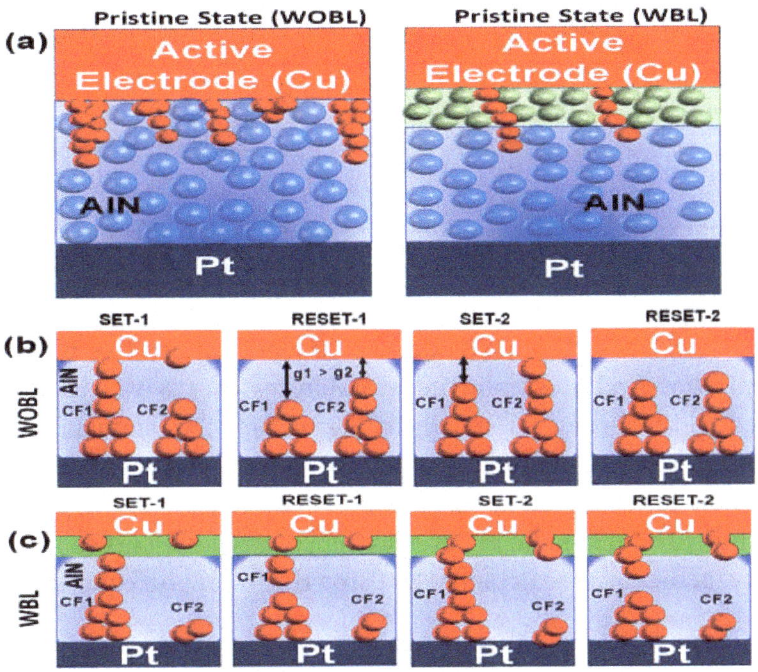

Figure 4. Schematic of (a) WOBL and WBL devices in a pristine state. Set and reset the switching mechanism of (b) WOBL and (c) WBL devices.

The endurance plots for both devices are shown in **Figure** 5a,b for 500 DC cycles. The HRS/LRS were evaluated at a read voltage of 0.2 V. The WBL device has shown a stable and consistent memory window as compared to the WOBL device. To further evaluate the nonvolatile operation of RRAM, the data retention characteristic behavior at HRS/LRS was evaluated at room temperature (RT) as well as the elevated temperature of 120 °C as displayed in **Figure** 5c,d. Both WOBL and WBL devices exhibited high retention of 10^4 s at RT. Nonetheless, at 120 °C, the HRS of the WOBL device degrades after 10^3 s due to temperature-assisted metal diffusion into the switching layer.

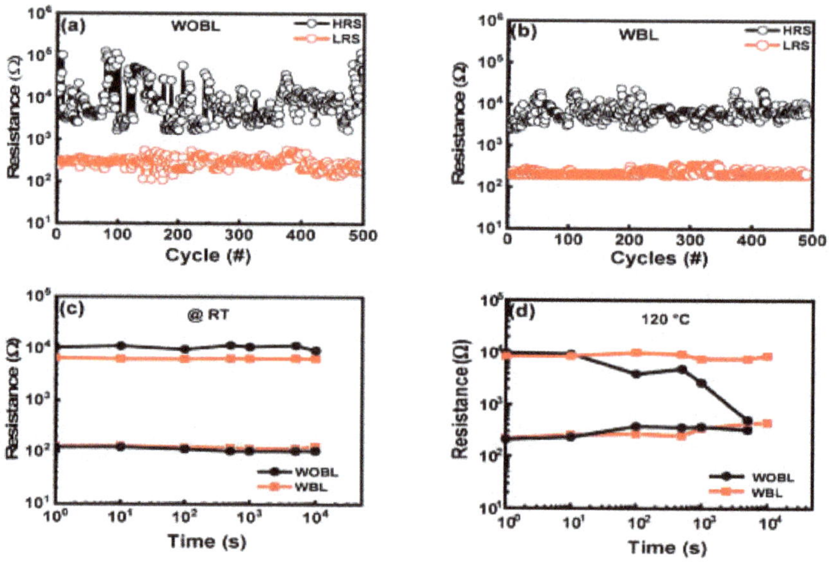

Figure 5. DC endurance behavior of (a) WOBL and (b) WBL devices. Data retention characteristics of WOBL and WBL devices were measured at (c) room temperature and (d) 120 °C.

The gradual conductance switching rather than abrupt switching upon application of a pulse is an important requirement for the memristive device to behave as an artificial synapse in neuromorphic hardware. To evaluate the synaptic characteristics of the WBL device, 50 identical positive pulses of 0.8 V/1 μs for potentiation and 50 similar negative pulses of -0.8 V/1 μs for depression were applied to the device as shown in **Figure 6**a. The conductance increased gradually from 400 to 800 μS during the potentiation period, and similar progressive decrementing behavior was observed during the depression period. **Figure 6**b shows the normalized potentiation/depression curves from which the non-linearity factor can be calculated by using the following Equations (4)–(6) [38]:

$$G_P = B\{1 - e^{(\frac{P}{A})}\} + G_{min} \tag{4}$$

$$G_D = -B\left\{1 - e^{(\frac{P-P_{max}}{A})}\right\} + G_{max} \tag{5}$$

$$B = \frac{G_{max} - G_{min}}{1 - e^{\frac{-P_{max}}{A}}} \tag{6}$$

where G, G_{max}, G_{min}, and P_{max} denote the conductance, maximum conductance, minimum conductance, and maximum number of pulses, respectively.

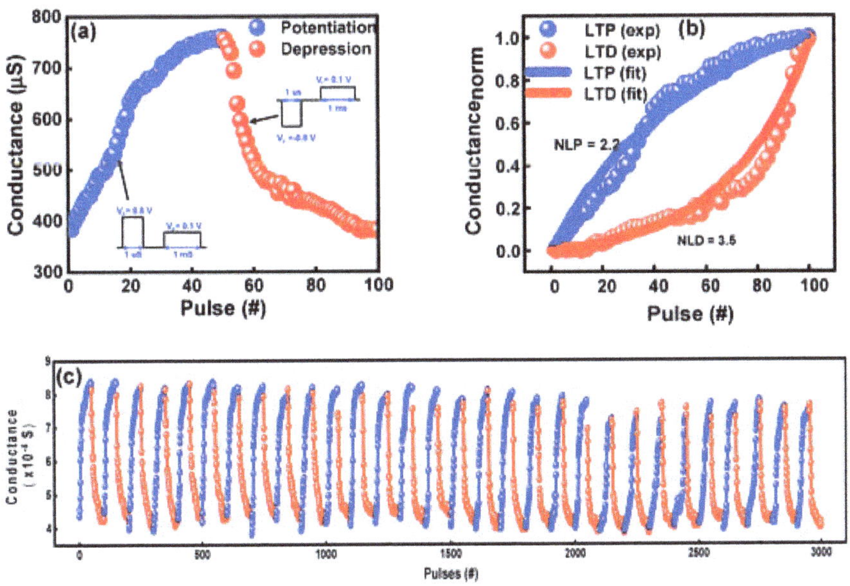

Figure 6. (a) Synaptic potentiation/depression characteristics measured by applying 50 identical 0.8 V/1 μs potentiation pulses and 50 identical -0.8 V/1 μs depression pulses. (b) Normalized conductance with nonlinearity measurement. (c) Thirty repeated potentiation/depression cycles of the WBL device.

The non-linearities calculated for potentiation and depression are 2.2 and 3.5, respectively. **Figure 6**c displays the long-term potentiation/depression (LTP/LTD) for 30 epochs. The device exhibits seamless and unvarying changes in conductance when transitioning between consecutive pulses, which is advantageous for improving the learning precision of neuromorphic systems. The device's long-term potentiation (LTP) and long-term depression (LTD) features are well-suited for pattern recognition applications in neural networks. Neurons propagate impulses across synapses, while the learning process modifies the weight value of the synaptic device. The WBL gadget exhibits a multilayer conductance state that demonstrates improved linearity and enables rapid pattern recognition accuracy. The aforementioned findings validate the potential of utilizing AlN CBRAM with WNx barrier layer as a synaptic device in the field of neuromorphic computing.

4. CONCISE SUMMARY

This study showcases the enhanced stability of switching and the exceptional ability to retain data, as well as the extremely linear 30 epoch long-term potentiation/long-term depression synaptic functions achieved by including the WNx barrier layer into the AlN memristor. The observed outcome can be ascribed to the incorporation of a 7 nm WNx barrier layer, which hinders the generation of multiple conductive connections by restricting the diffusion of Cu into the switching layer. The regulation of Cu migration in the host layer significantly enhanced the consistency of switching, both from cycle to cycle and from device to device. The CBRAM device demonstrated longer DC endurance for 500 cycles and stable data retention for 10^3 s at 120 °C. The current transport in the switching layer was used to explain the ohmic and ionic conduction mechanisms of the non-volatile CBRAM cell during set/reset operation. Ultimately, the recurrent long-term potentiation (LTP) and long-term depression (LTD) synaptic activities are illustrated using a consistent pulse scheme.

At a Glance

This study examines the impact of the tungsten nitride (WNx) diffusion barrier layer on the resistive switching function of the aluminum nitride (AlN) based conductive bridge random access memory. The WNx barrier layer restricts the movement of Cu ions in the AlN switching layer, hence regulating the creation of metallic conductive filaments in the host layer. The gadget functioned at a significantly low voltage, with a Vset of 0.6 V and a Vreset of 0.4 V. The introduction of a barrier layer resulted in a considerable reduction in both spatial and temporal switching variability. When barrier layer devices are used, the worst-case coefficient of variations (σ/μ) for HRS and LRS are 33% and 18%, respectively. In contrast, when the barrier layer is not present, the worst-case coefficient of variations are 167% and 33% for HRS and LRS, respectively. When a barrier layer is present, the device demonstrates data retention for over 104 seconds at a temperature of 120 °C. However, when there is no barrier layer, the device stops functioning after only 103 seconds. The device exhibited synaptic activity characterized by long-term potentiation/depression (LTP/LTD) over a period of 30 epochs. This was achieved by applying a series of identical tailored pulses, each lasting 1 μs, as a stimulus.

REFERENCES

1. Waser, R.; Dittmann, R.; Staikov, G.; Szot, K. Redox-Based Resistive Switching Memories—Nanoionic Mechanisms, Prospects, and Challenges. *Adv. Mater.* **2009**, *21*, 2632–2663. **[Google Scholar] [CrossRef]**

2. Wang, Z.; Wu, H.; Burr, G.W.; Hwang, C.S.; Wang, K.L.; Xia, Q.; Yang, J.J. Resistive switching materials for information processing. *Nat. Rev. Mater.* **2020**, *5*, 173–195. **[Google Scholar] [CrossRef]**

3. Cha, J.-H.; Yang, S.Y.; Oh, J.; Choi, S.; Park, S.; Jang, B.C.; Ahn, W.; Choi, S.-Y. Conductive-bridging random-access memories for emerging neuromorphic computing. *Nanoscale* **2020**, *12*, 14339–14368. **[Google Scholar] [CrossRef]**

4. Choi, S.; Tan, S.H.; Li, Z.; Kim, Y.; Choi, C.; Chen, P.-Y.; Yeon, H.; Yu, S.; Kim, J. SiGe epitaxial memory for neuromorphic computing with reproducible high performance based on engineered dislocations. *Nat. Mater.* **2018**, *17*, 335–340. **[Google Scholar] [CrossRef] [PubMed]**

5. Liu, Y.; Gao, J.; Wu, F.; Tian, H.; Ren, T.L. The Origin of CBRAM With High Linearity, On/Off Ratio, and State Number for Neuromorphic Computing. *IEEE Trans. Electron. Devices* **2021**, *68*, 2568–2571. **[Google Scholar] [CrossRef]**

6. Pan, F.; Gao, S.; Chen, C.; Song, C.; Zeng, F. Recent progress in resistive random access memories: Materials, switching mechanisms, and performance. *Mater. Sci. Eng. R Rep.* **2014**, *83*, 1–59. **[Google Scholar] [CrossRef]**

7. Valov, I.; Waser, R.; Jameson, J.R.; Kozicki, M.N. Electrochemical metallization memories—Fundamentals, applications, prospects. *Nanotechnology* **2011**, *22*, 254003. **[Google Scholar] [CrossRef] [PubMed]**

8. Sun, W.; Gao, B.; Chi, M.; Xia, Q.; Yang, J.J.; Qian, H.; Wu, H. Understanding memristive switching via in situ characterization and device modeling. *Nat. Commun.* **2019**, *10*, 3453. **[Google Scholar] [CrossRef] [Green Version]**

9. Yu, S.; Wong, H.S.P. Compact Modeling of Conducting-Bridge Random-Access Memory (CBRAM). *IEEE Trans. Electron Devices* **2011**, *58*, 1352–1360. **[Google Scholar] [CrossRef]**

10. Ilyas, N.; Li, D.; Li, C.; Jiang, X.; Jiang, Y.; Li, W. Analog Switching and Artificial Synaptic Behavior of Ag/SiOx:Ag/TiOx/p++-Si Memristor Device. *Nanoscale Res. Lett.* **2020**, *15*, 30. **[Google Scholar] [CrossRef]**

11. Ohno, T.; Hasegawa, T.; Tsuruoka, T.; Terabe, K.; Gimzewski, J.K.; Aono, M. Short-term plasticity and long-term potentiation mimicked in single inorganic synapses. *Nat. Mater.* **2011**, *10*, 591–595. **[Google Scholar] [CrossRef] [PubMed]**

12. Yan, X.; Qin, C.; Lu, C.; Zhao, J.; Zhao, R.; Ren, D.; Zhou, Z.; Wang, H.; Wang, J.; Zhang, L.; et al. Robust Ag/ZrO2/WS2/Pt Memristor for Neuromorphic Computing. *ACS Appl. Mater. Interfaces* **2019**, *11*, 48029–48038. **[Google Scholar] [CrossRef] [PubMed]**

13. Lim, S.; Sung, C.; Kim, H.; Kim, T.; Song, J.; Kim, J.J.; Hwang, H. Improved Synapse Device With MLC and Conductance Linearity Using Quantized Conduction for Neuromorphic Systems. *IEEE Electron Device Lett.* **2018**, *39*, 312–315. **[Google Scholar] [CrossRef]**

14. Sun, B.; Guo, T.; Zhou, G.; Ranjan, S.; Jiao, Y.; Wei, L.; Zhou, Y.N.; Wu, Y.A. Synaptic devices based neuromorphic computing applications in artificial intelligence. *Mater. Today Phys.* **2021**, *18*, 100393. **[Google Scholar] [CrossRef]**

15. Sun, B.; Zhou, G.; Yu, T.; Chen, Y.; Yang, F.; Zhao, Y. Multi-factor-controlled ReRAM devices and their applications. *J. Mater. Chem. C* **2022**, *10*, 8895–8921. **[Google Scholar] [CrossRef]**

16. Lu, W.; Jeong, D.S.; Kozicki, M.; Waser, R. Electrochemical metallization cells—Blending nanoionics into nanoelectronics? *MRS Bull.* **2012**, *37*, 124–130. **[Google Scholar] [CrossRef]**

17. Moon, K.; Lim, S.; Park, J.; Sung, C.; Oh, S.; Woo, J.; Lee, J.; Hwang, H. RRAM-based synapse devices for neuromorphic systems. *Faraday Discuss.* **2019**, *213*, 421–451. **[Google Scholar] [CrossRef]**

18. Celano, U.; Goux, L.; Belmonte, A.; Opsomer, K.; Detavernier, C.; Jurczak, M.; Vandervorst, W. Conductive filaments multiplicity as a variability factor in CBRAM. In Proceedings of the 2015 IEEE International Reliability Physics Symposium, Monterey, CA, USA, 19–23 April 2015; pp. MY.11.11–MY.11.13. **[Google Scholar]**

19. Mohanty, S.K.; Chen, Y.-Y.; Yeh, P.-H.; Horng, R.-H. Thermal Management of GaN-on-Si High Electron Mobility Transistor by Copper Filled Micro-Trench Structure. *Sci. Rep.* **2019**, *9*, 19691. **[Google Scholar] [CrossRef] [Green Version]**

20. Choi, B.J.; Torrezan, A.C.; Strachan, J.P.; Kotula, P.G.; Lohn, A.J.; Marinella, M.J.; Li, Z.; Williams, R.S.; Yang, J.J. High-Speed and Low-Energy Nitride Memristors. *Adv. Funct. Mater.* **2016**, *26*, 5290–5296. **[Google Scholar] [CrossRef] [Green Version]**

21. Mohanty, S.K.; Reddy, P.K.; Prasad, O.K.; Wu, C.H.; Chang, K.M.; Lin, J.C. Interface Engineering for 3-Bit per Cell Multilevel Resistive Switching in AlN Based Memristor. *IEEE Electron Device Lett.* **2021**, *42*, 1770–1773. **[Google Scholar] [CrossRef]**

22. Belmonte, A.; Kim, W.; Chan, B.T.; Heylen, N.; Fantini, A.; Houssa, M.; Jurczak, M.; Goux, L. A Thermally Stable and High-Performance 90-nm Al_2O_3\Cu-Based 1T1R CBRAM Cell. *IEEE Trans. Electron Devices* **2013**, *60*, 3690–3695. **[Google Scholar] [CrossRef]**

23. Kuo, C.C.; Chen, I.C.; Shih, C.C.; Chang, K.C.; Huang, C.H.; Chen, P.H.; Chang, T.C.; Tsai, T.M.; Chang, J.S.; Huang, J.C. Galvanic Effect of Au–Ag Electrodes for Conductive Bridging Resistive Switching Memory. *IEEE Electron. Device Lett.* **2015**, *36*, 1321–1324. **[Google Scholar] [CrossRef]**

24. Wang, J.; Li, L.; Huyan, H.; Pan, X.; Nonnenmann, S.S. Highly Uniform Resistive Switching in HfO2 Films Embedded with Ordered Metal Nanoisland Arrays. *Adv. Funct. Mater.* **2019**, *29*, 1808430. **[Google Scholar] [CrossRef]**

25. Yeon, H.; Lin, P.; Choi, C.; Tan, S.H.; Park, Y.; Lee, D.; Lee, J.; Xu, F.; Gao, B.; Wu, H.; et al. Alloying conducting channels for reliable neuromorphic computing. *Nat. Nanotechnol.* **2020**, *15*, 574–579. **[Google Scholar] [CrossRef]**

26. Holloway, K.; Fryer, P.M.; Cabral, C.; Harper, J.M.E.; Bailey, P.J.; Kelleher, K.H. Tantalum as a diffusion barrier between copper and silicon: Failure mechanism and effect of nitrogen additions. *J. Appl. Phys.* **1992**, *71*, 5433–5444. **[Google Scholar] [CrossRef] [Green Version]**

27. Ting, C.Y.; Wittmer, M. The use of titanium-based contact barrier layers in silicon technology. *Thin Solid Film.* **1982**, *96*, 327–345. **[Google Scholar] [CrossRef]**

28. Farahani, M.M.; Turner, T.E.; Barnes, J.J. Evaluation of Titanium as a Diffusion Barrier Between Aluminum and Silicon for 1.2 μm CMOS Integrated Circuits. *J. Electrochem. Soc.* **1987**, *134*, 2835–2845. **[Google Scholar] [CrossRef]**

29. Ono, H.; Nakano, T.; Ohta, T. Diffusion Barrier Effects Against Cu of W-N Layer Formed by Electron Cyclotron Resonance Plasma Nitridation on W Layer. *Jpn. J. Appl. Phys.* **1995**, *34*, 1827–1830. **[Google Scholar] [CrossRef]**

30. Hsieh, T.-E.; Lin, Y.-C.; Li, F.-M.; Shi, W.-C.; Huang, Y.-X.; Lan, W.-C.; Chin, P.-C.; Chang, E.Y. GaN High-Electron-Mobility Transistor with WN x /Cu Gate for High-Power Applications. *J. Electron. Mater.* **2015**, *44*, 4700–4705. **[Google Scholar] [CrossRef]**

31. Takeyama, M.; Noya, A. Preparation of WN_x Films and Their Diffusion Barrier Properties in Cu/Si Contact Systems. *Jpn. J. Appl. Phys.* **1997**, *36*, 2261–2266. **[Google Scholar] [CrossRef]**

32. Asgary, S.; Hantehzadeh, M.R.; Ghoranneviss, M.; Boochani, A. Thermal stability of amorphous tungsten/tungsten nitride synthesis using HFCVD as a diffusion barrier for copper. *Appl. Phys. A* **2016**, *122*, 518. **[Google Scholar] [CrossRef]**

33. Panda, D.; Simanjuntak, F.M.; Chandrasekaran, S.; Pattanayak, B.; Singh, P.; Tseng, T.Y. Barrier Layer Induced Switching Stability in Ga:ZnO Nanorods Based Electrochemical Metallization Memory. *IEEE Trans. Nanotechnol.* **2020**, *19*, 764–768. **[Google Scholar] [CrossRef]**

34. Iraji-zad, A.; Ahadian, M.M.; Vashaei, Z. Room temperature diffusion of Cu in vanadium pentoxide thin films. *J. Phys. D Appl. Phys.* **2002**, *35*, 1176–1182. **[Google Scholar] [CrossRef]**

35. Kang, Y.; Liu, T.; Potnis, T.; Orlowski, M.K. Composite Cu/VO and VO/Cu Nanofilaments in Cu/Ta2O5/Pt Devices. *ECS Solid State Lett.* **2013**, *2*, Q54–Q57. **[Google Scholar] [CrossRef]**

36. Kwon, K.-C.; Song, M.-J.; Kwon, K.-H.; Jeoung, H.-V.; Kim, D.-W.; Lee, G.-S.; Hong, J.-P.; Park, J.-G. Nanoscale CuO solid-electrolyte-based conductive-bridging-random-access-memory cell operating multi-level-cell and 1selector1resistor. *J. Mater. Chem. C* **2015**, *3*, 9540–9550. **[Google Scholar] [CrossRef]**

37. Seung, H.-M.; Kwon, K.-C.; Lee, G.-S.; Park, J.-G. Flexible conductive-bridging random-access-memory cell vertically stacked with top Ag electrode, PEO, PVK, and bottom Pt electrode. *Nanotechnology* **2014**, *25*, 435204. **[Google Scholar] [CrossRef]**

38. Wu, W.; Wu, H.; Gao, B.; Yao, P.; Zhang, X.; Peng, X.; Yu, S.; Qian, H. A Methodology to Improve Linearity of Analog RRAM for Neuromorphic Computing. In Proceedings of the 2018 IEEE Symposium on VLSI Technology, Honolulu, HI, USA, 18–22 June 2018; pp. 103–104. [**Google Scholar**]

11

Chapter

AN EXAMINATION OF CERTAIN CHARACTERISTICS AND THE APPLICATION OF THE TWIST MAP IN THE CONTEXT OF THE FINITE FRENKEL-KONTOROVA MODEL

1. INTRODUCTION: AN OVERVIEW

The Frenkel-Kontorova model (FK) was introduced in 1938 to examine the presence of dislocations in a linear arrangement of particles [1]. The FK model has served as a prominent example in the discussion of several elements of solid-state physics for the past century. One can partition a collection of particles into a subsystem consisting of one-dimensional interacting elements and a remaining substrate. The latter interacts with the extracted subsystem through a potential. Electronic applications for Wigner electrons or Josephson junctions are of particular interest, as indicated by references [2,3,4,5,6,7]. Additional examples include charge-density wave conductors [8,9,10], charge transport in solids and on crystal surfaces [11], magnetic or ferro- and antiferromagnetic domain walls [12], magnetic superlattices [13], superconductivity [14,15], and vortex matter [16,17,18], among others.

The FK model for a chain of N particles (called atoms) at positions ui on an axis has the energy **[1,19]**

$$U(\mathbf{u}) = \sum_{i=1}^{N} \lambda\, V(u_i) + \sum_{i=1}^{N-1} \frac{k}{2}(u_{i+1} - u_i - a_0)^2 . \qquad (1)$$

The substrate potential function is the usual one

$$V(u_i) = 1 - \cos\left(\frac{2\pi}{a_s} u_i\right) .$$

The value as represents the period of this potential, whereas λ denotes its amplitude. The second term refers to the spring term that accounts for the nearest neighbor forces acting between the atoms. The parameter k represents its quantity. The chain's equilibrium distance, ao, would remain unchanged in the absence of the potential V, specifically when λ=0. Typically, the two periods *ao* and *as* exhibit dissimilarities [20]. Occasionally, the chain's equilibrium distance, *ao*, is absent [21,22,23].

At first, one has to assume that a_o=0 is used. One treats the FK model as a different one with the pure harmonic potential of the spring forces between the nearest neighbors but often with an additional linear tensile force of amount μ

$$\sum_{i=1}^{N-1}\left[\frac{k}{2}(u_{i+1}-u_i)^2-\mu\,(u_{i+1}-u_i)\right]. \qquad (2)$$

Note that the alternating tensile force of form (2) reduces to two summands

$$-\mu\,(u_N-u_1)\,.$$

If μ=0, then putting all *ui*=0 (or putting all *ui*=na_s with an integer *n*) gives zero energy [**24,25,26**]. This is the absolute minimal value on a global scale, yet it is considered to be a simple and unchallenging one. Starting with a finite chain that has no zero distances between the atoms, and using the parameter values from reference [21] for $\lambda \le k$ and μ=0, performing a minimization would yield a simple conclusion. In Appendix A, we note that when λ is much larger than *k*, additional stationary states can exist when *ao* is equal to zero and when μ is equal to zero. Nevertheless, these states do not represent global minima.

The chapter primarily focuses on the utilization of the twist map on a finite FK chain. In Section 2, we review the equations for a stable configuration and present a counterargument against utilizing these equations for the twist map application. In Section 3, we provide a comprehensive explanation of the concept of tensile force. Sections 4 and 5 of our discussion focus on the boundary conditions of a finite FK chain. We also provide examples to illustrate the wrong outcomes of the twist map. Additional minor considerations include the energy of the FK chain and the potential issue of the illogical relationships between the two periodicities of the FK model. Our remedies incorporate linkages to an endless FK chain [21].

2. EQUATIONS FOR STATIONARY STATES

At first, it seems that the gap of form (2) does not matter, because for the minimum search for the inner atoms, *ui*, holds an equation, where the a_o disappears

$$\frac{\partial U(\mathbf{u})}{\partial u_i}=\frac{2\pi\lambda}{a_s}\sin\left(\frac{2\,\pi u_i}{a_s}\right)+k[2\,u_i-u_{i+1}-u_{i-1}]=0,\ i=2,..,N-1. \qquad (3)$$

Note that the linear tensile force with amount μ also disappears [**27**]. However, the a_o as well as the μ do not disappear at all; they emerge in the two boundary equations for the first and the last atom of the chain, u_1 and u_N [**19,28**]

$$\frac{\partial U(\mathbf{u})}{\partial u_1} = \frac{2\pi\lambda}{a_s}\sin\left(\frac{2\pi u_1}{a_s}\right) - k[u_2 - u_1 - a_o] + \mu = 0 \tag{4}$$

and

$$\frac{\partial U(\mathbf{u})}{\partial u_N} = \frac{2\pi\lambda}{a_s}\sin\left(\frac{2\pi u_N}{a_s}\right) + k[u_N - u_{N-1} - a_o] - \mu = 0\,. \tag{5}$$

Equations (3)–(5), for all atoms i=1,...N, form a system of coupled equations, where we have to include the boundary conditions Equations (**4**) and (**5**). To select a single 'solution' of one or a small number of them will usually not give a correct stationary state of the FK chain.

In contrast to this simple remark, many researchers propose to use the 'twist map', which successively calculates one u_{i+1} if the former u_i and u_{i-1} are known—an assumption that we cannot make. If one does not know the solution, one also does not know the location of the two special atoms at the start of the map [**19**]. In real calculations, the twist map is started at an arbitrary pair of two points in a finite region [**21**]. Of course, to start the twist map at arbitrary point pairs (see the M_i in Figure 1 of ref. [**21**]) provides nice pictures, including the excursion to chaos. We claim that many calculations of twist maps do not use the correct initial values; for example, see Example 1 and **Section 6** below.

Example 1.

A simple example is a chain with N = 3 atoms, with the parameters a_s=2π, k=2.5, λ=1, *and* a_o=2a_s/3. *The equilibrium structures of a minimum are obtained by the NMinimize procedure of Mathematica 13.0, starting with a chain with* u_i=a_o(i−1) *for i*=1,..,3, *as shown on the left of* ***Figure 1.***

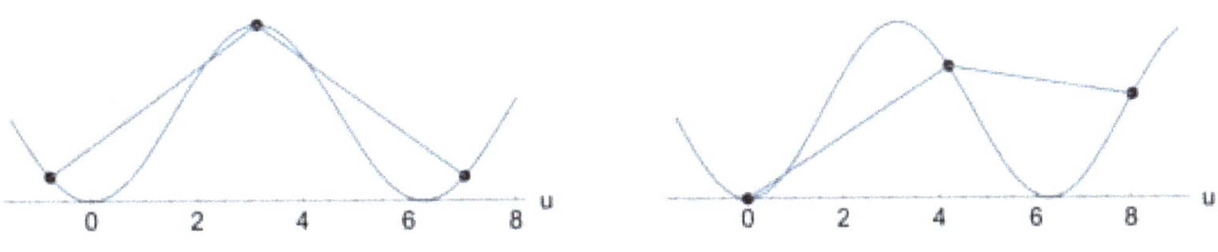

Figure 1. (Left): Minimum solution for an FK chain with N=3 atoms. (Right): With an arbitrary start of u_1=0, we obtain u_2=a_0=2a_s/3, and we obtain with Equation (3) the shown point u_3. However, the structure is not a minimum, see text. Note that the atoms are shifted on the site-up potential to help the imagination.

Using the twist map of Equations (4) and (3) starting at an arbitrary point, for example, with u_1=0, we obtain with Equation (**4**) u_2=a_0=2a_s/3, and we obtain with Equation (**3**) the structure on the right hand side of **Figure 1**. It does not look like a minimum. The left minimum has an energy of 2.756 units, but this special twist map 'solution' has an energy of 2.826 units. Its gradient (0, 0, 0.118)T is not a zero vector.

The reason for the deviation of the gradient from zero is that the third Equation (**5**) for i=N=3 is not fulfilled for the obtained u_3. Only when one begins with u_1 of a stationary structure is the third equation fulfilled. Further arguments to understand the result are given in ref. [19]. We conclude with the following theorem.

Theorem 1.

*Using the twist map at an arbitrary start point is useless. In the general case, it will not provide a stationary structure of the FK chain. The reason is that Equation (**5**) is not fulfilled, in the general case. This is not restricted to the N=3 chain.*

The twist map fulfills the demand of equilibrium only from step to step. Only when the start pair belongs to the correct minimum do the results (under numerical problems, see below) build a correct minimum structure. In the system, one can fulfill some equations—with false values, which vice versa compensate—however, one cannot assume that the other equations obtain the correct solution.

3. A REMARK ON THE LENGTH OF THE CHAIN

We treat the energy of the chain structure (1). When the spring distance of the chain, a_o, and the period of the site-up potential, a_s, are equal, and if μ=0, or if a_0=0 and μ/k=a_s, or if a_0+μ/k=a_s, then an equilibrium chain exists in its ground state with all particles at the bottom of the site-up potential, with energy zero for every particle, and thus with zero energy for the chain in its ground state. (This case was originally treated [**1**] for an infinite chain.)

In any other case, an infinite chain needs an infinite energy in the sum, which is impossible. Strangely, the authors of ref. [29] stated that "...one does not hope that the sum of the energies converges." Nevertheless, they worked with the divergent sum. Tong et al. [30] treated a finite segment of the infinite incommensurate FK chain without the boundary conditions. Without further explanation, the problem was circumscribed by the word 'formal sum' [31,32].

A way out, avoiding the infinite energy of an infinite chain, was proposed [33] by dividing the energy of the chain by N, the number of atoms. So, one has a finite energy per atom [34,35], and one can apply the optimization per atom at u_i, discussed in **Section 2** [36].

4. THE MEANING OF THE TENSILE FORCE

If λ=0, we see with Equations (**4**) and (**5**) that the equilibrium separation of the chain becomes

$$a_0 + \mu/k \,. \tag{6}$$

In addition, if one uses the ansatz of form (2) with a_0=0, then the value of the tensile force divided by the value of the spring force, μ/k, becomes the placeholder of the a_0 in the original FK model. One can put [37,38]

$$l_0 = \mu/k \quad \rightarrow \quad \text{new} \quad a_0 \,.$$

The meaning of the tensile force in (2) is to include an equilibrium separation of the chain, l_0 for a_0. One can ask why one needs the tensile force.

On the other hand, one can ask for the action of the tensile force, if its amount μ increases. It has the form

$$-\mu\ \mathbf{f}^T\ \mathbf{u}$$

with the special form of the force vector

$$\mathbf{f}^T \ = \ (-1, 0, \ldots, 0, 1)\,.$$

At μ=0, we may be in an original minimum (with a_0=0 or not). Increasing the amount of μ, we can optimize another minimum. The curve, which describes the consecutive series of minima, is a Newton trajectory; it was named the curve of the force-displaced stationary points [19,39,40,41,42,43,44]. Newton trajectories connect stationary points of an index difference of one [45] on the original potential energy surface. It is a quite interesting property that the special form of the f vector acts as a change in the orginal distance, a_0, of the atoms of the FK chain, see value (6), and not only as the tilding of the potential of the first and the last atom of the chain. The FK chain is similar to an accordion. If one pulls the ends, all of the bellows relax. Newton trajectories with force f can be used to calculate many kinds of solitones of the FK chain and, thus, intermediate minima of the potential energy surface and saddle points with an increasing index [3,6,19,46,47,48].

5. THE MEANING OF THE FREE BOUNDARY CONDITIONS

Now, we discuss the boundary conditions (BCs). If one has free BCs, then we cannot start with an arbitrary 'left' BC, because the minimization will probably result in different BC in the end. So, the steps of an arbitrary twist map start in a nebula. The developments by some researchers [21,49,50,51] (to name a few), which use the twist map, ignore that the chain will find another minimum structure at the boundary, in comparison to an arbitrary twist map result [19].

Of course, if one starts the twist map with the correct u_1, u_2 of a minimum structure or of another stationary state of the FK chain, then one can regain this stationary structure with the result of the twist map, at least for the first section of the structure; see the example below. However, we do not know where the stationary structure of an FK chain begins if the BCs are free.

Example 2.

The example is again the chain with parameters a_s=2π, k=2.5, λ=1, and a_o=2a_s/3, for N = 3, 4, 8–10, 20, 47, 74, 99, 500, 1000, and N=5000 atoms. The equilibrium structures, probably minima, are obtained by the NMinimize procedure of Mathematica 13.0 starting with a chain with u_i=a_o(i–1) for i=1,...,N. The first atom of the chain at u_1 changes its place with N, as well as the average distance

$$\tilde{a} = \frac{u_N - u_1}{N-1}, \tag{7}$$

see ***Table 1****. The numbers in* ***Table 1*** *are rational numbers, because they are the result of a computer calculation. We do not know whether the character of the limit for N to ∞ is rational or irrational. The two values possibly converge for large N against defined values around u_1 = 0.369 and a_o. However, we do not know the exact limit of $\tilde{a}$.*

Table 1. Position of the first atom of a minimum of an FK chain and its average distance.

N	u_1	ã/2π	N	u_1	ã/2π
3	-0.769	0.6224	47	0.390	0.6700
4	0.117	0.6542	74	0.370	0.6688
8	0.751	0.6801	99	0.358	0.6621
9	-0.341	0.6386	500	0.368	0.6670
10	0.239	0.6582	1000	0.369	0.6665
20	0.632	0.6736	5000	0.369	0.6667

We show in **Figure 2** a minimum of an FK chain with *N*=500 atoms. The central region represents the winding number, the ratio 2/3 of a_o/a_s; however, the two boundary regions show the large influence of the free BCs. A swing in at the left boundary needs ≈50 atoms, and an analogous transient process needs ≈50 atoms at the right boundary. One could imagine that an infinite chain with the 2/3 ratio could be represented by the central piece of **Figure 2**. However, a cut anywhere relaxes the strain at the border [20], and the shown boundary pattern will emerge again. It also holds for *N*=1000 and *N*=5000, where we obtain similar pictures.

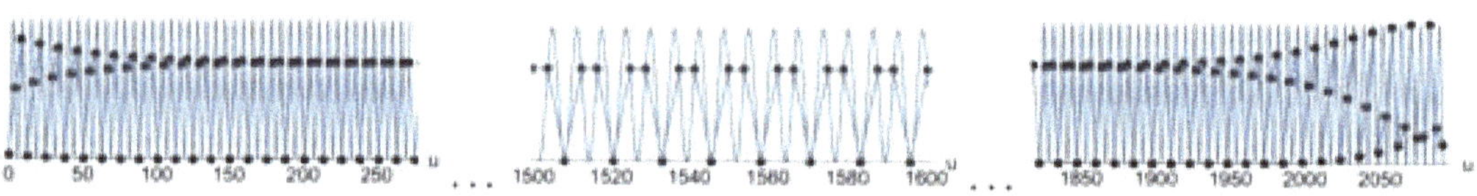

Figure 2. Minimum solution for a fixed a_o, see text. FK chain with *N*=500 atoms, with the first 66 atoms on the left and the last 66 atoms on the right. The center is an enlarged piece with the regular pattern representing the rational relation to a_s sorting three atoms in two wells.

The structures of the FK chain at the reported minima are mirror symmetrical up to a length of $N = 46$; see the left panel of **Figure 1**. Beginning with $N = 47$, the symmetry is broken, and the different form at the two boundaries emerges, as in **Figure 2**.

In **Figure 3**, we show the corresponding twist map representation of the first 66 atoms for N=500. There, we used the map $\Phi_i=u_i(\mathrm{mod}\ a_s)$ for the minimization result. Consecutive pairs of atoms are shown, (Φ_i, Φ_{i+1}), (Φ_{i+1}, Φ_{i+2}), (Φ_{i+2}, Φ_{i+3}),... and so on. The dashed triangle is the start set of points, and the red triangle shows the periodic stable cycle of the central region of the FK chain. The swing in at the left boundary of the chain is the path from the dashed to the red triangle. Note that the points of the inner cycle form a kind of a unit cell defined by the used winding number, 2/3, but they are not free of tension. This is demonstrated by the relaxation at the boundaries.

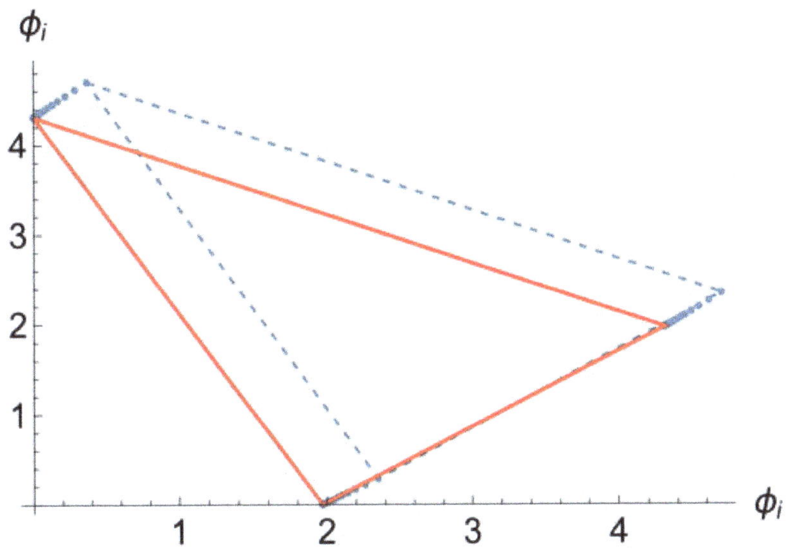

Figure 3. Twist map representation of the first 66 atoms of a long chain. The dashed line is the start triangle of points, and the red line is the quasi-stable cycle of three atoms in one cell of two troughs, as in the central region of Figure 2. (The calculation was a minimization.)

6. THE TWIST MAP FOR A FINITE FK CHAIN

In the right side of **Figure 4**, we depict the result of the twist map used with the two correct initial points, $u1$=0.369 and $u2$=4.701, of the optimized structure. The calculation was performed with Mathematica 13.0 with the usual accuracy. The result fits the true structure up to $N \approx 300$ atoms. Then, the twist map 'breaks out', and at $N = 353$ the former right BC appears. Its way out emerges by numerical effects, which can accumulate over the consecutive iterations.

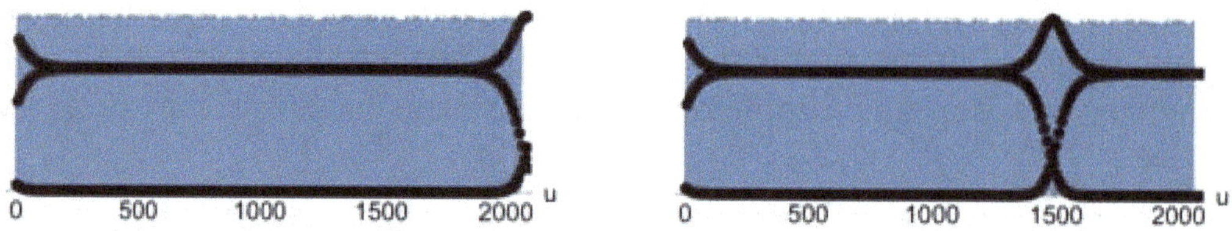

Figure 4. (Left): FK chain, optimized minimum with N=500 atoms. (Right): The result of the twist map with the two points of the minimum, u_1 and u_2, at the start. It repeats the left BC; however, it 'finishes' at N = 353 with a right BC; then, the calculation is reflected.

The twist map must include the handling of the tension of the central quasi-unit cells. The N=4 minimum shown on the left of **Figure 5**, with a u_1 value of 0.117, is two times 'shorter' than the putative unit cell of 4π, see **Table 1**. The distances of the minimized structure also change from atom to atom in the central part. It only looks like a unit cell. In reality, a unit cell does not exist.

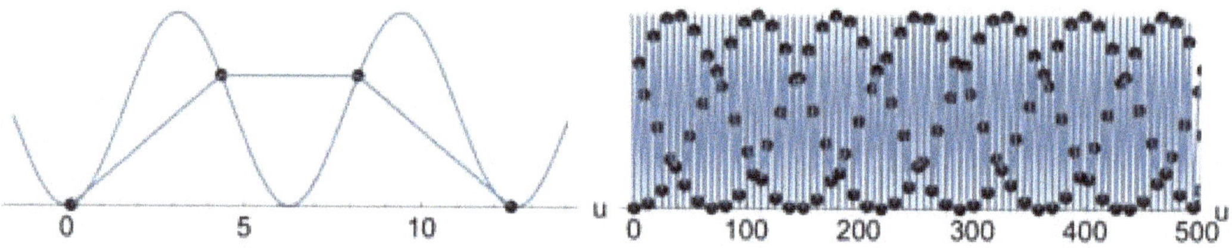

Figure 5. (Left): FK chain minimum for N=4 atoms; it is not the unit cell of Figure 2, center; however, it is shorter, see Table 1. (Right) The result of a twist map calculation with the two points, u_1=0 and u_2=$2a_s/3$, of a putative unit cell at the start. It does not result in the quasi-unit cells of Figure 2.

The result for an infinite chain [**21**] does not hold here, where one has a unit cell if a_o/a_s is a rational number. There is no period of the kind

$$u_i+3 = u_i + 2\,a_s\text{, but it holds } u_i+3 = ui + 2\,a_s + \epsilon i$$

with variable values of ϵ_i. The smallest ϵ_i we find at the central region of the chain with $\epsilon_{244}\approx 2\times 10^{-11}$.

The fact that even the twist map calculation by Equation (**3**) finds (though at a false place) a correct upper boundary structure, as the direct optimization does, is very interesting. One has to consider that the quasi-perfect cyclic behavior in the central region is not the behavior of a limit cycle, in contrast to the propositions of ref. [36].

Next, we show a twist map calculation in the right panel of **Figure 5** and in **Figure 6**. The start is the putative unit cell with u_1=0 and u_2=$4\pi/3$. However, the twist map does not find the structure of consecutive unit cells. It finds that the 3-cycle of **Figure 2** and **Figure 3** rotates and does not form a correct stable cycle, at least not for three atoms. See the twist map representation in Figure 6.

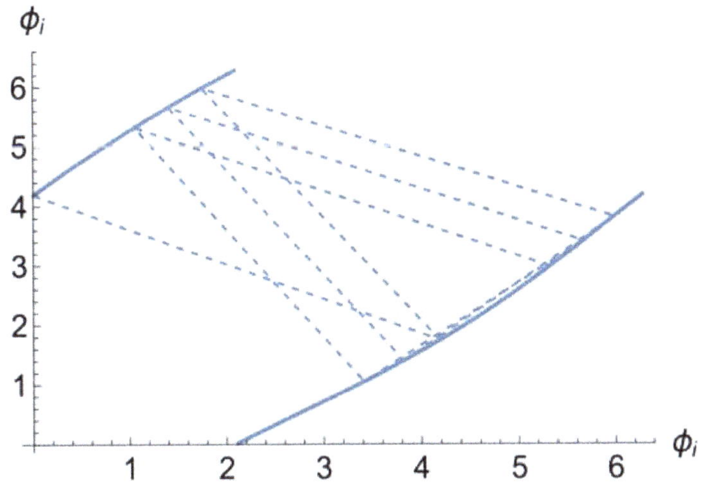

Figure 6. Twist map representation of a twist map calculation with u_1=0 and u_2=4π/3. The dashed line represents the start sequence of points.

7. THE DISAPPEARENCE OF INCOMMENSURABILITIES

For a given length of a finite FK chain with N atoms for λ=0,

$$L_O = (N-1)\,a_O,$$

one can determine an integer M such that the chain of length L_o covers up M troughs of the site potential. If λ is set correctly, the FK chain will fit into the M troughs forming a structure of a minimum with an average separation $\tilde{a}$, with Equation (7). It holds independently of whether the numbers, L_o, a_o, $\tilde{a}$, and a_s, are rational or irrational. One can cite ref. [52] "...a finite chain on a periodic substrate will always be locked because of its free ends." The relation of N to M is in every case a commensurable ratio between two integers [53]; see Example 3.

Example 3.

We treat an example of a chain with a putative 'irrational' winding number [54] of a_o=0.873a_s and the other parameters a_s=2π, k=2.5, and λ=1, for N=500. The equilibrium structure, probably a minimum, is obtained by the NMinimize procedure of Mathematica starting with a chain with u_i=a_o(i-1) for i=1,...,N. ***Figure*** *7 shows the structure and the corresponding twist map representation.*

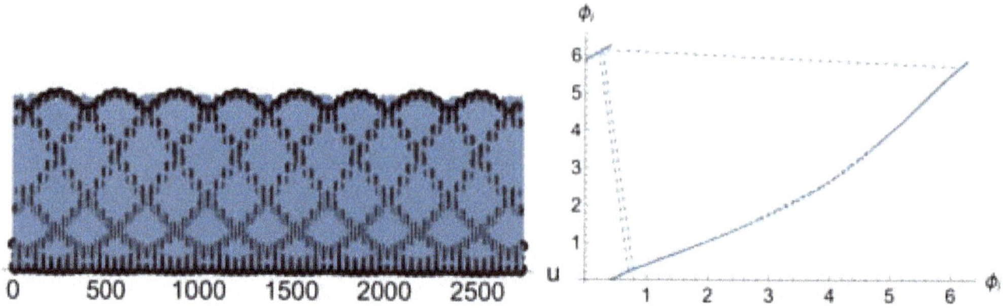

Figure 7. (Left): Optimized FK chain minimum for N=500 atoms with winding number 0.873. (Right): Twist map representation of the result. The dashed line is the start sequence of nine points plus the first step of the next cycle. It is a quasi-cyclic result.

The structure seems not to be very erratic. The twist map shows a quasi-cycle of nine points. However, the winding number 0.873 is not fully irrational.

8. REMARK ON THE AUBRY TRANSFORMATION OF A FINITE CHAIN

There should be a value λC, where a qualitative change in the behavior of the infinite FK chain occurs [21,55,56]. For a finite chain, there is no such value. If $\lambda>0$, and $k>0$, then the action of the site-up potential will fix the chain in a minimum, if we start a minimization for the 'relaxed' chain with distances a_0; see Section 7.

The putative Aubry-phase transition demonstrated by Figure 6 in ref. [54] is an incorrect interpretation. The first eigenvalue of the second derivatives of the potential of the chain is the frequency for a collective movement of the chain. For an unpinned chain, it has to be zero; thus, the potential has to be flat. However, in the example used in [54], the first eigenvalue is greater than zero. Thus, the potential is a minimum in both treated cases, $\lambda>\lambda_C$ and $\lambda<\lambda_C$, and the chain is pinned. Of course, the pinning is small if the λ is small. The chain can collectively vibrate; however, it cannot move.

For the interpretation of the experiment [57] with N = 1–5 atoms, we assume that the reported effect is for other reasons.

9. DISCUSSION

Our comprehension of the extensive chronicle of the twist map is as follows: In a finite FK chain, the value of parameter $a0$ is known, but the exact location of the minimum is uncertain. We do not know the values of u_1, u_N, and $\tilde{a}$. Thus, we cannot start the twist map.

In order to rectify this situation, one must establish a maximum limit, denoted as "l," for the average distance. Then, one can begin by selecting any two supposed points, denoted as $u1$ and $u2$, as the initial values. It is hypothesized that any random pair of initial values is conceivable. Figure 1 clearly demonstrates that the outcome is not advantageous for an initial finite FK chain.

We note that many researchers have treated finite chains [9,33,35,54,57,58,59, 60,61,62,63,64,65] (to name a few); however, they use a useless contrast of rational with irrational numbers in the finite FK model.

The procedure described in reference [66] organizes the FK chain in a 'commensurable' manner within the site's upward potential, creating a positive contrast.

10. CONCISE SUMMARY

We are surprised by researchers' attention to the twist map in the field of the FK model. For a finite FK chain, the twist map theory can only be a supplement; otherwise, important physics conclusions of the model are lost.

At a Glance

We analyze the twist map, specifically focusing on its use in the finite Frenkel-Kontorova model. We elucidate the significance of the tensile force in several proposed models. We prove that the utilization of the twist map in the finite FK model is erroneous, since it fails to account for the essential boundary requirements.

Abbreviations

The following abbreviations are used in this manuscript:

BC	Boundary condition
FK	Frenkel-Kontorova

Appendix A

For k=0 and μ=0, the FK model shortens to the pure site-up potential without spring forces. Any arbitrary distribution of the atoms at the bottom of the siteup potential is possible.

For $\lambda \gg \lambda_c > k > 0$ in the FK model, further minimum structures are possible under a_0=0, aside from the zero solution discussed in **Section 1.** λ_c depicts a critical value. Note that every small k>0 prevents the length of the FK chain from becoming too large.

In **Figure A1**, we depict a case for N=8, λ=10, k=1, and μ=0. The structure has two 'slightly stretched' springs, three 'stretched' springs, and two 'overstretched' springs, which are possible because of the much smaller k parameter. Note that most of the atoms are not at the minima of the site-up potential, as claimed for large λ [34].

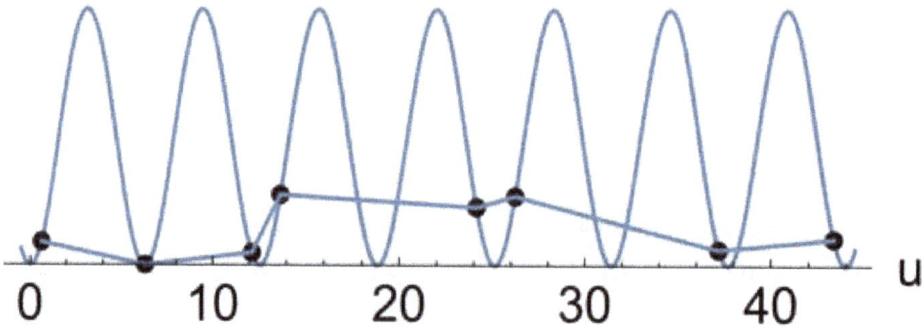

Figure A1. A nonzero FK minimum to potential (2).

REFERENCES

1. Kontorova, T.A.; Frenkel, Y.I. To the theory of plastic deformations and twinnings I. *Zh. Eksp. Teor. Fis.* **1938**, *8*, 89–95. (In Russian) **[Google Scholar]**
2. Lin, J.Y.; Smorodin, A.V.; Badrutdinov, A.O.; Konstantinov, D. Transport Properties of a Quasi-1D Wigner Solid on Liquid Helium Confined in a Microchannel with Periodic Potential. *J. Low Temp. Phys.* **2019**, *195*, 289–299. **[Google Scholar] [CrossRef] [Green Version]**
3. Quapp, W.; Lin, J.Y.; Bofill, J.M. The movement of a one-dimensional Wigner solid explained by a modified Frenkel-Kontorova model. *Eur. Phys. J. B* **2020**, *93*, 227. **[Google Scholar] [CrossRef]**
4. Duwel, A.E.; Watanabe, S.; Triás, E.; Orlando, T.P.; van der Zant, H.S.J.; Strogatz, S.H. Discreteness-induced resonances and ac voltage amplitudes in long one-dimensional Josephson junction arrays. *J. Appl. Phys.* **1997**, *82*, 4661–4668. **[Google Scholar] [CrossRef] [Green Version]**
5. Ustinov, A.V. Solitons in Josephson-junction arrays. *Phys. D* **1998**, *123*, 315–329. **[Google Scholar] [CrossRef]**
6. Quapp, W.; Bofill, J.M. Description of Shapiro steps on the potential energy surface of a Frenkel-Kontorova model Part II: Free boundaries. *Eur. Phys. J. B* **2021**, *94*, 64. **[Google Scholar] [CrossRef]**
7. Quapp, W.; Bofill, J.M. Description of zero field steps on the potential energy surface of a Frenkel-Kontorova model for annular Josephson junction arrays. *Eur. Phys. J. B* **2021**, *94*, 105. **[Google Scholar] [CrossRef]**
8. Grüner, G. The dynamics of charge-density waves. *Rev. Mod. Phys.* **1988**, *60*, 1129–1181. **[Google Scholar] [CrossRef]**
9. Garcia-Mata, I.; Zhirov, O.V.; Shepelyansky, D.L. The Frenkel-Kontorova model with cold trapped ions. *Eur. Phys. J. D* **2007**, *41*, 325–330. **[Google Scholar] [CrossRef] [Green Version]**

10. Pouget, J.P. The Peierls instability and charge density wave in one-dimensional electronic conductors. *Comptes Rendus Phys.* **2016**, *17*, 332–356. **[Google Scholar]** **[CrossRef]** **[Green Version]**

11. Braun, O.; Dauxois, T.; Paliy, M.V.; Peyrard, M. Mobility and diffusivity in a generalized Frenkel-Kontorova model. *Phys. Rev. B* **1996**, *54*, 321. **[Google Scholar]** **[CrossRef]**

12. Ekomasov, E.G.; Murtazin, R.R.; Bogomazova, O.B.; Gumerov, A.M. One-dimensional dynamics of domain walls in two-layer ferromagnet structure with different parameters of magnetic anisotropy and exchange. *J. Magn. Magn. Mater.* **2013**, *339*, 133–137. **[Google Scholar]** **[CrossRef]**

13. Trallori, L. Magnetic superlattices, classical spin chains, and the Frenkel-Kontorova model. *Phys. Rev. B* **1998**, *57*, 5923. **[Google Scholar]** **[CrossRef]**

14. Que, W.; Walker, M.B. Generalized Frenkel-Kontorova model for structural modulations in bismuth high-Tc superconductors and related compounds. *Phys. Rev. B* **1992**, *46*, 14772. **[Google Scholar]** **[CrossRef]** **[PubMed]**

15. Dinsmore, R.C., III; Bae, M.H.; Bezryadin, A. Fractional order Shapiro steps in superconducting nanowires. *Appl. Phys. Lett.* **2008**, *93*, 192505. **[Google Scholar]** **[CrossRef]** **[Green Version]**

16. Besseling, R.; Niggebrugge, R.; Kes, P.H. Transport properties of vortices in easy flow channels: A Frenkel-Kontorova study. *Phys. Rev. Lett.* **1999**, *82*, 3144–3147. **[Google Scholar]** **[CrossRef]** **[Green Version]**

17. Kokubo, N.; Besseling, R.; Vinokur, V.M.; Kes, P.H. Mode Locking of Vortex Matter Driven through Mesoscopic Channels. *Phys. Rev. Lett.* **2002**, *88*, 247004. **[Google Scholar]** **[CrossRef]** **[Green Version]**

18. Iaroshenko, O.; Rybalko, V.; Vinokur, V.M.; Berlyand, L. Vortex phase separation in mesoscopic superconductors. *Sci. Rep.* **2013**, *3*, 1758. **[Google Scholar]** **[CrossRef]** **[Green Version]**

19. Quapp, W.; Bofill, J.M. Newton Trajectories for the Frenkel-Kontorova model. *Mol. Phys.* **2019**, *117*, 1541–1558. **[Google Scholar]** **[CrossRef]** **[Green Version]**

20. Kochendörfer, A.; Seeger, A. Theorie der Versetzungen in eindimensionalen Atomreihen. I. Periodisch angeordnete Versetzungen. *Zeitschr. Phys.* **1950**, *127*, 533–550. **[Google Scholar]** **[CrossRef]**

21. Aubry, S. The twist map, the extended Frenkel-Kontorova model and the devil's staircase. *Phys. D* **1983**, *7*, 240–258. **[Google Scholar]** **[CrossRef]**

22. Hobart, R.; Celli, V. A Solution to the Frenkel-Kontorova Dislocation Model. *J. Appl. Phys.* **1962**, *33*, 60–62. **[Google Scholar]** **[CrossRef]**

23. Hu, B.; Shi, J. Nonanalytic twist maps and Frenkel-Kontorova models. *Phys. D* **1994**, *71*, 23–38. **[Google Scholar] [CrossRef] [Green Version]**

24. Aubry, S. Defectibility and frustration in incommensurate structures: The devil's stair case transformation. *Ferroelectrics* **1980**, *24*, 53–60. **[Google Scholar] [CrossRef]**

25. Guo, Y.; Qu, Z.; Zhang, Z. Lyapunov stability and precise control of the frictional dynamics of a one-dimensional particle array. *Phys. Rev. B* **2006**, *73*, 094118. **[Google Scholar] [CrossRef]**

26. Theodorakopoulos, N.; Peyrard, M.; MacKay, R.S. Nonlinear Structures and Thermodynamic Instabilities in a One-Dimensional Lattice System. *Phys. Rev. Lett.* **2004**, *93*, 258101. **[Google Scholar] [CrossRef] [Green Version]**

27. Aubry, S. Devils staircase and order without periodicity in classical condensed matter. *J. Phys. Fr.* **1983**, *44*, 147–162. **[Google Scholar] [CrossRef] [Green Version]**

28. Frank, F.C.; van der Merwe, J.H. One-Dimensional Dislocations. I. Static Theory. *Proc. R. Soc. Lond. A* **1949**, *198*, 205–216. **[Google Scholar]**

29. Blass, T.; de la Llave, R. The Analyticity Breakdown for Frenkel-Kontorova Models in Quasi-periodic Media: Numerical Explorations. *J. Stat. Phys.* **2013**, *150*, 1183–1200. **[Google Scholar] [CrossRef] [Green Version]**

30. Tong, P.; Li, B.; Hu, B. Wave transmission, phonon localization, and heat conduction of a one-dimensional Frenkel-Kontorova chain. *Phys. Rev. B* **1999**, *59*, 8639. **[Google Scholar] [CrossRef] [Green Version]**

31. Aubry, S.; MacKay, R.S.; Baesens, C. Equivalence of uniform hyperbolicity for symplectic twist maps and phonon gap for Frenkel-Kontorova models. *Phys. D* **1992**, *56*, 123–134. **[Google Scholar] [CrossRef]**

32. Salahshoor, H.; de la Llave, R. A Numerical Investigation of the Pinning Phenomenon in Quasi-Periodic Frenkel-Kontorova Model Under an External Force. *J. Stat. Phys.* **2018**, *173*, 398–410. **[Google Scholar] [CrossRef]**

33. Hu, B.; Li, B.; Zhao, H. Mode-locking of incommensurate phase by quantum zero point energy in the Frenkel-Kontorova model. *Europhys. Lett.* **2001**, *53*, 342. **[Google Scholar] [CrossRef]**

34. Griffiths, R.B.; Chou, W. Effective Potentials: A New Approach and New Results for One-Dimensional Systems with Competing Length Scales. *Phys. Rev. Lett.* **1986**, *56*, 1929. **[Google Scholar] [CrossRef] [PubMed]**

35. Sharma, S.R.; Bergersen, B.; Joos, B. Aubry transition in a finite modulated chain. *Phys. Rev. B* **1984**, *29*, 6335–6340. **[Google Scholar] [CrossRef]**

36. Baesens, C.; MacKay, R.S. Gradient dynamics of tilted Frenkel-Kontorova models. *Nonlinearity* **1998**, *11*, 949–964. **[Google Scholar] [CrossRef]**
37. Aubry, S.; Abramovici, G. Chaotic trajectories in the standard map. The concept of anti-intergability. *Phys. D* **1990**, *43*, 199–219. **[Google Scholar] [CrossRef]**
38. Aubry, S.; Le Daeron, P.Y. The diskrete Frenkel-Kontorova model and its extensions I. Exact results for the ground-states. *Phys. D* **1983**, *8*, 381. **[Google Scholar] [CrossRef]**
39. Avdoshenko, S.M.; Makarov, D.E. Reaction Coordinates and Pathways of Mechanochemical Transformations. *J. Phys. Chem. B* **2015**, *120*, 1537–1545. **[Google Scholar] [CrossRef]**
40. Bofill, J.M.; Ribas-Ariño, J.; García, S.P.; Quapp, W. An Algorithm to Locate Optimal Bond Breaking Points on a Potential Energy Surface. *J. Chem. Phys.* **2017**, *147*, 152710. **[Google Scholar] [CrossRef] [Green Version]**
41. Quapp, W. Can We Understand the Branching of Reaction Valleys for more than two Degrees of Freedom? *J. Math. Chem.* **2015**, *54*, 137–148. **[Google Scholar] [CrossRef]**
42. Quapp, W.; Bofill, J.M. A contribution to a theory of mechanochemical pathways by means of Newton trajectories. *Theor. Chem. Acc.* **2016**, *135*, 113. **[Google Scholar] [CrossRef]**
43. Quapp, W.; Bofill, J.M. Mechanochemistry on the Müller-Brown Surface by Newton Trajectories. *Int. J. Quant. Chem.* **2018**, *118*, e25522. **[Google Scholar] [CrossRef]**
44. Quapp, W.; Bofill, J.M.; Ribas-Ariño, J. Towards a Theory of Mechanochemistry, Simple Models from the Very Beginnings. *Int. J. Quant. Chem.* **2018**, *118*, e25775. **[Google Scholar] [CrossRef] [Green Version]**
45. Hirsch, M.; Quapp, W. Reaction Channels of the Potential Energy Surface: Application of Newton Trajectories. *J. Mol. Struct. Theochem* **2004**, *683*, 1–13. **[Google Scholar] [CrossRef] [Green Version]**
46. Quapp, W.; Bofill, J.M. A Model for a Driven Frenkel-Kontorova Chain. *Eur. Phys. J. B* **2019**, *92*, 95–117. **[Google Scholar] [CrossRef]**
47. Quapp, W.; Bofill, J.M. Sliding Paths for Series of Frenkel-Kontorova Models—A Contribution to the Concept of 1D-superlubricity. *Eur. Phys. J. B* **2019**, *92*, 193. **[Google Scholar] [CrossRef]**
48. Quapp, W.; Bofill, J.M. Description of Shapiro steps on the potential energy surface of a Frenkel-Kontorova model Part I: Chain in a box. *Eur. Phys. J. B* **2021**, *94*, 66. **[Google Scholar] [CrossRef]**

49. Bichoutskaia, E.; Heggie, M.I.; Lozovik, Y.E.; Popov, A.M. Multi-Walled Nanotubes: Commensurate-Incommensurate Phase Transition and NEMS Applications. *Fuller. Nanotubes Carbon Nanostruct.* **2006**, *14*, 131–140. **[Google Scholar] [CrossRef]**

50. Black, R.C.; Satija, I.I. Universal phase diagram in the generalized Frenkel-Kontorova model. *Phys. Rev. B* **1991**, *44*, 4089–4096. **[Google Scholar] [CrossRef]**

51. Vanossi, A.; Bechinger, C.; Urbakh, M. Structural lubricity in soft and hard matter systems. *Nat. Commun.* **2020**, *11*, 4657. **[Google Scholar] [CrossRef] [PubMed]**

52. Stoyanov, S.; Müller-Krumbhaar, H. Resonance-induced Cluster-mobility: Dynemics of a finite Frenkel-Kontorova model. *Surf. Sci.* **1985**, *159*, 49–68. **[Google Scholar] [CrossRef]**

53. Quapp, W.; Bofill, J.M. Comment on 'Out-of-equilibrium Frenkel-Kontorova model' (A. Imparato, 2021, J. Stat. Mech. 013214). *J. Stat. Mech.* **2022**, *2022*, 013204. **[Google Scholar] [CrossRef]**

54. Muñoz, J.M.; Sawant, R.; Maffei, A.; Wang, X.; Barontini, G. Realizing the Frenkel-Kontorova model with Rydberg-dressed atoms. *Phys. Rev. A* **2020**, *102*, 043308. **[Google Scholar] [CrossRef]**

55. Axel, F.; Aubry, S. Polarisation and transition by breaking of analyticity in a one-dimensional model for incommensurate structures in an electric field. *J. Phys. A* **1987**, *20*, 487. **[Google Scholar] [CrossRef]**

56. Peyrard, M.; Aubry, S. Critical behaviour at the transition by breaking of analyticity in the discrete Frenkel-Kontorova model. *J. Phys. C Solid State Phys.* **1983**, *16*, 1593–1608. **[Google Scholar] [CrossRef]**

57. Bylinskii, A.; Gangloff, D.; Counts, I.; Vuletić, V. Observation of Aubry-type transition in finite atom chains via friction. *Nat. Mater.* **2016**, *15*, 717–721. **[Google Scholar] [CrossRef]**

58. Babushkin, A.Y.; Abkaryan, A.K.; Dobronets, B.S.; Krasikov, V.S.; Filonov, A.N. The Ground State of the Frenkel-Kontorova Model. *Phys. Solid State* **2016**, *58*, 1834–1845. **[Google Scholar] [CrossRef]**

59. Ma, Y.; Wang, J.; Xu, X.; Wei, Q.; Kais, S. A density-matrix renormalization group Study of one-dimensional incommensurate quantum Frenkel-Kontorova model. *J. Phys. Soc. Jpn.* **2014**, *83*, 094605. **[Google Scholar] [CrossRef] [Green Version]**

60. Mikheikin, I.D.; Kuznetsov, M.Y.; Makhonina, E.V.; Pervov, V.S. Defects in Inorganic Suprastructures with Incommensurate Structural Elements: The Static Frenkel-Kontorova Model for Finite Systems. *Dokl. Phys. Chem.* **2001**, *376*, 52–55. **[Google Scholar]** **[CrossRef]**

61. Nasilowski, R. An unorthodox analysis of the Frenkel-Kontorova model. *Phys. A* **1996**, *230*, 266–284. **[Google Scholar]** **[CrossRef]**

62. Novak, I.; Truskinovsky, L. Segmentation in cohesive systems constrained by elastic environments. *Phil. Trans. R. Soc. A* **2017**, *375*, 20160160. **[Google Scholar]** **[CrossRef]** **[PubMed]**

63. Wang, C.L.; Tekić, J.; Duan, W.S.; Shao, Z.G.; Yang, L.P. Existence and stability of the resonant phenomena in the dc- and ac-driven overdamped Frenkel-Kontorova model with the incommensurate structure. *Phys. Rev. E* **2011**, *84*, 046603. **[Google Scholar]** **[CrossRef]** **[PubMed]**

64. Zhirov, O.V.; Lages, J.; Shepelyansky, D.L. Thermoelectricity of cold ions in optical lattices. *Eur. Phys. J. D* **2019**, *73*, 149. **[Google Scholar]** **[CrossRef]** **[Green Version]**

65. Zhirov, O.V.; Lages, J.; Shepelyansky, D.L. Thermoelectricity Modeling with Cold Dipole Atoms in Aubry Phase of Optical Lattice. *Appl. Sci.* **2020**, *10*, 2090. **[Google Scholar]** **[CrossRef]** **[Green Version]**

66. Thomas, A.; Leoni, T.; Siri, O.; Becker, C.; Unzog, M.; Kern, C.; Puschnig, P.; Zeppenfeld, P. A one-dimensional high-order commensurate phase of tilted molecules. *Phys. Chem. Chem. Phys.* **2022**, *24*, 9118. **[Google Scholar]** **[CrossRef]** **[PubMed]**

12

Chapter

ADVANCEMENTS IN NON-CADMIUM AND ORGANIC QUANTUM DOTS FOR OPTOELECTRONIC APPLICATIONS, SPECIFICALLY PHOTODETECTOR DEVICES

1. INTRODUCTION TO NANOMATERIALS

Nanomaterials have attracted significant interest in recent years because of their captivating and unique properties that exceed the constraints of larger materials. A substance that has at least one dimension of less than 100 nm can be classified as a nanomaterial [1,2]. Materials that have dimensions larger than the nanoscale are referred to as 3D materials or bulk materials. Researchers utilize various classifications based on the size of nanomaterials. When the size of two dimensions exceeds the nanoscale, they are referred to be 2D nanomaterials. Examples of such materials are quantum wells and nanolayers, which are heterostructures consisting of two barrier layers enclosing a single thin well layer. The layer, consisting of electrons and holes, has a thickness of around 100 Å. Consequently, the electrons and holes begin to exhibit wave-like behavior. Practically, the standing waves that occur in a direction that is perpendicular to the layers of this structure represent the allowed states. The system is quantized due to the presence of standing waves that can only exist at specified frequencies. Thin quantum well stacked semiconductor devices enable the observation and manipulation of numerous quantum mechanical processes. Advanced epitaxial crystal growth techniques can achieve precise production of them. A material is classified as a 1D nanomaterial if one of its dimensions falls within the nanoscale range, while the other two dimensions are greater than the nanoscale. Examples of 1D nanomaterials include nanorods and nanowires [3,4]. 1D nanomaterials are utilized in the field of solid-state electronics and serve as diagnostic tools in the realm of medical sciences. They are capable of operating at a single degree of freedom. This allows for the use of nanowires in applications that require electrical conduction [5]. A substance is classified as a 0D material if all of its dimensions are less than 10 nm, such as nanoparticles and quantum dots (QDs). As the dimensions decrease and the region becomes highly restricted, it is also known as a dot. Quantum dots (QDs) are formed when the dimensions of a material decrease, causing them to exhibit atomic-like behavior and display intriguing characteristics. An electron-hole pair is generated from an energy input, such as UV light, and is held together by Coulombic forces [6]. Quantum dots (QDs) have a dimension that varies between 2 and 10 nm, which is equivalent to a range of 10 to 50 atoms. Additionally, they include a number of electrons ranging from 100 to 1000. The majority of material systems, such as metals, insulators, and semiconductors, have electrical or optical properties that vary with size in the quantum size range. The change of the energy band gap of semiconductors is highly significant both in terms of its fundamental importance and its technical applications [7]. Semiconductors with highly adjustable energy band gaps [8] are being used as the materials for the upcoming generation of flat panel displays, photovoltaic, and optoelectronic devices. Quantum dots, also known as QDs, are minuscule particles made of semiconducting material that have the ability to be employed in photodetectors and various other applications. They usually have a size of just a few nanometers. QDs can serve as a photosensitive material, also known as a photoconductive material, in a photodetector. An instance of a photodetector that utilizes quantum dots is a quantum dot infrared photodetector (QDIP) [9,10]. QDIPs employ a layer of quantum dots (QDs) as the active substance, which absorbs the incoming infrared light and produces an electric current [11]. Empirical evidence has confirmed that they exhibit commendable stability, rapid response times, and heightened sensitivity. Another instance is a photodetector that utilizes quantum dots to detect UV light. The detector utilizes QDs as the photoactive material, which has a high sensitivity and rapid response time for detecting UV light [12,13]. Figure 1 illustrates the various categories of materials categorized by their dimensions.

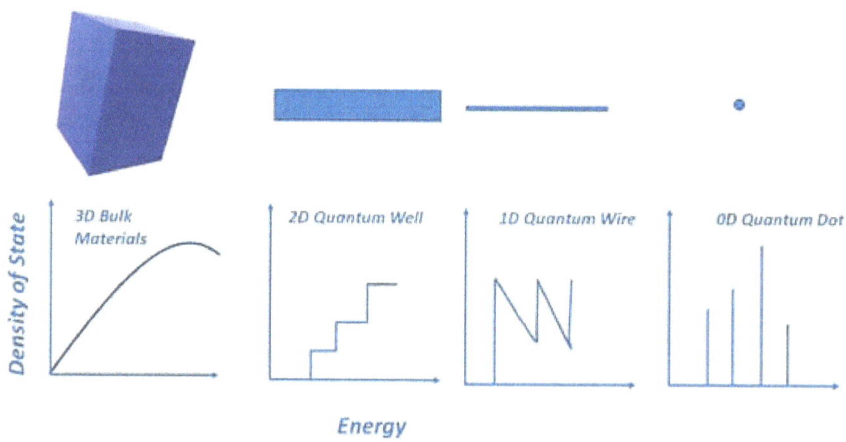

Figure 1. Schematic of nanomaterials based on types.

Another instance is a hybrid photodetector that utilizes quantum dots (QDs) in conjunction with conventional semiconductor materials like silicon [14,15]. These devices utilize both silicon and quantum dots (QDs) to create a highly responsive and efficient photodetector. Quantum dots show unique optical and electrical properties that have the potential to improve the performance of photodetectors. As a result, they are considered a promising material for use in these devices [16]. While the original observation of quantum dots (QDs) like cadmium raised concerns about their hazardous nature, their application is now restricted [17]. The semiconductor materials often employed for the fabrication of photodetectors based on quantum dots (QDs) encompass cadmium sulfide (CdS), lead sulfide (PbS), indium arsenide (InAs), and organic quantum dots [18]. These quantum dots (QDs) have unique optical and electrical properties, making them suitable for use in photodetector sensors. Cadmium sulfide (CdS) is the most thoroughly studied material for photodetector sensors [19]. Due to their high quantum yield and absorption coefficient, they have the ability to efficiently transform light into an electrical signal, enabling them to absorb a significant amount of light. The features of CdS QDs make them well-suited for ultraviolet and infrared photodetectors [20]. However, CdS possesses numerous drawbacks. Cadmium is a toxic heavy metal that can lead to severe health problems if inhaled. It is crucial to take appropriate measures to safely dispose of gadgets containing cadmium. CdS-based quantum dots are non-biodegradable and have the potential to impact the ecosystem [21]. Cadmium sulfide (CdS) is a pricier option compared to other quantum dot materials. Consequently, scientists are exploring alternative quantum dot materials that can be utilized for photodetection purposes [22]. Lead sulfide (PbS) QDs are the second most often utilized quantum dots (QDs) in photodetection applications. Infrared photodetectors greatly benefit from these materials due to their specific band gap, which renders them very sensitive to infrared light [23]. Indium arsenide (InAs) quantum dots, known for their strong absorption of near-infrared light, are a promising candidate for near-infrared photodetectors in sensor applications [24,25]. The optimal choice of QD material depends on the desired wavelength range of the light to be detected and the exact specifications of the device [26]. The scope of this review is limited to nanomaterials that do not include cadmium and their applications in the field of electronics. QDs, like light-emitting diodes, can also be used for electron-hole recombination, making them potentially useful in this field [27]. Quantum dots find uses in diverse disciplines such as electronics, biology, and renewable energy. Quantum dots have practical applications in electronics, such as the production of exceptionally efficient solar cells, LED lights, and various other gadgets. Quantum dots are employed as imaging agents in biomedicine to investigate cells and tissues throughout the body [28]. Quantum dots are utilized in renewable energy systems to generate hydrogen fuel from water by the utilization of sunlight [29].

2. OPTICAL AND ELECTRICAL PROPERTIES OF QUANTUM DOTS

Quantum dots exhibit distinct energy levels for electrons. When quantum dots (QDs) receive energy that is higher than their band gap, it leads to the creation of an exciton, which is a pair of electrons and holes. If the diameter of the nanomaterial is lower than its exciton Bohr radius, which is the typical physical distance between the electron and hole, the electron and hole pair is contained [30]. The confinement of an electron and hole pair is referred to as the quantum confinement effect. This effect results in the formation of a discrete packet of energy, rather than a continuous one [31,32]. Photoluminescence is a captivating phenomenon displayed by quantum dots (QDs). QDs exhibit the ability to absorb photons with higher energy levels and subsequently emit photons with lower energy levels. Upon absorption, the electron transitions to an excited state. Upon recombination with a pair, it releases light of reduced energy. This process occurs within a very short duration of time, specifically in femtoseconds [33]. Photoluminescence is mostly determined by the size of materials rather than their nature [34]. As the size of the quantum dot shrinks, the band gap of the quantum dot increases, resulting in a shift of the emission wavelength towards a shorter wavelength. This phenomenon is referred to as the blue shift and corresponds to emission in the visible range. Nevertheless, the emission can also manifest in alternative segments of the electromagnetic spectrum, such as ultraviolet or infrared. Quantum dots (QDs) possess remarkable optical characteristics, including enhanced photostability, elevated molar extinction, and a high quantum yield (the ratio of released photons to absorbed photons). Quantum dots (QDs) can be visualized using several techniques such as confocal microscopy, fluorescence spectroscopy, total internal reflection microscopy, and others [35,36].

QDs also have fascinating electrical properties, such as when sufficient energy is absorbed by the QD electron excited from the valence band to the conduction band while leaving the hole (empty state) in the valence band. This hole can be considered a positive mobile charge in the valence band. QDs possess a discrete energy state like an electron of a single atom [37]. Typically, a photon can only form one exciton and excite one electron across the band gap of fluorescent materials, with extra energy being lost as heat. However, multiple exciton generation is possible in QDs, which can be produced simultaneously, increasing the efficiency with which the nanocrystals convert energy [38].

Quantum yield measures the efficiency with which a system converts absorbed light into a different form of energy, such as heat or electricity. In the case of quantum dots, it refers to the efficiency with which the dots convert absorbed light into emitted light [39,40].

The equation for quantum yield is typically expressed as the ratio of the number of photons emitted by the quantum dots to the number of photons absorbed by the dots. This ratio can be expressed as follows:

Quantum yield = Number of photons emitted Number of photons absorbed Quantum yield = Number of photons emittedNumber of photons absorbed

The quantum yield is a crucial factor in determining the characteristics of quantum dots (QDs). The confinement characteristics exhibited by quantum dots' core/shell architectures are intricately linked to the materials used for the core and shell [41]. For example, when the energy band gap of the shell material is larger than that of the core material, the electron-hole pair is restricted to the core. Conversely, when the energy band gap of the core material is larger, the electron-hole pair is confined to the shell [42]. Quantum dots (QDs), including carbon quantum dots, exhibit high sensitivity to the presence of additional charges, such as electrons or holes, on their surfaces and in their surroundings. These charges have the ability to alter the photoluminescence and absorption wavelength of the QDs. The surface of carbon quantum dots has a functional group that has the ability to absorb and capture photons. This absorption process can result in a change in the emission wavelength of the carbon quantum dots [43]. Quantum dots are utilized in many applications, including solar cells [44], lasers [45], and photodetectors [46], due to their intriguing characteristics.

3. QUANTUM DOT LASERS

Semiconductor spherical nanocrystals (colloidal quantum dots), monolayers, and nanowires have found usage in the laser industry. CDs offer wide electromagnetic spectrum emissions from infrared to the visible range due to their size tunability, optical transition, and solution processibility. Compared to conventional semiconductors, QD-based semiconductors are easier to manipulate in size and shape [47,48]. Due to the unique absorption and emission properties of QDs, it is possible to produce lasing from a single exciton. A QD laser provides even spacing, low threshold amplification, and continuous wave operation. The shell absorbs the energy and acts as an antenna when the core–shell QDs are excited by photons having energies above the energy band gap. These QD outpace Auger recombination, so the threshold for lasing operation decreases [45].

4. QUANTUM DOTS FOR SOLAR CELLS

For more than two decades, semiconductor quantum dots (QDs) have attracted much attention due to the optoelectronic properties based on zero dimensions. Through the QD band energy level, colloidal quantum dots and self-assembled quantum dots improve the photogeneration of carriers for photovoltaic applications [49]. By expanding the range of photoexcitation, theoretically, it is possible to achieve maximum thermodynamic efficiency although the mechanism of carrier transport and carrier collection also contribute to its efficiency. Self-assembled and colloidal quantum dots are the two most-used types of QDs in solar cell applications [50]. The growth of solar cell wafers is also crucial for their efficiency, including surface passivation and well-controlled heterointerface [**51**].

Future risks include rising temperatures caused by global warming, shrinking fossil fuel supplies, and ever-increasing energy needs, which will require scientists and researchers to continuously work on developing efficient and affordable renewable energy alternatives [52]. Sunlight, which is abundantly available worldwide and can be converted into electricity or other forms of energy without polluting the environment, is one of the most commonly used renewable energy sources. Photovoltaic systems are based on converting sunlight into electricity, another valuable form of energy [53]. Photovoltaic device usage to produce electrical energy has increased, but obstacles still need to be overcome before these devices can significantly contribute to satisfying the world's energy needs [54].

Generally, there are three types of solar cells: traditionally, silicon-based monocrystalline and polycrystalline solar cells have been used [55]. The second type is thin-film solar cells, which are less expensive due to the low cost of absorbing material compared to silicon. These include copper gallium indium diselenide, cadmium telluride, and organic solar panels [56]. The third type, perovskite solar cells, uses hybrid organic–inorganic materials as the light-harvesting active layer. Perovskite materials, such as all-inorganic cesium lead halides and methylammonium lead halides, are cheap and straightforward to manufacture. Carrier mobility, doping density, trap density, and diffusion length in films are crucial factors that affect the performance of photovoltaic devices [57].

Junwei Yang et al. [58] reported synthesizing CdTe core QD-based solar cells. They used the aqueous route to synthesize QDs. CdTe core QDs were first synthesized and then deposited on a TiO_2 photoanode. An increase in absorption spectrum is observed for CdTe core QDs compared to CdTe cells. The power conversion efficiency of QD-based solar cells is increased up to 22% compared to CdTe/CdS plain solar cells. The suppressed charge recombination is also confirmed in QD-based solar cells. The authors also reported that thickness plays an important role in quantum efficiency. With optimization of $CdSe_xS_{1-x}$ shell thickness, about 7.24% conversion efficiency can be achieved, while Jun Du reported that Zn–Cu–In–Se alloyed QDs increase the absorption efficiency when deposited on TiO_2 film electrode, and reported conversion efficiency up to 11.66% [44].

5. NANOMATERIALS FOR PHOTODETECTION

Nanomaterials have properties that are suitable for photodetection, especially two-dimensional nanomaterials Two-dimensional (2D) materials have emerged as promising candidates for photodetectors due to their unique electrical and optical properties. In particular, their ultra-thin structure allows for efficient light absorption and carrier transport, leading to high sensitivity and fast response times. Two examples of 2D materials that are commonly used for photodetectors are:

Graphene is a single layer of carbon atoms arranged in a honeycomb lattice. It has excellent electrical conductivity and high carrier mobility, which makes it an attractive material for photodetectors [59]. When light is absorbed by graphene, it generates electron–hole pairs, which the graphene's high carrier mobility can efficiently collect. Graphene photodetectors have shown high responsivity and fast response times, making them useful for optical communications and imaging applications, while graphene hybrid composite materials with metallic nanomaterials and inorganic nanomaterials can increase photosensitivity [60].

Transition metal dichalcogenides (TMDs): TMDs are a family of 2D materials consisting of a transition metal atom sandwiched between two chalcogen atoms. They have a direct bandgap, which allows for efficient light absorption and emission [61]. TMDs have shown high sensitivity to light in the visible and near-infrared regions, making them useful for applications such as biomedical imaging and environmental monitoring. In addition, TMDs can be easily integrated with other electronic components, making them promising for on-chip optoelectronics [62]. 2D materials have shown great potential for photodetector applications due to their unique electrical and optical properties. Graphene and TMDs are just two examples of 2D materials that are being actively researched for their photodetection capabilities. As research in this area continues, it is likely that even more promising 2D materials will be discovered and developed for photodetector applications.

Infrared light detection is essential for biological sensing, spectroscopy, and in-depth 3D imaging. At the same time, semiconductors such as silicon, Cds, HgCdTe, and HgS are used for photodetectors. However, these materials have a few drawbacks, such as toxicity, low responsivity, and high cost of fabrication. The imaging performance of the quantum well photodetector, which is comparable to that of the mercury cadmium telluride (HgCdTe) detector, has recently attracted a lot of interest [63].

Following progress in the nanomaterials industry, it has been found that they have some advantages for photodetection over conventional materials, and the most-used nanomaterials are quantum well.

Quantum well infrared photodetectors (QWIP) absorb electrons through electronic intersubband polariton in a quantum well [64]. To be used for infrared detection, the energy difference between the first and second quantized states of the quantum well infrared photodetector's quantum wells must match the energy of the incoming infrared photons. III-V direct band gap semiconductor gallium arsenide (GaAs) is usually used to make QWIPs [65].

Therefore, research is also being carried out to create high-performance photodetectors using these nanomaterials. This review article focuses on quantum dot photodetectors [66]. They have been used in applications such as gas sensors and medical images, etc. Quantum well infrared detectors also have some limitations, such as the intersubband transition in the conduction band due to infrared photon polarization not being allowed in quantum wells [67,68]. This limitation can be overcome by using grating. However, they have a few more drawbacks, such as higher dark current, low quantum yield, lower responsivity, and detection level, and require cryogenic temperature to operate, so these limitations can be fulfilled by quantum dot photodetectors that have similar properties to quantum wells and also have some advantages due to the confinement of electrons in three dimensions [69]. QDs have reduced dependence on temperature for carrier transport, and carrier lifetime is also 10–100 times longer, which means that the electron remains excited for a more extended period, which causes a reduction in the dark current and increases sensitivity. QD-based detectors can absorb normal incident light without grating [70].

Similar to a quantum well, a quantum dot exhibits detection through the confinement-induced intraband photoexcitation of electrons. A photocurrent is generated by attracting the electrons in the collector with an electric field generated by the applied bias.

Colloidal QDs are low-cost materials, due to solution-based fabrication into semiconductor wafers and tunable band gap material, that are used as light emitters in flat panel displays. QDs are a cheap alternative to III–V group-based materials for imaging and sensing [71]. Research is being already carried out to make it possible to place QDs in a complementary metal–oxide–semiconductor (CMOS) circuit and increase the responsivity of QDs while decreasing the low noise equivalent power. If QDs can emit electrons across a wide range of the electromagnetic spectrum, they will be more practically efficient [72].

Infrared detectors are mainly categorized based on the material used for detection. Some of these are already extensively used in industry, including direct band gap semiconductor alloys such as HgCdTe and InSb, majority carrier Type-I superlattices, e.g., GaAs/AlGaAs QWIPs, and extrinsic semiconductors, e.g., Si: Ga,

Ge: Hg. The fourth type is silicon Schottky barriers such as PtSi, and IrS. At the same time, the other two types include high-temperature superconductors and QD-based photodetectors, e.g., InAs/GaAs-based detectors [73,74]. The QD-based photodetector will be discussed in this review. Examples of quantum dots for photodetection are given below.

6. LEAD(II) SULFIDE (PBS) QUANTUM DOTS

Lead(II) sulfide (PbS) QDs are special semiconductors with distinctive optoelectronic characteristics. Their strong absorption coefficient and band gap of 0.35 eV make them suitable for photovoltaic, photoconductor, and light-emitting diode applications [75,76]. PbS quantum dots are advantageous for sensing and imaging applications because they also have a high quantum yield and a narrow emission spectrum. Furthermore, their band gap can be modified by varying the quantum dot size, enabling further improvement of their optoelectronic characteristics [60,77].

One of these characteristics is their high absorption coefficient, which enables effective light absorption over a broad range of wavelengths. This is because PbS quantum dots have a significant quantum confinement effect, which confines the electrons to a limited area and causes them to have a high absorption coefficient [78].

The high responsivity of PbS quantum dots is a crucial characteristic that makes them effective as photodetectors which means more transformation of absorbed photons into an electrical signal [59,79]. Due to their high electron–hole recombination rate and rapid electron mobility, PbS quantum dots have a high responsiveness that enables the effective conversion of absorbed photons into electrical current. PbS quantum dots are also ideally suited for applications with high sensitivity and low noise because of their excellent signal-to-noise ratio and low dark current [80,81]. Therefore, lead sulfide (PbS) quantum dots (QDs) are considered potential material options for next-generation light, affordable, and flexible photodetectors due to their wide tunable band gaps, high absorption coefficients, and simple solution synthesis methods.

Alberto Maulu et al. synthesized PbS QDs and then modified their surface properties by using 3-mercaptopropionic acid (MPA) and tetrabutylammonium iodide (TBAI) [82]. It was observed that modification with this ligand significantly alters the properties of QDs. MPA-based QDs showed a decrease in dark current and low noise ratio, which causes the enhancement of detective and photosensitivity compared to TBAI-based QDs. MPA-modified QDs have functional group (Pb–OH bonds), while TBAI has a functional group (Pb–I bond), which is the primary reason for the Pb enhancement properties.

QDs used the trap sensitization method to show photoconductivity, increasing the responsivity but decreasing the response time for low optical power. At medium power incidents, light, when traps are filled with both types of photodetector, shows similar types of properties.

Urvashi Bothra et al. investigated the effect of ligands on PbS properties. They showed that a PbX_2 [X = I, Br])-based photodetector decreases the trap density compared to a PbS-based device, which increases the performance of (PbS) PbX_2-based devices [83].

Jin Beom Kwon et al. fabricated short-wave infrared (SWIR) photodetectors based on PbS QDs [84]. These infrared photodetectors minimize the harmful effect caused by the low wavelength band of SWIR. They fabricated a PbS photodetector with an absorbance peak at 1410 nm and used poly(3-hexylthiophene) (P3HT) as a conductive polymer with PbS QDs. Measurements were made on the properties of the synthesized PbS QDs, and the current–voltage (I–V) features of the PbS SWIR photodetectors. The maximal responsiveness of the PbS SWIR photodetector with P3HT optimization was found to be 2.26 times greater than that of the PbS SWIR photodetector without P3HT. Moreover, the former demonstrated a reduced operating voltage due to the high hole mobility and an adequate highest occupied molecular orbital level of P3HT. **Figure 2** shows the I–V characteristics of the PbS-based photodetector.

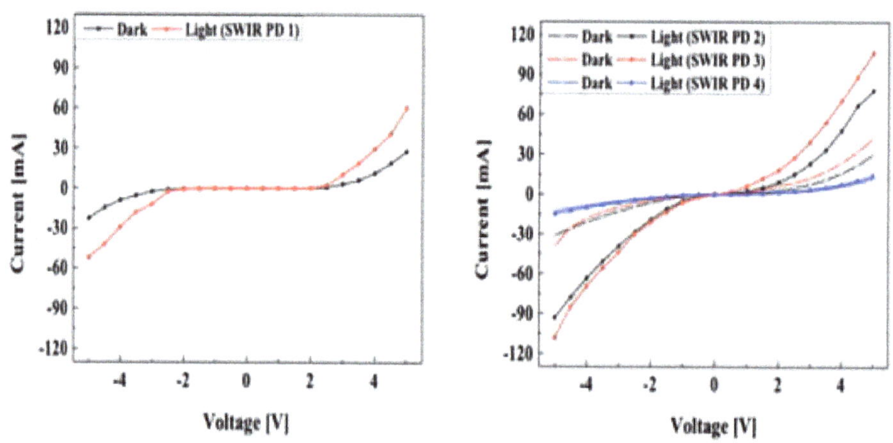

(a) (b)

Figure 2. I–V characteristics of (a) PbS SWIR photodetector without P3HT polymer and (b) PbS SWIR photodetector with P3HT polymer [84].

To compare their performance, they measured the I–V characteristics and responsiveness of the manufactured PbS SWIR photodetectors. The electron–hole pairs produced in the photoactive layer of the PbS SWIR photodetectors were removed from the electrodes by an external electric field when exposed to IR light. When the IR light source was turned off, the dark current was measured, and the light current was measured when the light source was turned on. The output power of the IR light source was 1 W/m^2, and the voltage sweep range was between -5 V and 5 V. The highest current difference between the dark current and light current and responsivity was recorded at -5 V in the case of the PbS SWIR photodetector without a P3HT layer. The largest current difference was 23.72 mA, and when this was divided by the input voltage, the corresponding responsivity was calculated as 2635.6 A/W. The PbS SWIR photodetector with a P3HT layer exhibited the best properties when applying 20 mg/mL of P3HT. The greatest current difference was confirmed to be 5891.1 A/W, which is 2.26 times greater than SWIRs, and the maximum current difference was 53.02 mA at -5V.

The drawbacks of current single-layer PbS-QDs photodetectors include excessive dark currents, slow light responses, and poor on–off ratios. It has been demonstrated that combining high-conducting graphene with organic materials and metal nanoparticles to create hybrid PbS-QDs devices can increase photoresponsivity. However, these hybrid devices bring other problems, and Zhenwei Ren et al. [80] synthesized the bilayer QDs device, which can be easily integrated into a flexible polyimide substrate. It increases the responsivity, signal-to-noise ratio, linear dynamic, and detection.

Moreover, this improvement in the device's properties is attributable to QDs' significant role. For example, the QDs junction accelerates carrier separation when the light is on. When the light is off, the QDs support the recombination of accumulated carriers in two films via the interface. Junction control is more dominant than trap control, which is the main determining factor in single-layer devices. They fabricated the bilayer photodetector using tetrabutylammonium iodide (TBAI)- and 1,2-ethanedithiol (EDT)-treated QDs. They compared it with a single-layer detector and observed that the bilayer photodetector enhanced optoelectronic properties. The bilayer PbS- TBAI/PbS-EDT devices have higher light current and detection efficiency, while the responsivity of PbS- TBAI/PbS-EDT is comparable to PbS-EDT devices due to the very close photocurrent.

It can be concluded that surface functional chemistry and modification of PbS with ligands improve the performance, and hybrid structure with modifications from different ligands can also improve the performance.

7. INDIUM ARSENIDE (INAS) QUANTUM DOTS

Indium arsenide (InAs) quantum dots (QDs) are also potential materials for photodetector sensors due to their distinct optical and electrical characteristics.

These QDs are especially beneficial for NIR photodetectors due to their high near-infrared (NIR) absorption coefficient [85]. NIR imaging is one possible use for InAs QD-based photodetectors. Due to InAs QDs' band gap, developing highly sensitive NIR imaging systems with them is possible. These gadgets are used in telecommunications for sensing, night vision, and medical imaging [86]. Deep-tissue imaging, optical communication, and covert illumination are just a few of the optoelectronics and biomedical uses for indium arsenide quantum dots, which typically emit in the near-infrared spectrum. Systems with bigger optical band gaps have yet to be produced, despite theory suggesting that further quantum confinement through size reduction could enable visible light emission.

NIR spectroscopy is another use for InAs QD-based photodetectors. Due to the NIR light detection capability, highly sensitive NIR spectroscopy devices can be made [87]. Additionally, the development of optoelectronic devices such as solar cells, LEDs, and lasers has utilized InAs QD-based photodetectors. These devices can increase the efficiency of solar cells and other optoelectronic devices by effectively converting light into electrical energy. Applications for these devices include bio-photonics, environmental monitoring, and chemical sensing [88,89].

Overall, because of their distinct optical and electrical characteristics, as well as their capacity to detect NIR light, InAs QDs have the potential to be utilized in a variety of photodetector applications. They might find use in spectroscopy, optoelectronic devices, and NIR imaging [90]. InAs quantum dots (QDs) might have a different band gap depending on their size and shape. Due to quantum confinement, the band gap of InAs QDs is typically between 0.35 and 0.4 eV, which is lower than that of bulk InAs (0.36 eV) [91], which is shown in **Figure 3.**

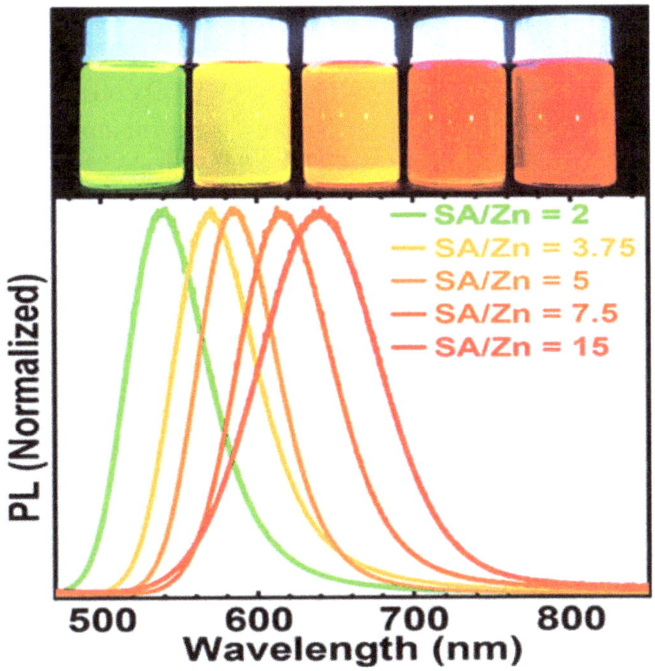

Figure 3. Photoluminescence spectra of In(Zn)As/ZnSe/ZnS QDs synthesized with different amounts of stearic acid (SA) ligand relative to the initial zinc precursor under UV illumination (Reproduced from [92], with permission from ACS Publications).

A low-temperature nanocluster synthesis technique was used by Daryl Darwan et al. to create highly luminous, visible-light-emitting In(Zn)As/ZnSe/ZnS QDs [**92**]. Each QD has an ultra-confined In(Zn)As nanocluster and emits light with a high photoluminescence quantum efficiency of 58% at tunable wavelengths between 538 and 640 nm. Theoretical calculations support the hypothesis that In(Zn)As nanoclusters are responsible for the infrared to visible spectral shift. This suggests that optical tuning in the visible region is also possible by utilizing confined semiconductor systems with a wide range of applications. The schematic is shown in **Figure 4**. Using an ultra-confinement technique, they were able to extend the spectrum coverage of InAs-based QDs into the visible range.

8. ORGANIC MATERIALS BASED ON QUANTUM DOTS

Organic materials, such as polymers or small organic compounds, are used to create organic quantum dots (OQDs), a particular kind of quantum dot. Due to their special optical and electronic characteristics, they could be used in photodetector sensors. Organic photovoltaics is one potential use for OQD-based photodetectors (OPVs) [93,94]. The active layer in OPVs, which transform light into electrical energy, can be made of OQDs. OQDs may absorb various wavelengths, which can enhance the efficiency of OPVs. They could also be used to make solar cells that are less expensive than ordinary inorganic solar cells. OQD-based photodetectors can also be used in flexible and wearable gadgets. OQDs offer thin, bendable, and solution-processable advantages, making them ideal for application in bendable and flexible products such as flexible screens and wearable technology [95].

Devices for sensing and bioimaging have also been created using OQDs. OQDs have the potential to be used in biosensing and biomedical imaging, as well as optical imaging. They are advantageous in these applications because of their high quantum yield, strong stability, and great sensitivity [**96**]. OQDs are a promising material for photodetector sensors overall because of their distinct optical and electronic characteristics and their potential for usage in a variety of applications, including solar cells, flexible and wearable devices, and bio-imaging and sensing equipment [**97**].

9. CARBON-BASED MATERIALS FOR PHOTODETECTION

Carbon materials, such as carbon nanotubes, graphene, and fullerenes, are used to synthesize carbon-based quantum dots (CQDs), a particular kind of quantum dot. Due to their extraordinary optical and electronic characteristics, they could be used in photodetector sensors [98].

The photodetection of diverse wavelength ranges is one potential use for CQD-based photodetectors. It has been demonstrated that carbon-based quantum dots have a broad absorption spectrum, making them appropriate for use in photodetectors that must detect a variety of wavelengths [**99**]. They are, therefore, very beneficial for multi-spectral imaging and sensing. High-speed optoelectronics is a further possible use for CQD-based photodetectors. Because of their rapid response times, carbon-based quantum dots can be used in high-speed optoelectronic systems, including optical communications and high-speed data transfer [100]. Devices enabling sensing and bioimaging have also been made with CQDs. The unique optical and electrical characteristics of carbon-based quantum dots make them suitable for biosensing and bioimaging applications, including in vivo imaging. They are helpful in various applications due to their high quantum yield, outstanding stability, and great sensitivity [101]. CQDs are a promising material for photodetector sensors overall because of their distinctive optical and electronic properties and their potential to be employed in a variety of applications, including multi-spectral imaging and sensing, high-speed optoelectronics, and bio-imaging and sensing devices [102].

Both graphene quantum dots (GQD) and carbon dots (CD) are being studied for potential application in photodetector sensors. Small, carbon-based nanoparticles known as CDs have been shown to exhibit robust fluorescence characteristics

[103]. On the other hand, GQDs, which are small graphene flakes, have been discovered recently to have fluorescence characteristics similar to those of CDs, with the added advantage of being more stable and simpler to functionalize. CDs and GQDs are promising candidates for photodetector sensors due to their efficiency at absorbing light and transforming it into an electrical signal [104].CDs' performance is negatively influenced by their inefficient near-infrared (NIR-I and NIR-II) excitation and emission. By resolving this, CDs can be utilized for in vivo bioimaging.

10. GRAPHENE QUANTUM DOTS (GQDS)

There may be applications for graphene quantum dots (GQDs), tiny graphene sheets, in photodetector sensors. Their unique electrical and optical features make them suitable for sensing technologies. GQDs have been found to possess strong fluorescence properties, effectively absorbing light and turning it into an electrical signal [105]. GQDs are extremely stable and may be functionalized with various chemical groups, enabling precise control over their optical and electrical characteristics. They are beneficial for sensing applications requiring specificity and sensitivity [106].

Graphene, in particular, is a 2D monolayer of sp^2 hybrid carbon. At the Dirac point, graphene's conduction band meets with its valence band. This causes a linear dispersion relationship, making it a zero-band gap semiconductor with zero density of electronic states and an incredibly high room temperature carrier mobility [**107**]. The broad spectrum response and quick response of graphene are perfect photoelectric qualities due to its zero-band gap semiconducting nature. Although graphene quantum dots have lesser absorption and responsiveness, researchers are primarily focusing on hybridizing graphene with high-absorption materials to enhance photodetection capability like quantum dots [108].

GQDs are also ideally suited for photodetectors and other optoelectronic devices due to their outstanding charge transfer characteristics. Different photodetector designs, including p–n junction, Schottky, and photoconductive ones, can incorporate GQDs [109]. They are helpful for various sensing applications because they can detect a wide range of wavelengths, from visible to infrared.

Molahalli Vandana et al. fabricated a UV-based photodetector by utilizing graphene QDs with polypyrrole polymer [110]. They used two different amounts of graphene, 20 and 40 mL in polypyrrole, labeled as PGC2 and PGC4, while single graphene QDs (GQDs) were also utilized. The active layer of ITO electrode was illuminated with UV light of 265 and 355 nm for about 200 s. The graphene QDs are an excellent light absorber and electron donor to increase the carrier concentration. **Figure 4** compares the photocurrent of GQDs, PGC2, and PGC4.

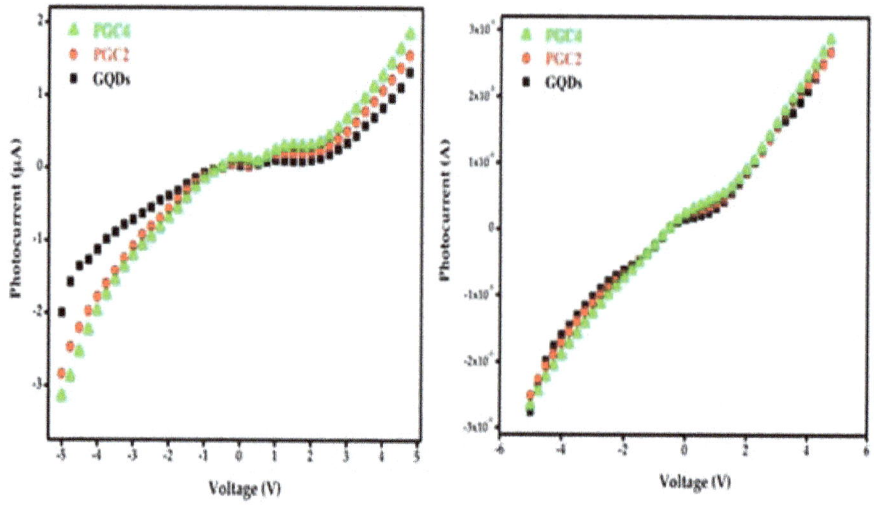

Figure 4. Comparison between the photocurrents of three photodetector devices under illumination [110].

PGC2 and PGC4 show responsivity of about 0.33 μA/W and 0.36 μA/W and corresponding photocurrent of 2.65 μA and 2.9 μA when illuminated at wavelength 265 nm, while a responsivity of 1.93 and 2.33 μA/W and photocurrent of 15.5 μA and 18.7 μA was observed for PGC2 and PGC4 when illuminated with UV light of 355 nm. Based on the results, which show that the PGC4 photodiode had the highest 18.7μA photocurrent, it can be concluded that the quantum dots in the polymer boosted electron mobility due to the high electrical properties of the nanocomposite.

11. CARBON DOTS

Carbon QDs are nanomaterials made of carbon, similar to substances such as nanodiamonds, fullerenes, carbon nanotubes, graphene, and graphene oxide. Small, water-soluble carbon nanoparticles known as carbon dots (CDs) have been shown to have potential uses in photodetector sensors [111]. CDs are a desirable material for use in sensing technologies due to their distinctive electrical and optical characteristics. Strong fluorescence qualities have been discovered in them, making them efficient at absorbing light and transforming it into an electrical signal [112]. CDs' high quantum yield and outstanding stability have been found to make them appropriate for use in photodetectors [113]. The precise control of their optical and electrical properties is made possible by their simplicity in synthesis and functionalization. As a result, they are advantageous for sensing applications that demand specificity and sensitivity [114]. CDs' superior charge transport characteristics have also been discovered, making them suitable for photodetectors and other optoelectronic devices. Many photodetector designs, including p-n junction, Schottky, and photoconductive ones, can incorporate CDs [115]. They are helpful for various sensing applications because they can detect a wide range of wavelengths, from visible to infrared. Additionally, CDs are non-toxic and biocompatible, which makes them advantageous in medicinal applications such as cancer therapy and bioimaging. Additionally, CDs can be functionalized with different biomolecules to enable the detection of certain biomolecules and biological activities [116].

They have gained attention due to their environmentally friendliness, chemical stability, and good conductivity, making them useful in various fields such as biomedical and biotechnological research, solar cells, light-emitting devices, imaging, electrochemical studies, and electrochemiluminescence studies [**117**]. Carbon dots are a class of nanoparticles composed mainly of carbon and oxygen, while some amino and carboxyl functional groups are also present on the surface [118]. These nanodots have unique optical, electrical, and chemical properties that make them highly versatile and helpful in various applications. Some of the critical properties of carbon dots include: Size: Carbon dots are nanoparticles with diameters ranging from 1 to 10 nanometers. Their small size allows them to display unique properties and quantum confinement effects, which are helpful for a number of applications. These properties make carbon dots valuable for various applications, including sensors, catalysts, and energy storage and conversion. Luminescence: Carbon dots can emit light, making them useful for applications such as biosensors and imaging [119]. Chemical stability: Carbon dots are highly stable in many different chemical environments, including basic and acidic conditions. Their stability makes them useful for sensing and imaging applications. Conductivity: Since CDs have excellent electrical conductivity, carbon dots can be used in photovoltaic and energy storage applications [120]. Biocompatibility: Due to their non-toxicity and biocompatibility, carbon dots can be used in biological systems without damaging healthy cells. Surface functionalization: The surface of carbon dots can be modified with different chemical groups, enabling them to be useful for specific purposes [121]. Optical properties: Carbon dots can absorb light in the ultraviolet and near-infrared parts of the spectrum, making them useful for specific applications. For example, their strong absorbance in these spectrum regions can be utilized in solar cells to absorb and convert sunlight into electricity and in LED lighting to provide energy-efficient and long-lasting light sources [122]. CDs can emit light when excited by an external energy source, such as electricity or heat. The emission wavelength, or the specific range of wavelengths that the light emitted by CDs falls within, can be manipulated by changing the size and composition of the CDs. This can be achieved by using different starting materials or synthesis methods while synthesizing CDs. Solubility: Since they are highly soluble in a wide range of solvents, including water, carbon dots are simple to handle and incorporate into different systems [123]. High surface area: The small size of carbon dots results in a high surface area, which makes them ideal for applications such as catalysis and drug delivery [123].

One crucial property of CDs is their ability to emit light when excited by light, also known as photoluminescence. The photophysical responses of CDs, including their light absorption and emission, are influenced by the isolated network of sp^2 carbon bonds in their structure [123]. In contrast, extended networks of carbon bonds, as found in carbon nanotubes, graphite, and graphene, do not interact with light as strongly. This may be due to the rapid recombination of photogenerated electrons and holes without the emission of radiation. CDs typically absorb light over a broad range of wavelengths [124]. They can absorb light over a wide range of wavelengths, but the specific wavelength at which they emit light is determined by the population of a particular energy level or "domain" and the surface properties of the quantum dots. This phenomenon, known as excitation-dependent emission, is a characteristic of CDs arising from these nanomaterials' surface state. Essentially, the surface of the quantum dots plays a crucial role in determining the specific wavelength at which they will emit light when excited by an external energy source [125].

Carbon dots have a high absorption coefficient, which implies they can absorb a lot of light, in addition to their fluorescence and quenching qualities. Due to their ability to transform light into energy that may be utilized to treat a range of medical problems, they are valuable in phototherapy applications [126].

K. Sarkar et al. [127] show the use of carbon dots (CDs) synthesized from Carica Papaya through a facile synthesis method as a potential broadband photodetector in a hybrid (organic–inorganic) heterostructure with a silicon wafer. The engineered CDs' maximum detectivity and responsivity are around 1 A W^{-1} and 2×10^{12} Jones, respectively. The device structure ensures a promising photodetector technology comparable to silicon process technology through a cheap and simple fabrication process. Additional optimization effort and defect engineering can speed up switching.

Sergii Kalytchuk et al. demonstrated the synthesis of purple-emissive CDs utilizing folic acid as the only precursor in a simple and reproducible process [128]. The CDs have a high photoluminescence quantum yield of 54.6% and emit light at a wavelength of 390 nm, which is independent of the excitation wavelength and integrates the emission color spectrum of intensely glowing carbon dots to purple. They employed CDs to increase the silicon photodetector's UV range sensitivity. It was found that the photoresponsivity of the silicon photodetector in the UV range of 0.8 to 2.5 mA/W was enhanced by 203.8% with CDs' integration.

Di Li et al. [129], by modifying the surface of CDs with functional groups of sulfoxide/carbonyl, changed the optical band gap of CDs. They proposed a mechanism for improved NIR emission, which can be seen in **Figure 5**. The CDs have a layered structure, and the outer layers have bonding sites. The interaction between the S = O/C = O functional group and layers increased surface oxidation, which caused the reduction in the lowest unoccupied orbital (LUMO). As a result, these functional groups disrupt CDs' organized structure. Due to the increased surface oxidation, the LUMO orbitals are lower in the outer layers compared to the inner layers and cause the NIR emission of CDs.

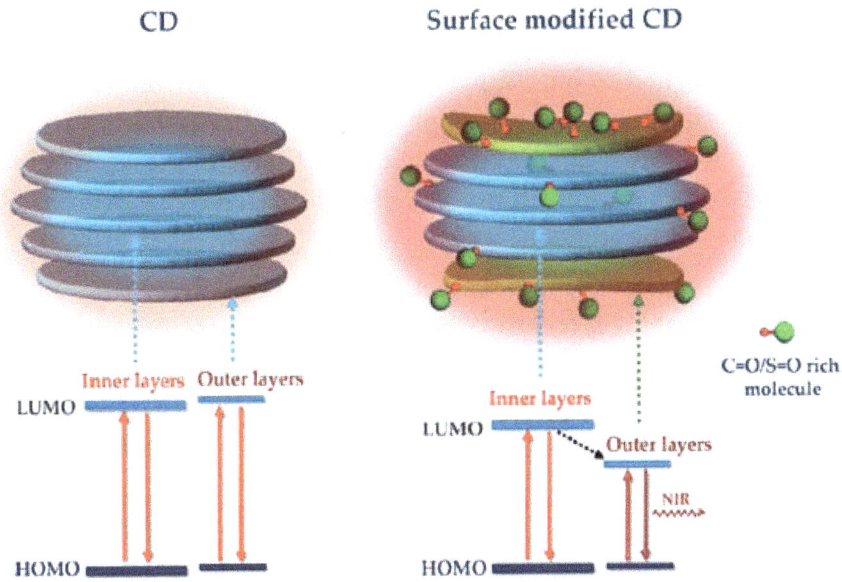

Figure 5. Non-treated CDs (left column) and CDs modified with S = O/C = O rich molecules (right column). The red (oxygen atom) and green double-bonded balls represent the C = O/S = O rich molecule (Reproduced from [129], with permission from John Wiley & Sons, Inc.).

12. ANALYSIS OF POTENTIAL FUTURE OUTCOMES AND FINAL REMARKS

This study addressed several optoelectronics applications that have been implemented as a result of breakthroughs in nanomaterial production. Scientists have employed quantum dots (QDs) in a range of technologies, such as solar cells, lasers, and photodetectors. Quantum dots exhibit distinct behavior in terms of light emission and absorption compared to bulk materials, mostly due to the quantum confinement effect. This phenomenon is particularly advantageous for applications involving lasers and solar panels. This paper provided a concise overview of several optoelectronic devices that are based on quantum dots (QDs). Quantum dots (QDs) used in photodetectors have promising prospects, including enhanced performance, lower expenses, and novel uses. Anticipated advancements in quantum dot photodetectors are projected to offer numerous advantages compared to existing photodetectors. For example, they have the potential to provide increased sensitivity, quicker response times, and reduced levels of noise. Due to their ability to be customized in terms of size and composition, they are expected to have greater flexibility in absorbing light across a wide range of wavelengths. This study discusses several factors that contribute to the superior usefulness of quantum dots (QDs) compared to standard photodetectors.

- Quantum dot-based photodetectors offer higher efficiency compared to traditional detectors, especially in the infrared range. This is because they can be designed to have a narrow absorption spectrum that matches the wavelength of the incoming light. On the other hand, organic quantum dots have a large surface area, allowing them to effectively absorb light and generate a significant number of excitons (electron-hole pairs) that are beneficial for photodetectors.
- Quantum dot-based photodetectors can be created using cost-effective processes that focus on providing practical solutions. Organic quantum dots (QDs) have the potential to reduce the cost of photodetectors, especially for applications that require large surface areas. Additionally, the ability to deposit organic QDs on flexible substrates allows for the development of wearable electronics and various other applications.
- QDs enable the construction of photodetectors with rapid response times, facilitating faster reaction times that are well-suited for high-speed applications.
- Quantum dots (QDs) have the ability to be integrated with other electronic components, resulting in the creation of hybrid devices that possess advanced and innovative characteristics. These could be integrated with complementary metal–oxide–semiconductor technologies to create cost-effective and high-performing bio-imaging systems.
- QDs possess greater versatility in integrating organic and inorganic substances, allowing for convenient modification of their characteristics to accommodate diverse requirements.

PbS quantum dots exhibit a high rate of electron-hole recombination and fast electron mobility, resulting in enhanced responsivity and improved efficiency in converting photons into electric current. The characteristics of PbS quantum dots can be altered through ligand reactions. The light currents and detections of devices including bilayer PbS quantum dots with various modification ligands, namely tetrabutylammonium iodide (TBAI) and 1,2-ethanedithiol (EDT), are significantly greater. This review study also examines indium arsenide (InAs) quantum dots

(QDs). It has been demonstrated that their high infrared absorption coefficient makes them effective photodetectors for infrared detection. They are also beneficial for imaging in the infrared spectrum. This paper examines the properties and applications of organic quantum dots, with a specific focus on graphene and carbon quantum dots. CDs possess unique electrical and optical characteristics that make them highly sought-after materials for sensing technologies. These entities have demonstrated robust fluorescence characteristics, efficiently capturing light and transforming it into an electric signal. Their ability to be easily synthesized and modified enables exact manipulation of their optical and electrical characteristics. Consequently, they are advantageous for photodetectors that necessitate specificity and sensitivity.

AT A GLANCE

Quantum dots (QDs) are nanomaterials with zero-dimensional (0D) characteristics, indicating that they exhibit charge confinement in all dimensions. This attribute has a significant influence on multiple applications. Metal-free organic quantum dots exhibit fascinating properties, such as the capacity to modify their bandgap according to their size, a high coefficient of optical absorption, the capability to alter the wavelengths of absorption and emission, and a cost-efficient production method. The characteristics of these materials originate from the quantum confinement of electrons, which occurs due to their extremely small size. This article will present a brief summary of the application of quantum dots (QDs) in solar cells and quantum dot lasers. Subsequently, it will conduct an in-depth examination of the use of quantum dots (QDs) in photodetectors. This topic will encompass various metallic compounds such as lead sulfide and indium arsenide, alongside nonmetallic materials like graphene and carbon nanotubes. The emphasis will be on scrutinizing their procedure of detection.

REFERENCES

1. Korotcenkov, G. Current trends in nanomaterials for metal oxide-based conductometric gas sensors: Advantages and limitations. part 1: 1D and 2D nanostructures. *Nanomaterials* **2020**, *10*, 1392. **[Google Scholar] [CrossRef] [PubMed]**
2. Daulbayev, C.; Sultanov, F.; Bakbolat, B.; Daulbayev, O. 0d 1d 2d Nanomater. Visible Photoelectrochem. Water Splitting. A Review. *Int. J. Hydrog. Energy* **2020**, *45*, 33325–33342. **[Google Scholar] [CrossRef]**
3. Fang, L.; Feng, J.J.; Shi, X.; Si, T.; Song, Y.; Jia, H.; Li, Y.; Li, H.-W.; Zhang, Q.J. 0D, 1D and 2D nanomaterials for visible photoelectrochemical water splitting. A review. *Chem. Commun.* **2019**, *55*, 10476–10479. **[Google Scholar] [CrossRef] [PubMed]**

4. Yang, T.; Liu, Y.; Wang, H.; Duo, Y.; Zhang, B.; Ge, Y.; Zhang, H.; Chen, W. Recent advances in 0D nanostructure-functionalized low-dimensional nanomaterials for chemiresistive gas sensors. *J. Mater. Chem. C* **2020**, *8*, 7272–7299. **[Google Scholar] [CrossRef]**

5. Garnett, E.; Mai, L.; Yang, P. Introduction: 1D nanomaterials/nanowires. *Chem. Rev.* **2019**, *119*, 8955–8957. **[Google Scholar] [CrossRef] [Green Version]**

6. Nozik, A.J. Quantum dot solar cells. *Phys. E: Low-Dimens. Syst. Nanostruct.* **2002**, *14*, 115–120. **[Google Scholar] [CrossRef]**

7. Bourlinos, A.B.; Stassinopoulos, A.; Anglos, D.; Zboril, R.; Karakassides, M.; Giannelis, E.P. Surface functionalized carbogenic quantum dots. *Small* **2008**, *4*, 455–458. **[Google Scholar] [CrossRef]**

8. Liu, X.; Zhang, Y.; Yu, T.; Qiao, X.; Gresback, R.; Pi, X.; Yang, D.J.P.; Characterization, P.S. Optimum quantum yield of the light emission from 2 to 10 nm hydrosilylated silicon quantum dots. *Part. Syst. Charact.* **2016**, *33*, 44–52. **[Google Scholar] [CrossRef]**

9. Martyniuk, P.; Rogalski, A. Quantum-dot infrared photodetectors: Status and outlook. *Prog. Quantum Electron.* **2008**, *32*, 89–120. **[Google Scholar] [CrossRef]**

10. Stiff-Roberts, A.D. Quantum-dot infrared photodetectors: A review. *J. Nanophotonics* **2009**, *3*, 031607. **[Google Scholar] [CrossRef]**

11. Liu, H.; Gao, M.; McCaffrey, J.; Wasilewski, Z.; Fafard, S. Quantum dot infrared photodetectors. *Appl. Phys. Lett.* **2001**, *78*, 79–81. **[Google Scholar] [CrossRef]**

12. Li, M.Y.; Yu, M.; Su, D.; Zhang, J.; Jiang, S.; Wu, J.; Wang, Q.; Liu, S. Ultrahigh responsivity UV photodetector based on Cu nanostructure/ZnO QD hybrid architectures. *Small* **2019**, *15*, 1901606. **[Google Scholar] [CrossRef] [PubMed]**

13. Praveen, S.; Veeralingam, S.; Badhulika, S. A Flexible Self-Powered UV Photodetector and Optical UV Filter Based on β-Bi2O3/SnO2 Quantum Dots Schottky Heterojunction. *Adv. Mater. Interfaces* **2021**, *8*, 2100373. **[Google Scholar] [CrossRef]**

14. Chen, G.; Liang, B.; Liu, X.; Liu, Z.; Yu, G.; Xie, X.; Luo, T.; Chen, D.; Zhu, M.; Shen, G. High-performance hybrid phenyl-C61-butyric acid methyl ester/Cd3P2 nanowire ultraviolet–visible–near infrared photodetectors. *ACS Nano* **2014**, *8*, 787–796. **[Google Scholar] [CrossRef]**

15. Dang, V.Q.; Trung, T.Q.; Kim, D.I.; Duy, L.T.; Hwang, B.U.; Lee, D.W.; Kim, B.Y.; Toan, L.D.; Lee, N.E. Ultrahigh responsivity in graphene–ZnO nanorod hybrid UV photodetector. *Small* **2015**, *11*, 3054–3065. **[Google Scholar] [CrossRef]**

16. Manga, K.K.; Wang, J.; Lin, M.; Zhang, J.; Nesladek, M.; Nalla, V.; Ji, W.; Loh, K. High-performance broadband photodetector using solution-processible PbSe–TiO_2–graphene hybrids. *Adv. Mater.* **2012**, *24*, 1697–1702. **[Google Scholar] [CrossRef]**

17. Banyai, L.; Gilliot, P.; Hu, Y.; Koch, S. Surface-polarization instabilities of electron-hole pairs in semiconductor quantum dots. *Phys. Rev. B* **1992**, *45*, 14136. **[Google Scholar] [CrossRef]**

18. Gandhi, S.A.; Sutariya, P.G.; Soni, H.N.; Chaudhari, D. Quantum dots: Application in medical science. *Int. J. Nano Dimens.* **2023**, *14*, 29–40. **[Google Scholar]**

19. Najm, N.I.; Hassun, H.K.; al-Maiyaly, B.K.H.; Hussein, B.H.; Shaban, A.H. Highly selective CdS: Ag heterojunction for photodetector applications. *AIP Conf. Proc.* **2019**, *2123*, 020031. **[Google Scholar]**

20. Hakami, J. Highly induced photosensing behavior of Erbium (Er), Yttrium (Y) and Terbium (Tb) doped nanostructured Cadmium Sulphide (CdS) thin films prepared by nebulizer spray pyrolysis method. *J. Alloys Compd.* **2022**, *924*, 166577. **[Google Scholar] [CrossRef]**

21. Habeebu, S.S.; Liu, J. Klaassen CDJT pharmacology, a. Cadmium-induced apoptosis in mouse liver. *Toxicol. Appl. Pharmacol.* **1998**, *149*, 203–209. **[Google Scholar] [CrossRef] [PubMed]**

22. Ning, X.; Lu, G. Photocorrosion inhibition of CdS-based catalysts for photocatalytic overall water splitting. *Nanoscale* **2020**, *12*, 1213–1223. **[Google Scholar] [CrossRef] [PubMed]**

23. McDonald, S.A.; Konstantatos, G.; Zhang, S.; Cyr, P.W.; Klem, E.J.; Levina, L.; Sargent, E. Solution-processed PbS quantum dot infrared photodetectors and photovoltaics. *Nat. Mater.* **2005**, *4*, 138–142. **[Google Scholar] [CrossRef]**

24. Kim, E.-T.; Madhukar, A.; Ye, Z.; Campbell, J. High detectivity InAs quantum dot infrared photodetectors. *Appl. Phys. Lett.* **2004**, *84*, 3277–3279. **[Google Scholar] [CrossRef]**

25. Marzin, J.-Y.; Gérard, J.-M.; Izraël, A.; Barrier, D.; Bastard, G. Photoluminescence of single InAs quantum dots obtained by self-organized growth on GaAs. *Phys. Rev. Lett.* **1994**, *73*, 716. **[Google Scholar] [CrossRef] [PubMed]**

26. Zhao, N.; Osedach, T.P.; Chang, L.-Y.; Geyer, S.M.; Wanger, D.; Binda, M.T.; Arango, A.C.; Bawendi, M.G.; Bulovic, V. Colloidal PbS quantum dot solar cells with high fill factor. *ACS Nano* **2010**, *4*, 3743–3752. **[Google Scholar] [CrossRef]**

27. Bretagnon, T.; Lefebvre, P.; Valvin, P.; Bardoux, R.; Guillet, T.; Taliercio, T.; Gil, B.; Grandjean, N.; Semond, F.; Damilano, B. Radiative lifetime of a single

electron-hole pair in Ga N/Al N quantum dots. *Phys. Rev. B* **2006**, *73*, 113304. **[Google Scholar] [CrossRef]**

28. Li, X.; Rui, M.; Song, J.; Shen, Z.; Zeng, H. Carbon and graphene quantum dots for optoelectronic and energy devices: A review. *Adv. Funct. Mater.* **2015**, *25*, 4929–4947. **[Google Scholar] [CrossRef]**
29. Litvin, A.; Martynenko, I.; Purcell-Milton, F.; Baranov, A.; Fedorov, A.; Gun'Ko, Y. Colloidal quantum dots for optoelectronics. *J. Mater. Chem. A* **2017**, *5*, 13252–13275. **[Google Scholar] [CrossRef]**
30. Li, S.-Y.; He, L. Recent progresses of quantum confinement in graphene quantum dots. *Front. Phys.* **2022**, *17*, 33201. **[Google Scholar] [CrossRef]**
31. Takagahara, T.; Takeda, K. Theory of the quantum confinement effect on excitons in quantum dots of indirect-gap materials. *Phys. Rev. B* **1992**, *46*, 15578. **[Google Scholar] [CrossRef] [PubMed]**
32. Wise, F. Lead salt quantum dots: The limit of strong quantum confinement. *Acc. Chem. Res.* **2000**, *33*, 773–780. **[Google Scholar] [CrossRef] [PubMed]**
33. Giessen, H.; Woggon, U.; Fluegel, B.; Mohs, G.; Hu, Y.; Koch, S.; Peyghambarian, N. Femtosecond optical gain in strongly confined quantum dots. *Opt. Lett.* **1996**, *21*, 1043–1045. **[Google Scholar] [CrossRef] [PubMed]**
34. Ruedas-Rama, M.J.; Orte, A.; Hall, E.A.; Alvarez-Pez, J.M.; Talavera, E. Quantum dot photoluminescence lifetime-based pH nanosensor. *Chem. Commun.* **2011**, *47*, 2898–2900. **[Google Scholar] [CrossRef] [PubMed]**
35. Mintz, K.J.; Zhou, Y.; Leblanc, R.M. Recent development of carbon quantum dots regarding their optical properties, photoluminescence mechanism, and core structure. *Nanoscale* **2019**, *11*, 4634–4652. **[Google Scholar] [CrossRef] [PubMed]**
36. Zhu, S.; Song, Y.; Zhao, X.; Shao, J.; Zhang, J.; Yang, B. The photoluminescence mechanism in carbon dots (graphene quantum dots, carbon nanodots, and polymer dots): Current state and future perspective. *Nano Res.* **2015**, *8*, 355–381. **[Google Scholar] [CrossRef]**
37. Benyettou, F.; Aissat, A.; Djebari, M.; Vilcot, J.R. Electrical properties of InAsP/Si quantum dot solar cell. *Int. J. Hydrog. Energy* **2017**, *42*, 19512–19517. **[Google Scholar] [CrossRef]**
38. Jolley, G.; Fu, L.; Lu, H.F.; Tan, H.H.; Jagadish, C.J. The role of intersubband optical transitions on the electrical properties of InGaAs/GaAs quantum dot solar cells. *Prog. Photovolt. Res. Appl.* **2013**, *21*, 736–746. **[Google Scholar] [CrossRef]**

39. Emerson, R. The quantum yield of photosynthesis. *Annu. Rev. Plant Physiol.* **1958**, *9*, 1–24. **[Google Scholar] [CrossRef]**

40. Weber, G.; Teale, F. Determination of the absolute quantum yield of fluorescent solutions. *Trans. Faraday Soc.* **1957**, *53*, 646–655. **[Google Scholar] [CrossRef]**

41. Hamood, R.; Abd El-sadek, M.; Gadalla, A.J.V. Facile synthesis, structural, electrical and dielectric properties of CdSe/CdS core-shell quantum dots. *Vacuum* **2018**, *157*, 291–298. **[Google Scholar] [CrossRef]**

42. Grieve, K.; Mulvaney, P.; Grieser, F. Synthesis and electronic properties of semiconductor nanoparticles/quantum dots. *Curr. Opin. Colloid Interface Sci.* **2000**, *5*, 168–172. **[Google Scholar] [CrossRef]**

43. Kouwenhoven, L.P.; Austing, D.; Tarucha, S. Few-electron quantum dots. *Rep. Prog. Phys.* **2001**, *64*, 701. **[Google Scholar] [CrossRef]**

44. Du, J.; Du, Z.; Hu, J.-S.; Pan, Z.; Shen, Q.; Sun, J.; Long, D.; Dong, H.; Sun, L.; Zhong, X. Zn–Cu–In–Se quantum dot solar cells with a certified power conversion efficiency of 11.6%. *J. Am. Chem. Soc.* **2016**, *138*, 4201–4209. **[Google Scholar] [CrossRef] [PubMed]**

45. Marko, I.P.; Adams, A.R.; Sweeney, S.J.; Mowbray, D.J.; Skolnick, M.S.; Liu, H.Y.; Groom, K. Recombination and loss mechanisms in low-threshold InAs-GaAs 1.3-/spl mu/m quantum-dot lasers. *IEEE J. Sel. Top. Quantum Electron.* **2005**, *11*, 1041–1047. **[Google Scholar] [CrossRef]**

46. Park, Y.-S.; Bae, W.K.; Pietryga, J.M.; Klimov, V.I. Auger recombination of biexcitons and negative and positive trions in individual quantum dots. *ACS Nano* **2014**, *8*, 7288–7296. **[Google Scholar] [CrossRef] [PubMed]**

47. Geiregat, P.; Van Thourhout, D.; Hens, Z. A bright future for colloidal quantum dot lasers. *NPG Asia Mater.* **2019**, *11*, 41. **[Google Scholar] [CrossRef] [Green Version]**

48. Ustinov, V.M.; Zhukov, A.E.; Zhokov, A.E.; Maleev, N.A.; Egorov, A.Y. *Quantum Dot Lasers*; Oxford University Press on Demand: Oxford, UK, 2003; Volume 11. **[Google Scholar]**

49. Semonin, O.E.; Luther, J.M.; Choi, S.; Chen, H.-Y.; Gao, J.; Nozik, A.J.; Beard, M.C. Peak external photocurrent quantum efficiency exceeding 100% via MEG in a quantum dot solar cell. *Science* **2011**, *334*, 1530–1533. **[Google Scholar] [CrossRef]**

50. Carey, G.H.; Abdelhady, A.L.; Ning, Z.; Thon, S.M.; Bakr, O.M.; Sargent, E.H. Colloidal quantum dot solar cells. *Chem. Rev.* **2015**, *115*, 12732–12763. **[Google Scholar] [CrossRef]**

51. Kamat, P.V. Quantum dot solar cells. The next big thing in photovoltaics. *J. Phys. Chem. Lett.* **2013**, *4*, 908–918. **[Google Scholar] [CrossRef]**

52. Sawadogo, W.; Abiodun, B.J.; Okogbue, E.C. Impacts of global warming on photovoltaic power generation over West Africa. *Renew. Energy* **2020**, *151*, 263–277. **[Google Scholar] [CrossRef]**

53. Green, M. Photovoltaic technology and visions for the future. *Prog. Energy* **2019**, *1*, 013001. **[Google Scholar] [CrossRef]**

54. Andreev, V.A.C.M.; Grilikhes, V.A.; Rumyantsev, V.D. *Photovoltaic Conversion of Concentrated Sunlight*; Wiley: New York, NY, USA, 1997. **[Google Scholar]**

55. Imamzai, M.; Aghaei, M.; Thayoob, Y.H.M.; Forouzanfar, M. A review on comparison between traditional silicon solar cells and thin-film CdTe solar cells. In Proceedings of the National Graduate Conference Nat-Grad, Kajang, Malaysia, 8–10 November 2012; pp. 1–5. **[Google Scholar]**

56. Brendel, R.J. Review of layer transfer processes for crystalline thin-film silicon solar cells. *Jpn. J. Appl. Phys.* **2001**, *40*, 4431. **[Google Scholar] [CrossRef]**

57. Rong, Y.; Hu, Y.; Mei, A.; Tan, H.; Saidaminov, M.I.; Seok, S.I.; McGehee, M.D.; Sargent, E.H.; Han, H.J.S. Challenges for commercializing perovskite solar cells. *Science* **2018**, *361*, eaat8235. **[Google Scholar] [CrossRef] [PubMed] [Green Version]**

58. Yang, J.; Zhong, X. CdTe based quantum dot sensitized solar cells with efficiency exceeding 7% fabricated from quantum dots prepared in aqueous media. *J. Mater. Chem. A* **2016**, *4*, 16553–16561. **[Google Scholar] [CrossRef]**

59. Ge, B.; Nie, C.; Shen, J. A hybrid photodetector of graphene/TiO2/inorganic PbS quantum dots for fast response. *Jpn. J. Appl. Phys.* **2022**, *61*, 040903. **[Google Scholar] [CrossRef]**

60. Kundu, B.; Özdemir, O.; Dalmases, M.; Kumar, G.; Konstantatos, G. Hybrid 2D-QD MoS2□PbSe Quantum Dot Broadband Photodetectors with High-Sensitivity and Room-Temperature Operation at 2.5 µm. *Adv. Opt. Mater.* **2021**, *9*, 2101378. **[Google Scholar] [CrossRef]**

61. Kagkoura, A.; Skaltsas, T.; Tagmatarchis, N. Transition-Metal Chalcogenide/ Graphene Ensembles for Light-Induced Energy Applications. *Chem. A Eur. J.* **2017**, *23*, 12967–12979. **[Google Scholar] [CrossRef]**

62. Jaiswal, K.; Girish, Y.R.; De, M.J. Group-VI-Chalcogenide-Based Nanomaterials in Photo/Thermal Organic Transformations. *Acc. Mater. Res.* **2022**, *3*, 1033–1048. **[Google Scholar] [CrossRef]**

63. Chatterjee, A.; Pendyala, N.B.; Jagtap, A.; Koteswara Rao, K.S.R. Uncooled Mid-Wave Infrared Focal Plane Array Using Band Gap Engineered Mercury

Cadmium Telluride Quantum Dot Coated Silicon ROIC. *E J. Surf. Sci. Nanotechnol.* **2019**, *17*, 95–100. **[Google Scholar] [CrossRef] [Green Version]**

64. Levine, B. Quantum-well infrared photodetectors. *J. Appl. Phys.* **1993**, *74*, R1–R81. **[Google Scholar] [CrossRef]**
65. Fox, M.; Ispasoiu, R.; Materials, P. *Quantum Wells, Superlattices, and Band-Gap Engineering*; Springer: Cham, Switzerland, 2017. **[Google Scholar]**
66. Phillips, J. Evaluation of the fundamental properties of quantum dot infrared detectors. *J. Appl. Phys.* **2002**, *91*, 4590–4594. **[Google Scholar] [CrossRef] [Green Version]**
67. Goossen, K.; Lyon, S. Performance aspects of a quantum-well detector. *J. Appl. Phys.* **1988**, *63*, 5149–5153. **[Google Scholar] [CrossRef]**
68. Rogalski, A. Quantum well photoconductors in infrared detector technology. *J. Appl. Phys.* **2003**, *93*, 4355–4391. **[Google Scholar] [CrossRef] [Green Version]**
69. Bhattacharya, P.; Ghosh, S.; Stiff-Roberts, A. Quantum dot opto-electronic devices. *Annu. Rev. Mater. Res.* **2004**, *34*, 1. **[Google Scholar] [CrossRef]**
70. Pan, D.; Towe, E.; Kennerly, S. Photovoltaic quantum-dot infrared detectors. *Appl. Phys. Lett.* **2000**, *76*, 3301–3303. **[Google Scholar] [CrossRef]**
71. Liu, M.; Yazdani, N.; Yarema, M.; Jansen, M.; Wood, V.; Sargent, E. Colloidal quantum dot electronics. *Nat. Electron.* **2021**, *4*, 548–558. **[Google Scholar] [CrossRef]**
72. Liu, J.; Liu, P.; Chen, D.; Shi, T.; Qu, X.; Chen, L.; Wu, T.; Ke, J.; Xiong, K.; Li, M. A near-infrared colloidal quantum dot imager with monolithically integrated readout circuitry. *Nat. Electron.* **2022**, *5*, 443–451. **[Google Scholar] [CrossRef]**
73. Brown, A.E. *Application of Mobility Spectrum Analysis to Modern Multi-Layered IR Device Material*; University of Illinois: Chicago, IL, USA, 2016. **[Google Scholar]**
74. Rogalski, A.; Chrzanowski, K.J.M.; Systems, M. Infrared devices and techniques (revision). *Metrol. Meas. Syst.* **2014**, *21*, 565–618. **[Google Scholar] [CrossRef]**
75. Hoyer, P.; Könenkamp, R. Photoconduction in porous TiO2 sensitized by PbS quantum dots. *Appl. Phys. Lett.* **1995**, *66*, 349–351. **[Google Scholar] [CrossRef]**
76. Wang, H.; Kubo, T.; Nakazaki, J.; Segawa, H. Solution-processed short-wave infrared PbS colloidal quantum Dot/ZnO nanowire solar cells giving high open-circuit voltage. *ACS Energy Lett.* **2017**, *2*, 2110–2117. **[Google Scholar] [CrossRef]**
77. Gao, J.; Perkins, C.L.; Luther, J.M.; Hanna, M.C.; Chen, H.-Y.; Semonin, O.E.; Nozik, A.J.; Ellingson, R.J.; Beard, M. N-Type transition metal oxide as a hole extraction layer in PbS quantum dot solar cells. *Nano Lett.* **2011**, *11*, 3263–3266. **[Google Scholar] [CrossRef] [PubMed]**

78. Liu, Y.; Wu, H.; Shi, G.; Li, Y.; Gao, Y.; Fang, S.; Tang, H.; Chen, W.; Ma, T.; Khan, I. Merging Passivation in Synthesis Enabling the Lowest Open-Circuit Voltage Loss for PbS Quantum Dot Solar Cells. *Adv. Mater.* **2022**, *35*, 2207293. **[Google Scholar] [CrossRef]**

79. Zhang, C.; Lian, L.; Yang, Z.; Zhang, J.; Zhu, H. Quantum confinement-tunable ultrafast charge transfer in a PbS quantum dots/WSe2 0D–2D hybrid structure: Transition from the weak to strong coupling regime. *J. Phys. Chem. Lett.* **2019**, *10*, 7665–7671. **[Google Scholar] [CrossRef]**

80. Wei, Y.; Ren, Z.; Zhang, A.; Mao, P.; Li, H.; Zhong, X.; Li, W.; Yang, S.; Wang, J. Hybrid organic/PbS quantum dot bilayer photodetector with low dark current and high detectivity. *Adv. Funct. Mater.* **2018**, *28*, 1706690. **[Google Scholar] [CrossRef]**

81. Kufer, D.; Nikitskiy, I.; Lasanta, T.; Navickaite, G.; Koppens, F.H.; Konstantatos, G. Hybrid 2D–0D MoS2–PbS quantum dot photodetectors. *Adv. Mater. Interfaces* **2015**, *27*, 176–180. **[Google Scholar] [CrossRef]**

82. Maulu, A.; Navarro-Arenas, J.; Rodríguez-Cantó, P.J.; Sánchez-Royo, J.F.; Abargues, R.; Suárez, I.; Martínez-Pastor, J. Charge transport in trap-sensitized infrared PbS quantum-dot-based photoconductors: Pros and cons. *Nanomaterials* **2018**, *8*, 677. **[Google Scholar] [CrossRef] [Green Version]**

83. Bothra, U.; Albaladejo-Siguan, M.; Vaynzof, Y.; Kabra, D. Impact of Ligands on the Performance of PbS Quantum Dot Visible–Near-Infrared Photodetectors. *Adv. Opt. Mater.* **2022**, *11*, 2201897. **[Google Scholar] [CrossRef]**

84. Kwon, J.B.; Han, M.; Jung, D.G.; Kong, S.H.; Jung, D.J. High sensitivity shortwave infrared photodetector based on PbS QDs using P3HT. *Nanomaterials* **2021**, *11*, 2683. **[Google Scholar] [CrossRef]**

85. Franke, D.; Harris, D.K.; Chen, O.; Bruns, O.T.; Carr, J.A.; Wilson, M.W.; Bawendi, M. Continuous injection synthesis of indium arsenide quantum dots emissive in the short-wavelength infrared. *Nat. Commun.* **2016**, *7*, 12749. **[Google Scholar] [CrossRef] [Green Version]**

86. Banin, U.; Cao, Y.; Katz, D.; Millo, O. Identification of atomic-like electronic states in indium arsenide nanocrystal quantum dots. *Nature* **1999**, *400*, 542–544. **[Google Scholar] [CrossRef]**

87. Kim, S.-W.; Zimmer, J.P.; Ohnishi, S.; Tracy, J.B.; Frangioni, J.V.; Bawendi, M.G. Engineering InAs x P1-x/InP/ZnSe III–V Alloyed Core/Shell Quantum Dots for the Near-Infrared. *J. Am. Chem. Soc.* **2005**, *127*, 10526–10532. **[Google Scholar] [CrossRef]**

88. Jalali, H.B.; De Trizio, L.; Manna, L.; Di Stasio, F. Indium arsenide quantum dots: An alternative to lead-based infrared emitting nanomaterials. *Chem. Soc. Rev.* **2022**, *51*, 9861–9881. **[Google Scholar] [CrossRef]**

89. Park, Y.; Jeong, S.; Kim, S. Medically translatable quantum dots for biosensing and imaging. *J. Photochem. Photobiol. C Photochem. Rev.* **2017**, *30*, 51–70. **[Google Scholar] [CrossRef]**

90. Fu, Y.; Ferdos, F.; Sadeghi, M.; Zhao, Q.; Wang, S.; Larsson, A.J.S. Strain and optical transitions in InAs quantum dots on (001) GaAs. *Superlattices Microstruct.* **2001**, *30*, 205–213. **[Google Scholar] [CrossRef]**

91. Dion, C.; Desjardins, P.; Shtinkov, N.; Robertson, M.; Schiettekatte, F.; Poole, P.; Raymond, S. Intermixing during growth of InAs self-assembled quantum dots in InP: A photoluminescence and tight-binding investigation. *Phys. Rev. B* **2008**, *77*, 075338. **[Google Scholar] [CrossRef] [Green Version]**

92. Darwan, D.; Lim, L.J.; Wang, T.; Wijaya, H.; Tan, Z. Ultra-Confined Visible-Light-Emitting Colloidal Indium Arsenide Quantum Dots. *Nano Lett.* **2021**, *21*, 5167–5172. **[Google Scholar] [CrossRef]**

93. Nguyen, H.T.; Das, R.; Duong, A.T.; Lee, S. Influence of quantum dot concentration on the opto-electronic properties of colloidal quantum-dots LEDs. *Opt. Mater.* **2020**, *109*, 110251. **[Google Scholar] [CrossRef]**

94. Wang, T.; Nie, C.; Ao, Z.; Wang, S.; An, T. Recent progress in gC 3 N 4 quantum dots: Synthesis, properties and applications in photocatalytic degradation of organic pollutants. *J. Mater. Chem. A* **2020**, *8*, 485–502. **[Google Scholar] [CrossRef]**

95. Weiss, E.A. Organic molecules as tools to control the growth, surface structure, and redox activity of colloidal quantum dots. *Acc. Chem. Res.* **2013**, *46*, 2607–2615. **[Google Scholar] [CrossRef]**

96. Li, J.; Zhu, J.-J. Quantum dots for fluorescent biosensing and bio-imaging applications. *Analyst* **2013**, *138*, 2506–2515. **[Google Scholar] [CrossRef] [PubMed]**

97. Vercelli, B.J.C. The role of carbon quantum dots in organic photovoltaics: A short overview. *Coatings* **2021**, *11*, 232. **[Google Scholar] [CrossRef]**

98. Alaghmandfard, A.; Sedighi, O.; Rezaei, N.T.; Abedini, A.A.; Khachatourian, A.M.; Toprak, M.S.; Seifalian, A.J. Recent advances in the modification of carbon-based quantum dots for biomedical applications. *Mater. Sci. Eng. C* **2021**, *120*, 111756. **[Google Scholar] [CrossRef] [PubMed]**

99. Namdari, P.; Negahdari, B.; Eatemadi, A.J.B. Synthesis, properties and biomedical applications of carbon-based quantum dots: An updated review. *Biomed. Pharmacother.* **2017**, *87*, 209–222. **[Google Scholar] [CrossRef] [PubMed]**

100. Xu, A.; Wang, G.; Li, Y.; Dong, H.; Yang, S.; He, P.; Ding, G. Carbon-based quantum dots with solid-state photoluminescent: Mechanism, implementation, and application. *Small* **2020**, *16*, 2004621. **[Google Scholar] [CrossRef] [PubMed]**

101. Luo, P.G.; Yang, F.; Yang, S.-T.; Sonkar, S.K.; Yang, L.; Broglie, J.J.; Liu, Y.; Sun, Y.-P. Carbon-based quantum dots for fluorescence imaging of cells and tissues. *RSC Adv.* **2014**, *4*, 10791–10807. **[Google Scholar] [CrossRef]**

102. Anilkumar, P.; Wang, X.; Cao, L.; Sahu, S.; Liu, J.-H.; Wang, P.; Korch, K.; Tackett Ii, K.N.; Parenzan, A.; Sun, Y.-P. Toward quantitatively fluorescent carbon-based "quantum" dots. *Nanoscale* **2011**, *3*, 2023–2027. **[Google Scholar] [CrossRef]**

103. Walther, B.K.; Dinu, C.Z.; Guldi, D.M.; Sergeyev, V.G.; Creager, S.E.; Cooke, J.P.; Guiseppi-Elie, A.J. Nanobiosensing with graphene and carbon quantum dots: Recent advances. *Mater. Today* **2020**, *39*, 23–46. **[Google Scholar] [CrossRef]**

104. Paulo, S.; Palomares, E.; Martinez-Ferrero, E.J.N. Graphene and carbon quantum dot-based materials in photovoltaic devices: From synthesis to applications. *Nanomaterials* **2016**, *6*, 157. **[Google Scholar] [CrossRef]**

105. Bacon, M.; Bradley, S.J.; Nann, T.J.P.; Characterization, P.S. Graphene quantum dots. *Part. Syst. Charact.* **2014**, *31*, 415–428. **[Google Scholar] [CrossRef]**

106. Tian, P.; Tang, L.; Teng, K.; Lau, S. Graphene quantum dots from chemistry to applications. *Mater. Today Chem.* **2018**, *10*, 221–258. **[Google Scholar] [CrossRef]**

107. Zhang, M.K.; Liu, W.D.; Gong, Y.P.; Liu, Q.; Chen, Z. Graphene/Quantum Dot Heterostructure Photodetectors: From Material to Performance. *Adv. Opt. Mater.* **2022**, *10*, 2201889. **[Google Scholar] [CrossRef]**

108. De Fazio, D.; Uzlu, B.; Torre, I.; Monasterio-Balcells, C.; Gupta, S.; Khodkov, T.; Bi, Y.; Wang, Z.; Otto, M.; Lemme, M.C. Graphene–quantum dot hybrid photodetectors with low dark-current readout. *ACS Nano* **2020**, *14*, 11897–11905. **[Google Scholar] [CrossRef] [PubMed]**

109. Konstantatos, G.; Badioli, M.; Gaudreau, L.; Osmond, J.; Bernechea, M.; De Arquer, F.P.G.; Gatti, F.; Koppens, F.H. Hybrid graphene–quantum dot phototransistors with ultrahigh gain. *Nat. Nanotechnol.* **2012**, *7*, 363–368. **[Google Scholar] [CrossRef] [PubMed]**

110. Vandana, M.; Devendrappa, H.; Padova, P.D.; Hegde, G.J.N. Polymer Nanocomposite Graphene Quantum Dots for High-Efficiency Ultraviolet Photodetector. *Nanomaterials* **2022**, *12*, 3175. **[Google Scholar] [CrossRef] [PubMed]**

111. Cao, L.; Yang, S.-T.; Wang, X.; Luo, P.G.; Liu, J.-H.; Sahu, S.; Liu, Y.; Sun, Y.-P. Competitive performance of carbon "quantum" dots in optical bioimaging. *Theranostics* **2012**, *2*, 295. **[Google Scholar] [CrossRef]**

112. Zhang, X.; Zeng, Q.; Xiong, Y.; Ji, T.; Wang, C.; Shen, X.; Lu, M.; Wang, H.; Wen, S.; Zhang, Y.J. Energy level modification with carbon dot interlayers enables efficient perovskite solar cells and quantum dot based light-emitting diodes. *Adv. Funct. Mater.* **2020**, *30*, 1910530. **[Google Scholar] [CrossRef]**

113. Hu, L.; Sun, Y.; Li, S.; Wang, X.; Hu, K.; Wang, L.; Liang, X.-J.; Wu, Y. Multifunctional carbon dots with high quantum yield for imaging and gene delivery. *Carbon* **2014**, *67*, 508–513. **[Google Scholar] [CrossRef]**

114. Meziani, M.J.; Dong, X.; Zhu, L.; Jones, L.P.; LeCroy, G.E.; Yang, F.; Wang, S.; Wang, P.; Zhao, Y.; Yang, L.; et al. Visible-light-activated bactericidal functions of carbon "Quantum" dots. *ACS Appl. Mater. Interfaces* **2016**, *8*, 10761–10766. **[Google Scholar] [CrossRef] [Green Version]**

115. Chen, X.; Yang, C.; Sun, H.; Ning, S.; Zhou, H.; Zhang, H.; Wang, S.; Feng, G.; Zhou, S. Enhanced photoresponsivity in carbon quantum dots-coupled graphene/silicon Schottky-junction photodetector. *Laser Phys. Lett.* **2019**, *16*, 076201. **[Google Scholar] [CrossRef]**

116. da Silva, J.C.E.; Gonçalves, H.M.R. Analytical and bioanalytical applications of carbon dots. *TrAC Trends Anal. Chem.* **2011**, *30*, 1327–1336. **[Google Scholar] [CrossRef]**

117. Chu, K.-W.; Lee, S.L.; Chang, C.-J.; Liu, L.J.P. Recent progress of carbon dot precursors and photocatalysis applications. *Polymers* **2019**, *11*, 689. **[Google Scholar] [CrossRef] [PubMed] [Green Version]**

118. Zhang, X.; Zhang, Y.; Wang, Y.; Kalytchuk, S.; Kershaw, S.V.; Wang, Y.; Wang, P.; Zhang, T.; Zhao, Y.; Zhang, H. Color-switchable electroluminescence of carbon dot light-emitting diodes. *ACS Nano* **2013**, *7*, 11234–11241. **[Google Scholar] [CrossRef] [PubMed]**

119. Kurian, M.; Paul, A. Recent trends in the use of green sources for carbon dot synthesis–A short review. *Carbon Trends* **2021**, *3*, 100032. **[Google Scholar] [CrossRef]**

120. Guo, R.; Li, L.; Wang, B.; Xiang, Y.; Zou, G.; Zhu, Y.; Hou, H.; Ji, X. Functionalized carbon dots for advanced batteries. *Energy Storage Mater.* **2021**, *37*, 8–39. **[Google Scholar] [CrossRef]**

121. Shabbir, H.; Tokarski, T.; Ungor, D.; Wojnicki, M. Eco Friendly Synthesis of Carbon Dot by Hydrothermal Method for Metal Ions Salt Identification. *Materials* **2021**, *14*, 7604. **[Google Scholar] [CrossRef] [PubMed]**

122. Li, X.; Zhao, S.; Li, B.; Yang, K.; Lan, M.; Zeng, L. Advances and perspectives in carbon dot-based fluorescent probes: Mechanism, and application. *Coord. Chem. Rev.* **2021**, *431*, 213686. **[Google Scholar] [CrossRef]**

123. De, B.; Karak, N. A green and facile approach for the synthesis of water soluble fluorescent carbon dots from banana juice. *RSC Adv.* **2013**, *3*, 8286–8290. **[Google Scholar] [CrossRef]**

124. Tepliakov, N.V.; Kundelev, E.V.; Khavlyuk, P.D.; Xiong, Y.; Leonov, M.Y.; Zhu, W.; Baranov, A.V.; Fedorov, A.V.; Rogach, A.L.; Rukhlenko, I.D. Sp2–sp3-Hybridized atomic domains determine optical features of carbon dots. *ACS Nano* **2019**, *13*, 10737–10744. **[Google Scholar] [CrossRef]**

125. Shi, W.; Han, Q.; Wu, J.; Ji, C.; Zhou, Y.; Li, S.; Gao, L.; Leblanc, R.M.; Peng, Z.J. Synthesis Mechanisms, Structural Models, and Photothermal Therapy Applications of Top-Down Carbon Dots from Carbon Powder, Graphite, Graphene, and Carbon Nanotubes. *Int. J. Mol. Sci.* **2022**, *23*, 1456. **[Google Scholar] [CrossRef]**

126. Li, W.-K.; Feng, J.-T.; Ma, Z.-Q. Nitrogen, sulfur, boron and flavonoid moiety co-incorporated carbon dots for sensitive fluorescence detection of pesticides. *Carbon* **2020**, *161*, 685–693. **[Google Scholar] [CrossRef]**

127. Sarkar, K.; Devi, P.; Lata, A.; Ghosh, R.; Kumar, P. Engineering carbon quantum dots for enhancing the broadband photoresponse in a silicon process-line compatible photodetector. *J. Mater. Chem. C* **2019**, *7*, 13182–13191. **[Google Scholar] [CrossRef]**

128. Kalytchuk, S.; Zdražil, L.; Scheibe, M.; Zbořil, R. Purple-emissive carbon dots enhance sensitivity of Si photodetectors to ultraviolet range. *Nanoscale* **2020**, *12*, 8379–8384. **[Google Scholar] [CrossRef] [PubMed]**

129. Li, D.; Jing, P.; Sun, L.; An, Y.; Shan, X.; Lu, X.; Zhou, D.; Han, D.; Shen, D.; Zhai, Y. Near-infrared excitation/emission and multiphoton-induced fluorescence of carbon dots. *Adv. Mater.* **2018**, *30*, 1705913. **[Google Scholar] [CrossRef]**

13 Chapter

OPTIMIZING THE PERFORMANCE OF FLIP-CHIP MICRO-LEDS BY RATIONAL DISTRIBUTED BRAGG REFLECTOR DESIGN

1. INTRODUCTION: AN OVERVIEW

Micro light-emitting diodes (micro-LEDs) are becoming acknowledged as a highly promising choice for developing improved display panels in the future. LEDs have numerous benefits such as superior luminosity, exceptional energy conservation, rapid reaction time, steadfastness, extended durability, and a diverse array of hues. Micro-LEDs has the necessary characteristics to effectively achieve both high dynamic range (HDR) and high resolution in display technology. The list [1,2,3,4,5,6,7,8,9]. Full-color micro-LED display panels can be realized by combining extremely efficient red, green, and blue micro-LEDs, sometimes referred to as RGB micro-LEDs [10,11,12,13]. However, to achieve the specified applications, it is imperative to significantly improve the external quantum efficiency (EQE) of RGB micro-LEDs. Many methods that improve the effectiveness of wide-area LEDs can also be utilized for micro-LEDs. The techniques employed in this study consist of flip-chip technology [6,14,15,16,17,18,19,20], surface roughening [21,22], chip shaping [23,24,25], and patterned sapphire substrate (PSS) [26,27,28,29]. Micro-LEDs often utilize flip-chip technology due to its advantages in terms of light extraction efficiency (LEE), heat dissipation, and current spreading.

The emission of light in the active zone of multiple quantum wells (MQWs) is widely recognized to occur in all directions. In flip-chip micro-LEDs, the primary source of light emission is anticipated to be from the top substrates towards the surrounding air. Nevertheless, a multitude of light produced in MQWs propagates downwards, but it does not contribute to the LEE. A bottom reflector is crucial for enhancing the Luminous Efficacy of a device by redirecting the light back towards the top substrates. The distributed Bragg reflector (DBR) is commonly employed as the bottom reflector due to its exceptional reflectivity within a specified wavelength range [30,31,32]. A typical distributed Bragg reflector (DBR) is constructed by layering dielectric materials with low refractive indices (such as SiO2) and high refractive indices (such as TiO2 and Ti3O5). This arrangement results in minimal absorption of visible light. The study found that five pairs of TiO2/SiO2 single-stack DBR, which were built for GaN-based green LEDs, had a reflective bandwidth of over 100 nm with a high reflectivity of more than 80% [31]. A dual-stack distributed Bragg reflector (DBR) was specifically engineered to achieve superior reflectivity and a wider reflective bandwidth compared to a single-stack DBR [33]. However, the double-stack DBR still exhibits significant angular dependence, which obstructs the advancement of LEE in flip-chip micro-LEDs. A complete-angle distributed Bragg reflector (DBR) consisting of 14 sub-DBRs, which were specifically designed for optimal performance in the blue, green, and red wavelength ranges, demonstrated a broad reflective bandwidth and reduced the impact of angular variations [34]. Nevertheless, the complete-angle distributed Bragg reflector (DBR) has a restricted reflective bandwidth specifically in the blue and green light wavelength range, making it unsuitable for red micro-LEDs. Thus, in order to enhance the efficiency of the RGB flip-chip micro-LEDs, it is necessary to use a distributed Bragg reflector (DBR) that has a broader range of reflectivity across the entire visible light spectrum and is less sensitive to changes in viewing angle. In order to address the limitations described before regarding DBRs, we suggest the implementation of a Ti3O5/SiO2 DBR with a wide reflected angle (WRA-DBR). The WRA-DBR consists of six sub-DBRs that are specifically designed for distinct central wavelengths, spanning from the blue light to the red light wavelength range. The WRA-DBR exhibits exceptional reflectivity across the whole range of visible light wavelengths and displays minimal deviation when the angle is altered. The WRA-DBR exhibits a mean reflectivity of up to 99.73% when light is incident at a perpendicular angle within the wavelength range of 400 nm to 700 nm. Furthermore, the WRA-DBR has an average reflectance of 97.93% when the incidence angle is 60°. The WRA-DBR exhibits a wider range of reflection frequencies and reduces the impact of angle dependency, as opposed to the normal double-stack DBR. A simulation was conducted utilizing the finite difference time domain (FDTD) approach to evaluate the effect of the WRA-DBR on the optical performance of RGB micro-LEDs. The results suggest that the WRA-DBR shows promise as a feasible choice for creating highly efficient RGB flip-chip micro-LEDs.

2. MODEL AND METHODS

The Distributed Bragg Reflector (DBR) is a configuration consisting of numerous layers that exploit the phenomenon of thin-film interference. The layers are composed of two types of dielectric materials, one with a high refractive index and the other with a low refractive index. The thickness of the dielectric layer can be calculated utilizing the formula [35]:

$$n_1 t_1 = n_2 t_2 = \lambda / 4, \tag{1}$$

The variables n1 and n2 represent the refractive indices of the high refractive layer and the low refractive layer, respectively. The variables t1 and t2 represent the thicknesses of the high refractive layer and the low refractive layer, respectively. λ represents the central wavelength.

Figure 1a shows the schematic illustration of the single-stack DBR. The single-stack DBR consists of *N* periods of layer 1 and layer 2. The refractive indices of layer 1 and layer 2 are n_1 and n_2, respectively. The thicknesses of layer 1 and layer 2 are t_1 and t_2, respectively. We consider the refractive indices of the incident medium and the substrate are n_0 and n_3, respectively. The reflection coefficient and reflectivity of the single-stack DBR can be calculated based on the transfer matrix method (TMM). From the TMM, the total transfer matrix *M* can be calculated according to the following equation [36,37]:

$$M = \begin{pmatrix} M_{11} & M_{12} \\ M_{21} & M_{22} \end{pmatrix} = D_0^{-1}\left[D_1 P_1 D_1^{-1} D_2 P_2 D_2^{-1}\right]^N D_3 \tag{2}$$

the reflection coefficient can be expressed as

$$r = \frac{M_{21}}{M_{11}}, \tag{3}$$

the dynamical matrix for the medium α can be expressed as [**36,37**]

$$D_\alpha = \begin{cases} \begin{pmatrix} 1 & 1 \\ n_\alpha \cos\theta_\alpha & -n_\alpha \cos\theta_\alpha \end{pmatrix} \text{for TE wave} \\ \begin{pmatrix} \cos\theta_\alpha & \cos\theta_\alpha \\ n_\alpha & -n_\alpha \end{pmatrix} \text{for TM wave} \end{cases}, \ \alpha = 0,\ 1,\ 2 \text{ and } 3 \tag{4}$$

the propagation matrix is given by

$$P_\alpha = \begin{pmatrix} \exp(i\phi_\alpha) & 0 \\ 0 & \exp(-i\phi_\alpha) \end{pmatrix},\ \alpha = 1,\ 2 \tag{5}$$

$$\phi_\alpha = \frac{2\pi}{\lambda} n_\alpha t_\alpha \cos\theta_\alpha,\ \alpha = 1,\ 2 \tag{6}$$

where λ is the wavelength of incident light, and θ_α is the incident angle of light in the dielectric layer calculated by Snell's law of refraction:

$$n_0 \sin\theta_0 = n_1 \sin\theta_1 = n_2 \sin\theta_2 = n_3 \sin\theta_3 \tag{7}$$

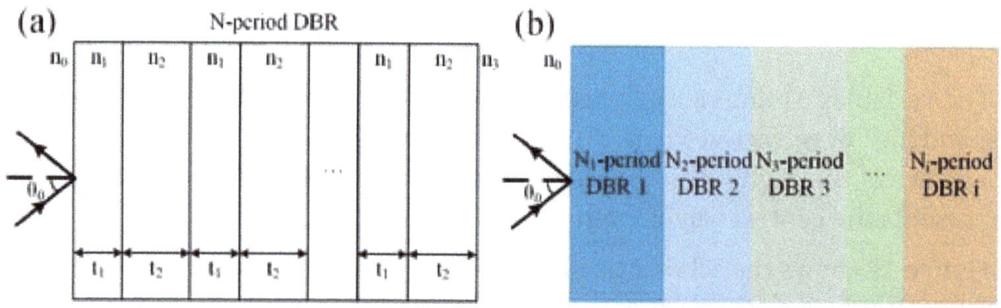

Figure 1. (a) Schematic illustration of a single-stack DBR consisting of N periods of layer 1 and layer 2. (b) Schematic illustration of DBR consisting of several sub-DBRs. Ni (i = 1, 2, 3...) is the periods of the sub-DBRs.

Then, the reflectivity of the single-stack DBR can be calculated by the following equation:

$$R = r^2 \tag{8}$$

Figure 1b shows the schematic illustration of the DBR consisting of several single-stack DBRs (sub-DBRs). N_i (i = 1, 2, 3...) is the periods of the sub-DBRs. The total transfer matrix of the DBR is

$$M = \begin{pmatrix} M_{11} & M_{12} \\ M_{21} & M_{22} \end{pmatrix} = D_0^{-1}\left[D_1P_1D_1^{-1}D_2P_2D_2^{-1}\right]^{N1}\cdots\left[D_1P_1D_1^{-1}D_2P_2D_2^{-1}\right]^{Ni}D_3 \tag{9}$$

Then, the reflectivity of the DBR can be calculated from the above-mentioned equations.

3. RESULTS AND DISCUSSION

Based on TMM, we investigate the angular dependence and reflective bandwidth of the double-stack Ti_3O_5/SiO_2 DBR. The double-stack Ti_3O_5/SiO_2 DBR is designed and modeled by TFCalc software. The double-stack DBR consists of two sub-DBRs, which are composed of 12 periods of Ti_3O_5/SiO_2. The refractive indices of Ti_3O_5 and SiO_2 in the simulation are fixed at 2.37 and 1.46, respectively. The central wavelengths of the two sub-DBRs are optimized for 465 nm and 630 nm, respectively. Therefore, the thicknesses of Ti_3O_5/SiO_2 dielectric layers are 48.97 nm/79.68 nm and 66.34 nm/108.00 nm, respectively. **Figure 2**a shows the schematic illustration of the double-stack DBR structure. **Figure 2**b shows the reflectance spectra of the double-stack DBR. The electroluminescent (EL) spectra of red, green, and blue micro-LEDs are also shown in **Figure 2**b. The peak wavelengths of red, green, and blue micro-LEDs are 630 nm, 520 nm, and 465 nm, respectively. It is clearly seen in **Figure 2**b that the reflectivity of the double-stack DBR is high within the whole visible light region when the incident angle of light is 0°. However, as the incident angle increases, the reflective bands exhibit a blueshift and the reflectivity decreases sharply in blue and green light wavelength regions. This result indicates that the double-stack DBR has a strong angular dependence.

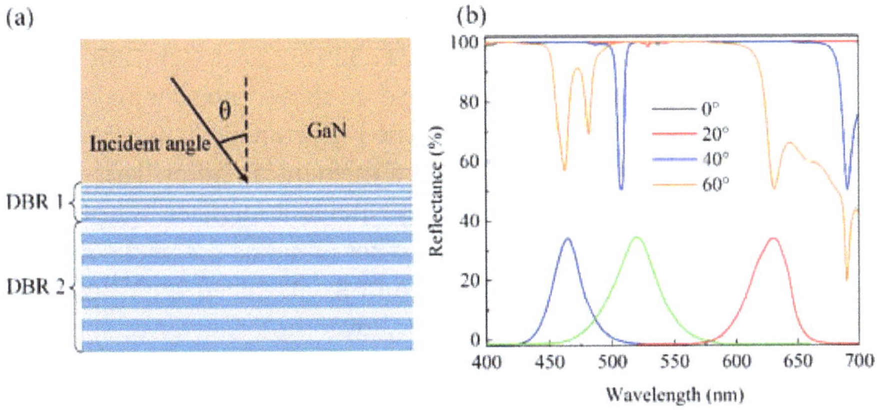

Figure 2. (a) Schematic illustration of double-stack DBR structure. (b) Reflectance spectra of the double-stack DBR as a function of incident angle.

To suppress the sharp decrease in the reflectivity in blue and green light wavelength regions, DBR III is designed by inserting a sub-DBR stack into the double-stack DBR. The inserted sub-DBR is optimized for a central wavelength at 520 nm, and the thickness of Ti_3O_5/SiO_2 of the inserted sub-DBR is 55.81 nm/90.82 nm. The total periods of DBR III are the same as that of double-stack DBR, which means that DBR III is composed of 24 periods of Ti_3O_5/SiO_2. Therefore, each sub-DBR of DBR III is made of eight periods of Ti_3O_5/SiO_2. **Figure 3**b shows the reflectance spectra of DBR III as a function of incident angle of light. After adding a sub-DBR, the sharp decrease of reflectivity of the double-stack DBR in the blue and green light wavelength region is alleviated, revealing that the added sub-DBR can broaden the reflective bandwidth. However, the reflectivity of DBR III in the red light wavelength region at a large incident angle of light remains relatively low.

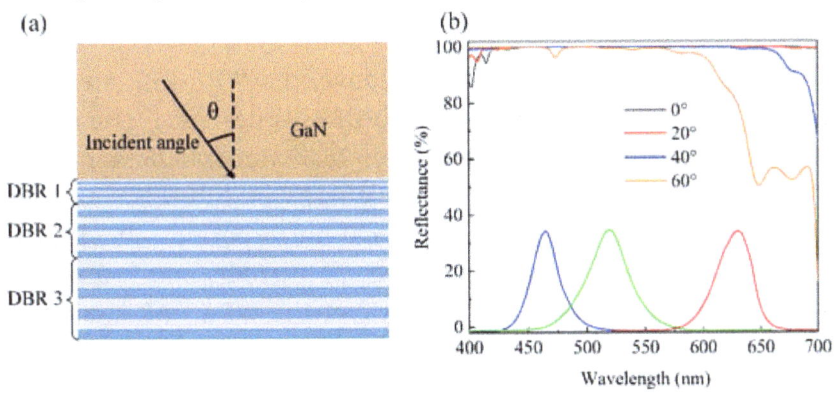

Figure 3. (a) Schematic illustration of DBR III structure. (b) Reflectance spectra of the DBR III as a function of incident angle.

To further improve the reflectivity of the DBR in the red light wavelength region, DBR IV with a combination of DBR III and another sub-DBR optimized for the central wavelength at 650 nm is designed, and the thickness of Ti_3O_5/SiO_2 of the inserted sub-DBR is 68.45 nm/111.38 nm. Each sub-DBR of DBR IV is composed of six periods of Ti_3O_5/SiO_2. **Figure 4**a shows the schematic illustration of the DBR IV structure. **Figure 4**b shows the reflectance of DBR IV. As shown in **Figure 4**b, the reflectivity is high when the incident angle of light is below 40°. However, the reflectivity of DBR IV at a large incident angle of light is not greatly improved in comparison with DBR III. To further enhance the performance of DBR IV, we investigate the influence of the sub-DBR optimized for different central wavelengths on the reflectivity. **Figure 4**c shows the reflectance spectra of DBR IV with the added sub-DBR optimized for different central wavelengths when the incident angle of light is 60°. The average reflectivity increases as the central wavelength of the sub-DBR increases from 650 nm to 730 nm. When the central wavelength of the inserted sub-DBR is 730 nm, the average reflectivity of DBR IV is larger than 90%. Therefore, we adopted the inserted sub-DBR optimized for the central wavelength at 730 nm.

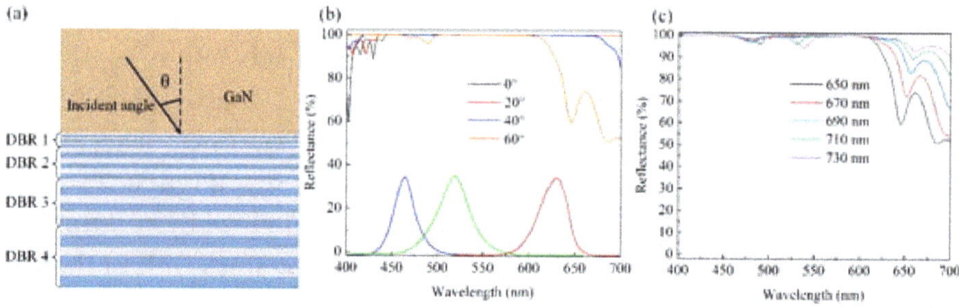

Figure 4. (a) Schematic illustration of DBR IV structure; (b) reflectance spectra of the DBR IV as a function of incident angle; (c) reflectance spectra of DBR IV with an added sub-DBR optimized for different central wavelengths at an incident angle of 60°.

To overcome the drawbacks outlined above, the WRA-DBR structure with a combination of DBR IV and another two sub-DBRs is designed. The thicknesses of Ti_3O_5/SiO_2 of the two sub-DBRs are 46.34 nm/75.39 nm and 78.98 nm/128.51 nm, respectively. Meanwhile, the central wavelengths of sub-DBRs and number of periods of Ti_3O_5/SiO_2 in the WRA-DBR are 440 nm/4 periods, 465 nm/4 periods, 520 nm/4 periods, 630 nm/5 periods, 730 nm/3 periods, and 750 nm/4 periods, respectively. **Figure 5**a shows the schematic of the WRA-DBR structure. **Figure 5**b shows the reflectance spectra of the WRA-DBR at different incident angles of light. Among all the mentioned DBRs, the WRA-DBR possesses the widest reflective bandwidth. It should be mentioned that the average reflectivity of the WRA-DBR at a normal incident angle of light is 99.73% within the wavelength region from 400 nm to 700 nm. The reflectivity of the WRA-DBR can reach up to 97.93% within the wavelength region from 400 nm to 700 nm when the incident angle of light is 60°, indicating that the WRA-DBR has alleviated the angular dependence.

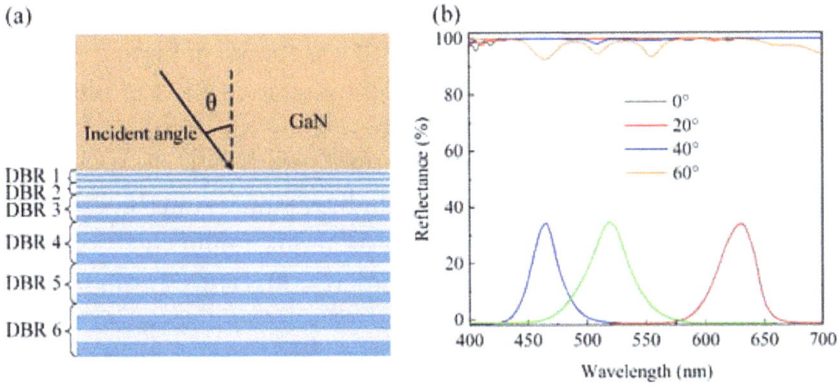

Figure 5. (a) Schematic illustration of the WRA-DBR structure. (b) Reflectance spectra of the WRA-DBR as a function of incident angle.

We investigate the influence of the double-stack DBR and the WRA-DBR on the performance of RGB flip-chip micro-LEDs using the finite-difference time-domain (FDTD) method. **Figure 6** shows the simulation models of the RGB flip-chip micro-LEDs with the double-stack DBR and the WRA-DBR. The optical parameters and the thicknesses of each layer of the RGB flip-chip micro-LEDs are shown in **Table 1**. The AlGaInP-based red micro-LED consists of ITO, p-GaP, p-AlInP, MQWs, n-AlGaInP, and n-GaAs. The GaN-based green and blue micro-LEDs are composed of ITO, p-GaN, MQWs, and n-GaN. The sidewalls of the RGB micro-LEDs are inclined and the inclination angle is 70° [38]. For simplicity, the metal electrode layer of the micro-LEDs is neglected. Considering the memory of the computer and computing time, the size of micro-LEDs is set to be 8 μm × 8 μm. The perfect matched layer (PML) is used to avoid the reflected electromagnetic wave at the boundary of the micro-LEDs. A 9 × 9 dipole source array is placed in the MQWs as a light source, and the space between dipoles is set to 0.8 μm. The emission wavelengths of the dipole sources in the RGB micro-LEDs are 630 nm, 520 nm, and 465 nm, respectively. A cross-sectional discrete Fourier transform (DFT) monitor is used to obtain the electric field distribution and a transmission box monitor is used to calculate the LEEs. **Figure** 7a–f show the electric field distributions of RGB flip-chip micro-LEDs with the double-stack DBR and WRA-DBR. In **Figure** 7a–f, region I is the bottom external space of micro-LEDs and region II is the top external space of the micro-LEDs. According to the electric field distributions shown in **Figure** 7a–f, it can be observed that there exists a weaker electric field intensity in the WRA-DBR in comparison with the electric field intensity in the double-stack DBR, indicating that the WRA-DBR has a higher reflectivity than the double-stack DBR. Therefore, more photons will be reflected upward to region II. Furthermore, compared to RGB flip-chip micro-LEDs with a double-stack DBR, stronger electric field intensity exists in the air of region II outside the RGB flip-chip micro-LEDs with the WRA-DBR, implying that the photons are more likely to be extracted from RGB flip-chip micro-LEDs with the WRA-DBR. Moreover, the LEEs of each face of the RGB micro-LEDs with the double-stack DBR and WRA-DBR are shown in **Figure** 7g–i. In **Figure** 7g, the LEEs of blue micro-LEDs with the double-stack DBR and blue micro-LEDs with the WRA-DBR are almost same. The reason for the similar LEEs in the blue micro-LEDs with the double-stack DBR and WRA-DBR is that the double-stack DBR and WRA-DBR both perform well in the blue light region. From the results of **Figure** 7h and **Figure** 7i, it can be observed that green and red micro-LEDs with the WRA-DBR exhibit higher LEEs than green and red micro-LEDs with the double-stack DBR.

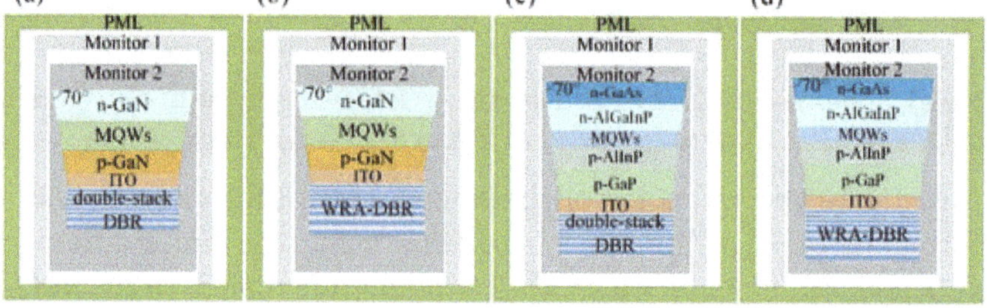

Figure 6. Simulation model of (a) blue/green micro-LED with double-stack DBR, (b) blue/green micro-LED with WRA-DBR, (c) red micro-LED with double-stack DBR, (d) red micro-LED with WRA-DBR.

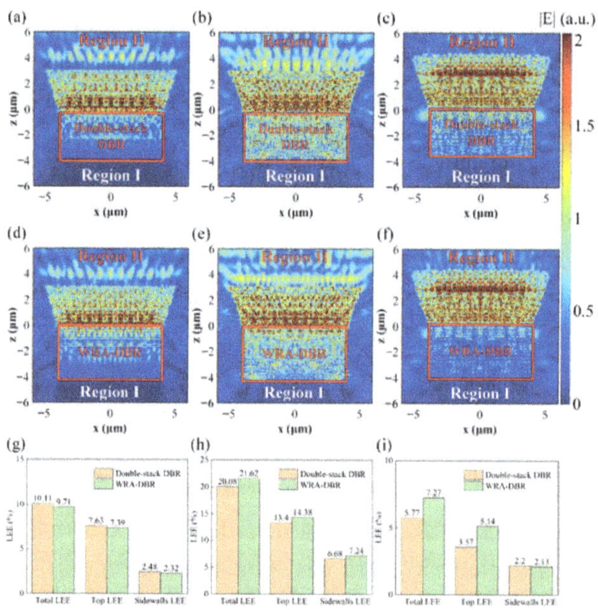

Figure 7. Simulated electric field distributions for a cross section of micro-LEDs: (a) blue, (b) green, (c) red micro-LEDs with double-stack DBR; (d) blue, (e) green, (f) red micro-LEDs with WRA-DBR. LEEs of each face (top and sidewalls) of (g) blue, (h) green, (i) red micro-LEDs with double-stack DBR and WRA-DBR.

Table 1. Optical parameters for each layer of AlGaInP-based red micro-LED and GaN-based blue and green micro-LEDs used in the FDTD simulations.

Red Micro-LED	Thickness (μm)	n	k	Green/Blue Micro-LED	Thickness (μm)	n	k
						Green (Blue)	Green (Blue)
p-GaP	2	3.322	0	p-GaN	0.5	2.38 (2.43)	5×10^{-3}(6×10^{-3})
p-AlInP	0.42	3.014	4×10^{-3}	n-GaN	2.5	2.38 (2.43)	5×10^{-3}(6×10^{-3})
n-AlGaInP	1.3	3.223	9×10^{-3}	ITO	0.28	1.89 (1.95)	4×10^{-3}(6×10^{-3})
n-GaAs	0.08	3.856	0.196	-	-	-	-
ITO	0.28	1.79	0	-	-	-	-

4. CONCISE SUMMARY

To summarize, we have developed a WRA-DBR that comprises six sub-DBRs. These sub-DBRs possess a high level of reflectivity and exhibit minimal angular dependency within the visible light wavelength range. The WRA-DBR has an average reflectivity of up to 99.73% when light is incident at a normal angle, within the wavelength range of 400 nm to 700 nm. In addition, the WRA-DBR demonstrates an average reflectivity of up to 97.93% at an incidence angle of 60°, indicating that it has minimal angular dependence. In addition, we are examining how the double-stack DBR and WRA-DBR affect the performance of RGB flip-chip micro-LEDs using the FDTD approach. The use of WRA-DBR in RGB flip-chip micro-LEDs results in a higher electric field intensity in the air outside the substrates compared to RGB flip-chip micro-LEDs with double-stack DBR. This suggests that the WRA-DBR can greatly enhance the performance of RGB flip-chip micro-LEDs. The results of our study show that the WRA-DBR has great potential for creating highly efficient RGB flip-chip micro-LEDs.

At a Glance

Due to its exceptional reflectivity, the distributed Bragg reflector (DBR) has found extensive application in flip-chip micro light-emitting diodes (micro-LEDs). Nevertheless, the traditional double-stack distributed Bragg reflector (DBR) exhibits a significant reliance on the angle of incidence and a limited range of wavelengths for reflection. In this study, we suggest the use of a wide reflected angle Ti3O5/SiO2 distributed Bragg reflector (WRA-DBR) for AlGaInP-based red and GaN-based green/ blue flip-chip micro-LEDs (RGB flip-chip micro-LEDs) as a solution to address the limitations of the double-stack DBR. The WRA-DBR, composed of six sub-DBRs, has a high level of reflectivity in the visible light wavelength range when the incidence angle of light falls between 0° and 60°. In addition, the impact of the WRA-DBR and double-stack DBR on the performance of RGB flip-chip micro-LEDs is analyzed using the finite-difference time-domain method. Due to the increased reflectivity and reduced angular dependence of the WRA-DBR, the RGB flip-chip micro-LEDs with this technology have a higher electric field intensity on the top side compared to RGB flip-chip micro-LEDs with the double-stack DBR. This suggests that micro-LEDs with the WRA-DBR can extract more photons.

REFERENCES

1. Huang, Y.M.; Chen, J.H.; Liou, Y.H.; Singh, K.J.; Tsai, W.C.; Han, J.; Lin, C.J.; Kao, T.S.; Lin, C.C.; Chen, S.C.; et al. High-Uniform and High-Efficient Color Conversion Nanoporous GaN-Based Micro-LED Display with Embedded Quantum Dots. *Nanomaterials* **2021**, *11*, 2696. **[Google Scholar]** **[CrossRef]** **[PubMed]**
2. Hsiang, E.L.; He, Z.; Huang, Y.; Gou, F.; Lan, Y.F.; Wu, S.T. Improving the Power Efficiency of Micro-LED Displays with Optimized LED Chip Sizes. *Crystals* **2020**, *10*, 494. **[Google Scholar]** **[CrossRef]**
3. Kishino, K.; Sakakibara, N.; Narita, K.; Oto, T. Two-dimensional multicolor (RGBY) integrated nanocolumn micro-LEDs as a fundamental technology of micro-LED display. *Appl. Phys. Express* **2019**, *13*, 014003. **[Google Scholar]** **[CrossRef]**
4. Gou, F.; Hsiang, E.L.; Tan, G.; Chou, P.T.; Li, Y.L.; Lan, Y.F.; Wu, S.T. Angular color shift of micro-LED displays. *Opt. Express* **2019**, *27*, A746–A757. **[Google Scholar]** **[CrossRef]**
5. Horng, R.H.; Chien, H.Y.; Tarntair, F.G.; Wuu, D.S. Fabrication and Study on Red Light Micro-LED Displays. *IEEE J. Electron Devices Soc.* **2018**, *6*, 1064–1069. **[Google Scholar]** **[CrossRef]**
6. Chang, K.P.; Tsai, Y.T.; Yen, C.C.; Horng, R.H.; Wuu, D.S. Structural design and performance improvement of flip-chip AlGaInP mini light-emitting diodes. *Semicond. Sci. Technol.* **2021**, *36*, 095008. **[Google Scholar]** **[CrossRef]**
7. Kim, T.K.; Islam, A.; Cha, Y.J.; Kwak, J.S. 32 × 32 Pixelated High-Power Flip-Chip Blue Micro-LED-on-HFET Arrays for Submarine Optical Communication. *Nanomaterials* **2021**, *11*, 3045. **[Google Scholar]** **[CrossRef]**
8. Huang, Y.; Hsiang, E.L.; Deng, M.Y.; Wu, S.T. Mini-LED, Micro-LED and OLED displays: Present status and future perspectives. *Light Sci. Appl.* **2020**, *9*, 105. **[Google Scholar]** **[CrossRef]**
9. Xiong, J.; Hsiang, E.L.; He, Z.; Zhan, T.; Wu, S.T. Augmented reality and virtual reality displays: Emerging technologies and future perspectives. *Light Sci. Appl.* **2021**, *10*, 216. **[Google Scholar]** **[CrossRef]**

10. Wu, Y.; Ma, J.; Su, P.; Zhang, L.; Xia, B. Full-Color Realization of Micro-LED Displays. *Nanomaterials* **2020**, *10*, 2482. **[Google Scholar] [CrossRef]**

11. Peng, D.; Zhang, K.; Chao, V.S.D.; Mo, W.; Lau, K.M.; Liu, Z. Full-Color Pixelated-Addressable Light Emitting Diode on Transparent Substrate (LEDoTS) Micro-Displays by CoB. *J. Display Technol.* **2016**, *12*, 742–746. **[Google Scholar] [CrossRef]**

12. Zhou, X.; Tian, P.; Sher, C.W.; Wu, J.; Liu, H.; Liu, R.; Kuo, H.C. Growth, transfer printing and colour conversion techniques towards full-colour micro-LED display. *Prog. Quantum Electron.* **2020**, *71*, 100263. **[Google Scholar] [CrossRef]**

13. Ding, K.; Avrutin, V.; Izyumskaya, N.; Özgür, Ü.; Morkoç, H. Micro-LEDs, a Manufacturability Perspective. *Appl. Sci.* **2019**, *9*, 1206. **[Google Scholar] [CrossRef]**

14. Zhou, S.; Liu, X.; Yan, H.; Chen, Z.; Liu, Y.; Liu, S. Highly efficient GaN-based high-power flip-chip light-emitting diodes. *Opt. Express* **2019**, *27*, A669–A692. **[Google Scholar] [CrossRef] [PubMed]**

15. Zhao, P.; Zhao, H. Analysis of light extraction efficiency enhancement for thin-film-flip-chip InGaN quantum wells light-emitting diodes with GaN micro-domes. *Opt. Express* **2012**, *20*, A765–A776. **[Google Scholar] [CrossRef] [PubMed]**

16. Jiang, P.; Peng, Y.; Mou, Y.; Cheng, H.; Chen, M.; Liu, S. Thermally stable multi-color phosphor-in-glass bonded on flip-chip UV-LEDs for chromaticity-tunable WLEDs. *Appl. Opt.* **2017**, *56*, 7921–7926. **[Google Scholar] [CrossRef]**

17. Konoplev, S.S.; Bulashevich, K.A.; Karpov, S.Y. From Large-Size to Micro-LEDs: Scaling Trends Revealed by Modeling. *Phys. Status Solidi Appl. Mater. Sci.* **2018**, *215*, 1700508. **[Google Scholar] [CrossRef]**

18. Liu, X.; Zhou, S.; Gao, Y.; Hu, H.; Liu, Y.; Gui, C.; Liu, S. Numerical simulation and experimental investigation of GaN-based flip-chip LEDs and top-emitting LEDs. *Appl. Opt.* **2017**, *56*, 9502–9509. **[Google Scholar] [CrossRef]**

19. Shchekin, O.B.; Epler, J.E.; Trottier, T.A.; Margalith, T.; Steigerwald, D.A.; Holcomb, M.O.; Martin, P.S.; Krames, M.R. High performance thin-film flip-chip InGaN–GaN light-emitting diodes. *Appl. Phys. Lett.* **2006**, *89*, 071109. **[Google Scholar] [CrossRef]**

20. Hu, H.; Tang, B.; Wan, H.; Sun, H.; Zhou, S.; Dai, J.; Chen, C.; Liu, S.; Guo, L.J. Boosted ultraviolet electroluminescence of InGaN/AlGaN quantum structures grown on high-index contrast patterned sapphire with silica array. *Nano Energy* **2020**, *69*, 104427. **[Google Scholar] [CrossRef]**

21. Tang, B.; Miao, J.; Liu, Y.; Zhou, S.; Xu, H.; Wan, H. Insights into the Influence of Sidewall Morphology on the Light Extraction Efficiency of Mini-LEDs. *IEEE Photon. J.* **2019**, *11*, 8200907. **[Google Scholar] [CrossRef]**

22. Guo, Y.; Zhang, Y.; Yan, J.; Xie, H.; Liu, L.; Chen, X.; Hou, M.; Qin, Z.; Wang, J.; Li, J. Light extraction enhancement of AlGaN-based ultraviolet light-emitting diodes by substrate sidewall roughening. *Appl. Phys. Lett.* **2017**, *111*, 011102. **[Google Scholar] [CrossRef]**
23. Zhuang, Z.; Iida, D.; Ohkawa, K. Effects of size on the electrical and optical properties of InGaN-based red light-emitting diodes. *Appl. Phys. Lett.* **2020**, *116*, 173501. **[Google Scholar] [CrossRef]**
24. Smith, J.M.; Ley, R.; Wong, M.S.; Baek, Y.H.; Kang, J.H.; Kim, C.H.; Gordon, M.J.; Nakamura, S.; Speck, J.S.; DenBaars, S.P. Comparison of size-dependent characteristics of blue and green InGaN microLEDs down to 1 μm in diameter. *Appl. Phys. Lett.* **2020**, *116*, 071102. **[Google Scholar] [CrossRef]**
25. Lu, S.; Zhang, Y.; Zhang, Z.H.; Zhu, B.; Zheng, H.; Tan, S.T.; Demir, H.V. High-Performance Triangular Miniaturized-LEDs for High Current and Power Density Applications. *ACS Photon.* **2021**, *8*, 2304–2310. **[Google Scholar] [CrossRef]**
26. Chao, S.H.; Yeh, L.H.; Wu, R.T.; Kawagishi, K.; Hsu, S.C. Novel patterned sapphire substrates for enhancing the efficiency of GaN-based light-emitting diodes. *RSC Adv.* **2020**, *10*, 16284–16290. **[Google Scholar] [CrossRef]**
27. Manley, P.; Walde, S.; Hagedorn, S.; Hammerschmidt, M.; Burger, S.; Becker, C. Nanopatterned sapphire substrates in deep-UV LEDs: Is there an optical benefit? *Opt. Express* **2020**, *28*, 3619–3635. **[Google Scholar] [CrossRef]**
28. Ooi, Y.K.; Zhang, J. Light Extraction Efficiency Analysis of Flip-Chip Ultraviolet Light-Emitting Diodes with Patterned Sapphire Substrate. *IEEE Photon. J.* **2018**, *10*, 8200913. **[Google Scholar] [CrossRef]**
29. Zhou, S.; Liu, S. *III-Nitride LEDs: From UV to Green*; Springer Nature: Singapore, 2022; pp. 91–95. **[Google Scholar]**
30. Wang, D.X.; Ferguson, I.T.; Buck, J.A. GaN-based distributed Bragg reflector for high-brightness LED and solid-state lighting. *Appl. Opt.* **2007**, *46*, 4763–4767. **[Google Scholar] [CrossRef]**
31. Huang, S.Y.; Horng, R.H.; Wang, W.K.; Wuu, D.S. GaN-Based Green Resonant Cavity Light-Emitting Diodes. *Jpn. J. Appl. Phys.* **2006**, *45*, 3433–3435. **[Google Scholar] [CrossRef]**
32. Zhi, T.; Tao, T.; Liu, B.; Yan, Y.; Xie, Z.; Zhao, H.; Chen, D. High Performance Wide Angle DBR Design for Optoelectronic Devices. *IEEE Photon. J.* **2021**, *13*, 8200206. **[Google Scholar] [CrossRef]**
33. Ding, X.; Gui, C.; Hu, H.; Liu, M.; Liu, X.; Lv, J.; Zhou, S. Reflectance bandwidth and efficiency improvement of light-emitting diodes with double-distributed

Bragg reflector. *Appl. Opt.* **2017**, *56*, 4375–4380. **[Google Scholar] [CrossRef] [PubMed]**

34. Shi, L.; Zhao, X.; Du, P.; Liu, Y.; Lv, Q.; Zhou, S. Enhanced performance of GaN-based visible flip-chip mini-LEDs with highly reflective full-angle distributed Bragg reflectors. *Opt. Express* **2021**, *29*, 42276–42286. **[Google Scholar] [CrossRef]**
35. Kats, M.A.; Blanchard, R.; Genevet, P.; Capasso, F. Nanometre optical coatings based on strong interference effects in highly absorbing media. *Nat. Mater.* **2013**, *12*, 20–24. **[Google Scholar] [CrossRef]**
36. Wu, C.J.; Chu, B.H.; Weng, M.T. Analysis of Optical Reflection in a Chirped Distributed Bragg Reflector. *J. Electromagn. Waves Appl.* **2009**, *23*, 129–138. **[Google Scholar] [CrossRef]**
37. Pérez, E.X. Design, Fabrication and Characterization of Porous Silicon Multilayer Optical Devices. Ph.D. Thesis, Universitat Rovira i Virgili, Tarragona, Spain, 2008. **[Google Scholar]**
38. Hu, X.; Cai, J.; Liu, Y.; Zhao, M.; Chen, E.; Sun, J.; Yan, Q.; Guo, T. Design of inclined omni-directional reflector for sidewall-emission-free micro-scale light-emitting diodes. *Opt. Laser Technol.* **2022**, *154*, 108335. **[Google Scholar] [CrossRef]**

14 Chapter

HYBRID QUANTUM DOTS SHOW GREAT POTENTIAL AS EFFECTIVE TOOLS FOR THERANOSTIC APPLICATIONS IN CANCER

1. INTRODUCTION: AN OVERVIEW

Cancer is the primary global cause of mortality, responsible for approximately 10 million fatalities in 2020 [1]. A proper diagnosis is essential for precise and efficient treatment, as each tumor requires specialized interventions such as surgery, radiation, and chemotherapy. Scientists, medics, and healthcare professionals are greatly concerned with developing new and effective treatments for cancer as well as detecting the disease at an early stage. Nanotechnology has been a leading area of research in various domains, including medicine, diagnostics, bioimaging, and other biomedical applications, for the past few decades [2]. Nanoparticles (NPs) are widely used as carrier systems for delivering drugs in the treatment of several types of cancer [3,4]. These carrier systems exhibit diverse shapes and dimensions, ranging from 1 to 100 nm, specifically for the purpose of delivering drugs in cancer treatment. These differences in physical and chemical properties can be utilized for numerous applications in cancer. Nanomaterials are utilized for targeted ultrasonic heating therapy [5,6,7], radiofrequency (RF) ablation [8,9,10], magnetic fluid hyperthermia [11,12,13], and magnetic particle imaging [14,15] in cancer, taking advantage of their special physicochemical features. Nanoparticles (NPs) are extensively utilized in various applications such as bioimaging, drug delivery systems[16,17,18], therapeutic agents for photodynamic therapy (PDT), photothermal therapy (PTT), regenerative medicine, and smart biomaterials[19,20,21],. Moreover, metallic nanoparticles (NPs), including gold [22], silver [23], platinum [24], and palladium [25], have been extensively studied as biocompatible NPs for the purpose of diagnosing and treating cancer. The use of imaging agents to treat various diseases, including different types of carcinomas, has generated interest in the theranostic approach. This approach involves the delivery of synthetic drugs, therapeutic peptides, and genes, and allows for both diagnosis and treatment to be carried out using a single system. Nanomaterials have excellent membrane penetration, biocompatibility, and multifunctionality owing to their small size and functionalization capabilities. The use of nanoparticles is very suitable for theranostic applications in various types of malignancies [26]. Quantum dots (QDs) are extremely tiny nanoparticles that can be either organic (such as carbon-QDs and graphene-QDs) or inorganic (such as zinc sulfide-QDs, cadmium telluride-QDs, and cadmium selenide-QDs). They typically have sizes ranging from 2 to 10 nm [27]. The small nano dimension of particles is advantageous for enhancing specific optical properties such as high brightness, high quantum yield, high extinction coefficient, intermittent fluorescence signals, and high stability against photobleaching [28]. These particles also possess electronic properties that are utilized in various biomedical applications [27,28,29]. The nanocrystal possesses a distinct energy band gap, which is necessary to stimulate the movement of an electron from a lower energy level in one electronic band to a higher energy level in another band. Due to their diminutive size, these semiconductor-based nanocarrier systems possess the ability to efficiently transfer electrons to a more elevated energy level, even under the influence of ultraviolet (UV) radiation. The diagnostic and therapeutic applications of quantum dots (QDs) are utilized in the management of several medical conditions, including cancer. This excitation strategy finally generates an electron-hole pair referred to as an exciton. When the exciton returns to its ground state, it emits energy in the form of a fluorescent photon [30]. The semiconducting nanocrystal, with a size ranging from 2 to 10 nm, began to exhibit characteristics similar to those of the bulk Bohr exciton radius, resulting in alterations to its electrical and optical properties [31]. The well-documented and understood phenomenon is the inverse correlation between the size of nanocrystals and the energy band gap. The inverse relationship states that when the nanocrystal size reduces, the energy band gap increases, resulting in a decrease in the excitation/emission wavelengths. By altering the size of the particle, the hue of the light emitted by the QDs when exposed to UV radiation can be modified. By manipulating the particle size and size distribution of the quantum dots (QDs), it is possible to obtain a broad spectrum of light absorption and produce emission spectra that are very symmetrical and narrow [32]. Various types of quantum dots (QDs) are synthesized using the bottom-up technique, wherein their precursor molecules are assembled into nanocrystals [33]. The manufacture of QDs can be classified into four main approaches: biotemplate-based synthesis, colloidal synthesis, biogenic synthesis, and electrochemical assembly [27,33]. The techniques used to prepare quantum dots (QDs) are depicted in Figure 1.

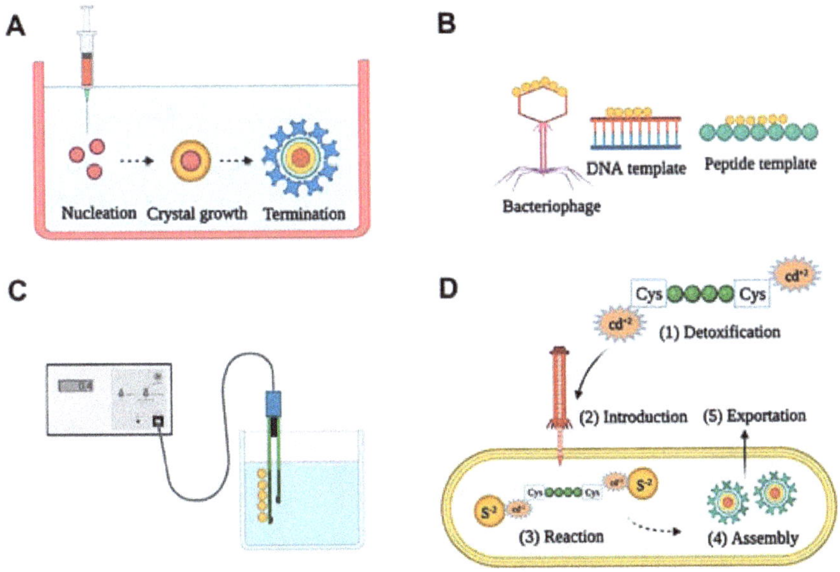

Figure 1. Schematic illustration of commonly used preparation approaches for QDs: (A) Colloidal synthesis approach. (B) Biotemplate-based synthesis approach. (C) Electrochemical assembly approach. (D) Biogenic synthesis approach. Reproduced from Abdellatif et al. [27], Dove Medical Press, 2022.

Carbon nanomaterials, including carbon nanotubes, carbon dots, graphene, and graphene derivatives, as well as other nanomaterials of organic and inorganic origin, are widely used due to their exceptional composition and outstanding inherent properties. These materials are highly valued for a variety of applications, such as fluorescence, fingerprinting, photocatalysis, electromagnetic shielding, and electric applications [3,34]. Nanomaterials have also had a practical impact in the field of biomedical engineering, aiding research in tissue engineering, drug delivery, biosensing, bioimaging, and cancer theranostic applications [35]. The system is modified with quantum dots (QDs) to create a hybrid nanoparticulate system known as hybrid QDs. These hybrid QDs have extremely small dimensions and are useful for both therapy and diagnosis in cancer, making them theranostic. These hybrid systems are employed to administer medications of synthetic/natural/biological origin while also serving as an imaging agent. This dual functionality is expected to enhance their concentration, particularly in malignant or tumor tissue. Therefore, these adaptable nanoparticulate systems, known as hybrid quantum dots (QDs), offer a broad range of applications in the detection and imaging of cancer, as well as the targeted administration of various types of treatments to malignant cells (as shown in Figure 2).

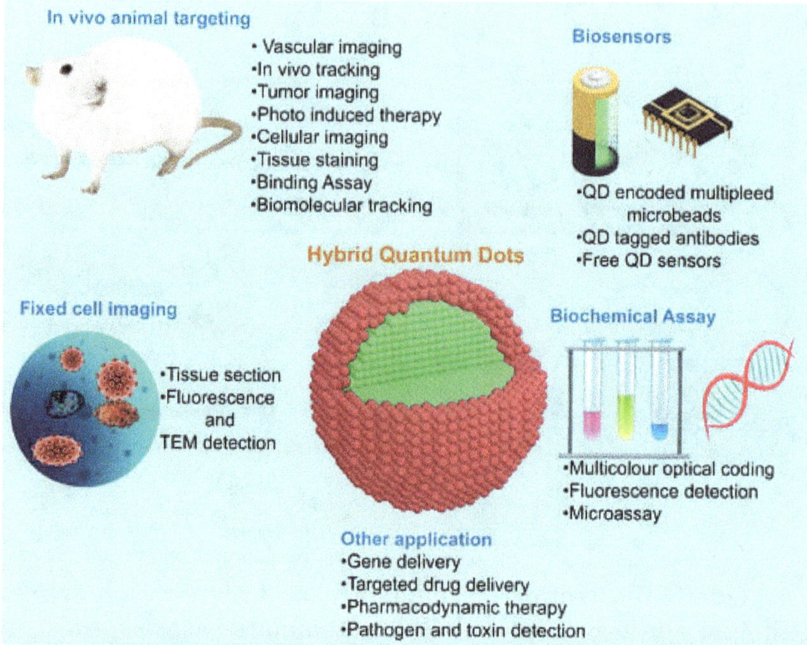

Figure 2. Illustration highlighting the utility of hybrid QDs as cancer theranostics for various applications.

To boost the effectiveness of drugs in treating cancer, one approach is to improve the EPR (increased permeability and retention) effect and address the obstacles posed by tumor heterogeneity. This can be achieved by constructing targeted hybrid quantum dots (QDs) with a stealthy nature [36]. Carbon-based porous particles, which are camouflaged with red blood cells and responsive to light, have the potential to selectively target and infiltrate tumor areas. Nanosponges that specifically target proteins and red blood cell (RBC) membranes enhance the efficiency of targeting and prolong the circulation half-life. The combination of porous carbon/silica and graphene quantum dots (QDs) forms a hybrid system that is both responsive to light and capable of delivering drugs to tumors. This system has potential applications in theranostics, as it may be used for both therapy and diagnosis. This information is supported by reference [37]. Cyclodextrins (CDs) are naturally occurring cyclic oligosaccharides that are soluble in water. They have hydrophilic exteriors and hydrophobic interiors, making them useful for delivering drugs. Their outside hydroxyl groups, both primary and secondary, can be easily altered, while their inner cavities, which are lipophilic, can be filled with a lipophilic component through the creation of an inclusion complex. These carriers are used to enhance the solubility of therapeutics or imaging agents that are not soluble in water. They are a promising system for delivering drugs in cancer treatment when combined with QDs for theranostic purposes. This is because they have many hydroxyl groups on their surface, which makes them suitable for this application [38]. This publication offers a comprehensive analysis and up-to-date progress on quantum dots (QDs) and their potential to enhance the effectiveness of therapeutic drugs and imaging techniques for the successful treatment of cancer. The article presented a comprehensive analysis of hybrid quantum dots as a cancer theranostics and highlighted the latest advancements in cancer theranostics using hybrid quantum dots, with a focus on research conducted within the past 5 years.

2. THE IMPORTANCE OF HYBRID QUANTUM DOT IN CANCER THERANOSTICS

The theranostic technique in cancer treatment is important because it allows for the simultaneous delivery of chemotherapeutics and photolytic chemicals. This approach is successful in penetrating deep into tumors and causing damage and inhibition when treated with a single irradiation. It can be employed to monitor the progress of cancer treatment by using integrated imaging agents [39]. Hybrid quantum dots (QDs) have been extensively studied for their theranostic potential in various cancer types. Nevertheless, the presence of heavy or inorganic elements in quantum dots (QDs) and their ability to generate reactive oxygen species (ROS) and exhibit significant surface responses have caused concerns over their toxicity. As a result, there is a need to modify or functionalize QDs for biomedical applications. Therefore, approaches have been developed to reduce the harmful effects of these substances and enhance their compatibility with living organisms by combining them with other substances (such as polymers, lipids, polysaccharides, proteins, etc.). This allows for targeted accumulation in tumor tissues while preventing accumulation in healthy tissues [40,41]. QDs are conjugated with biological molecules using hydrophilic surfactant shells that include reactive functional groups as COOH, NH2, or SH. Attachments can be achieved through several techniques, including adsorption, covalent bonding, electrostatic contact, and others [39,40,41]. According to reports, it has been seen to form chemical bonds with many biological molecules, such as biotin [42], folic acid [43,44], antibodies [45], and peptides [46]. Silanization is an effective approach for covalently coating hydroxyl-rich material surfaces with silica, which is a process that involves covering QDs with a layer of silica. Silanization enhances the cross-linking and chemical stability of ligand molecules. The terminal groups at the end of the silane shell can expose thiol, phosphonate, or methyl terminal ends, which can be used for coating quantum dots (QDs) and also enhance the material's biocompatibility. Silanization is preferred due of its lower toxicity compared to other ligands [47]. The thickness of the silica shell has the ability to regulate the reactivity of quantum dots to light. The QDs covered with silica were functionalized with amino, carboxyl, and epoxy groups and stabilized with PEG segments to evaluate their suitability. The modified quantum dots (QDs) effectively bound to antibodies and were employed as fluorescent markers in immunoassay detection [48]. A study conducted in living organisms has demonstrated that emissive silicon quantum dots (Si-QDs) undergo rapid biodegradation and generate non-toxic silicic acid, which may be excreted through urine [49]. The next section discusses the potential use of hybrid quantum dots (QDs) in cancer diagnosis/imaging and targeted delivery of therapeutic payload to tumor tissues.

2.1. Perspectives of QDs for Diagnostic/Imaging Utility

QDs in drug delivery may be utilized as therapeutic/imaging cargo that has photothermal and photodynamic features, making them excellent for bioimaging. Many clinically used photosensitizers (PSs) are not tumor-targeted; hence, they are treated with spatially controlled irradiation. After phototherapies, PS can increase reactive oxygen species (ROS) formation in healthy cells; hence, light exposure should be avoided to reduce skin photosensitivity. The hybrid QDs possess low toxicity and good biocompatibility, coupled with stable photoluminescence (PL), and therefore these are ideal candidates for both in vitro and in vivo bioimaging [40,41]. QDs by themselves are not as efficient as molecular PSs, but QDs can be used as antennae to improve light harvesting and energy transfer to molecular PSs because they absorb much light. NPs are commonly utilized for bioimaging, but their toxicity limits their utility. Because fluorescence imaging is very sensitive and has a good temporal and spatial resolution, hybrid QDs are a good choice for sensing and imaging cell targets. Hybrid QDs are chemically inert, dissolve well in water, are photostable, have a relationship between their optoelectronic properties and their shape and size, have fluorescence resonance energy transfer, high stability in physiological conditions, specific accumulation at target sites, are easy to modify on the surface [50], and have a high absorption coefficient because of hybridized C–C bonds. Therefore, these are good phototherapy chromophores.

Carbon QDs (C-QDs) synthesized and dispersed with excellent fluorescence, photostability, photobleaching resistance, and simple coupling with biological species [51]. C-QDs can carry Ce_6 and generate ROS. Using a 639 nm laser, water splitting produced oxygen and hydrogen in vivo. Increased oxygen yielded $^1O_2$ to improve PDT. C-QDs with specific cell targets can particularly detect malignant cells in different investigations. C-QDs conjugated with folic acid (FA) (C-dots-FA) to distinguish folate receptor (FR)-positive cancer cells from normal cells (FR-negative) by growing and analyzing NIH-3T3 and HeLa cells. Pheophytin (a natural, low-toxicity Mg-free chlorophyll derivative) was employed as a raw carbon source to synthesize C-QDs using a microwave technique [52]. QDs containing sulfur and nitrogen are used as PTT (photothermal therapy causes ease of cell death by protein denaturation and loosening of the cellular membrane by heating the tumor tissue exploiting irradiation of radiofrequency, ultrasound, microwaves, and magnetic fields, etc.) and PDT (photodynamic therapy utilizes a photosensitizer that absorbs light of a particular wavelength and produces oxygen-based molecular species to induce a cytotoxic effect) for cancers in animals [53,54]. PL and photoacoustic imaging [55,56,57,58] benefited from high photon conversion efficiencies. Passive targeting of QDs around cancer cells destroyed the tumor. Co-doped C-QDs had a strong photothermal conversion, optical and photoacoustic performance, and renal excretion [53,59]. N–O-CQDs with significant NIR absorption. Combining the biocompatible N-doped carbon dots (N-CDs) with folic acid, which possess a wide range of high-targeting capabilities (26 types of tumor cell lines) and alters the cellular metabolism leading to autophagy, results in a new targeted tumor therapy based on autophagy regulation [60]. Similarly, maleimide-terminated TTA1 aptamers complexed with CDs (TTA1–CDs), which is substantially expressed in HeLa and C6 (rat glioma cell line) but not in normal healthy CHO cells, exhibit a strong fluorescence along cancer cell membranes and minimal uptake in healthy cells [61].

Graphene quantum dots (GQDs) are one type of nanocarrier that has been seen in physics and chemical research due to their ultrasmall size, varied photoluminescence, and mechanical features [62]. Ultra-tiny GQDs exploiting imaging agents and labeling cell membranes are promising agents for drug transportation in cancer therapy because of their outstanding optical properties and transmembrane capabilities. The innate immune system and tumor heterogeneity continue to pose challenges to efficient tumor targeting and penetration; however, NIR irradiation, the energy created by photothermal conversion, can not only release therapeutic cargo but also burst the vesicle to suppress the tumor [63]. When NPs first enter the circulatory system, the innate immune system quickly recognizes them as foreign bodies and gets rid of them through the reticuloendothelial system and the mononuclear phagocyte system. This results in poor delivery efficiency. Because the tumor is a strong physiological barrier, only a small part of the dose injected gets to the deep tumor through the increased EPR effect, which helps particle accumulation. The high interstitial fluid pressure (IFP) and cancer-associated fibroblasts in tumors make it hard for therapeutic drugs to reach the perivascular cells of tumors [64,65]. Thus, in order to improve tumor therapy, it is crucial to create stealthy and permeable drug delivery systems for the efficient transportation of therapeutic agents.

For imaging or diagnostic applications, theranostic nanoplatforms must be robust enough to support them, have a superior cargo-loading and -releasing profile, and be able to do so. Hybridization between distinct NPs is a promising strategy because it can result in the accumulation of a wide range of chemical, physical, and biological properties inside a single complex. Because of their exceptional physical and chemical properties [32,66], GQDs have been put to use in a variety of biomedical settings. If GQD fluorescence could be made stable, it would greatly improve the efficiency of imaging in the life sciences. Due to their unique chemical, physical, and biological properties, graphene quantum dots (GQDs) and magnetic nanoparticles (MNPs) are two promising choices for use in these hybrids. Both magnetic resonance imaging (MRI) and computed tomography (CT) use contrast agents made of magnetic nanoparticles [67]. In addition to its use in biosensing and magnetic separation, this nanoparticle may also be put to use in hyperthermia therapy, thermo-ablation, targeted drug administration, and even bio-sensing. For example, magnetic nanoparticles could be added to GQDs to make even more complexes that could be used in medicine. The most commonly used magnetic nanoparticles are iron oxide NPs (usually Fe_2O_3 or Fe_3O_4), which can be classified as a pure metal, metal oxide, or magnetic nanocomposites [68]. Combining GQDs with other NPs, such as magnetic nanoparticles, could make them even better for use in biology.

Carbon quantum dots (CQDs), a novel kind of fluorescent carbon nanomaterial possessing the unique advantages of high stability, remarkable biocompatibility, easy synthesis and surface functionalization, and comparable optical characteristics, have been extensively studied, especially for bioimaging applications due to their tunable strong fluorescence emission property [69]. In light of the drawbacks of conventional chemotherapy, PTT has emerged as a viable option for treating cancer. Tumors can be heated from the inside out by injecting photothermal substances into the affected area or by targeting the tumor with other agents. These photothermal agents are designed to stimulate near-infrared (NIR) radiation and generate heat upon relaxation, killing cancer cells. Researchers have been investigating several facets of theranostic nanosystems [66] since it has been postulated that these systems, which meet both diagnostic and therapeutic needs, could be utilized to effectively eradicate cancer cells. It has been found that a wide range of organic, inorganic, organo-inorganic, and combinations can act as photothermal agents. Carbon nanomaterials, including carbon nanotubes, graphene, and graphene derivatives, have garnered significant attention due to their potential applications in fields as diverse as fingerprinting, photocatalysis, electromagnetic shielding, and electrics [70,71]. Due to their luminescence, versatile surface chemistry, easy cellular internalization, and high biocompatibility, CQDs are particularly promising in drug delivery. Additionally, the nano-formulation system also offers the possibility to increase drug solubility, bioavailability, and half-life. Although doxorubicin (DOX) is widely used for cancer treatment, it has many disadvantages, including a low EPR effect, low cellular internalization, and cytotoxicity to normal cells [72]. One approach to bypassing these problems is to use a multifunctional nanocarrier system for tumor-targeted drug delivery, which has the advantage of accumulating at the tumor site due to the increased EPR effect.

2.2. Perspectives of QDs for Therapeutic Utility

Different QD-based therapeutic systems for anticancer application have been widely explored in recent years [73,74]. Various studies have been conducted to investigate the potential of QDs for targeted drug delivery, PDT, PTT, and gene delivery in cancer treatment [75,76]. The QDs for cancer therapy have been investigated both at in vitro and in vivo levels. PDT is one of the most promising non-invasive cancer treatment approaches with limited side effects. It can be used alone or in combination with surgery, chemotherapy, or ionizing radiation to destroy undetected cancerous cells at the margins of resection. PDT uses photosensitizing drugs that are pharmacologically inactive until a particular light wavelength irradiates them in the presence of oxygen, which generates reactive oxygen species and induces cell death and tissue necrosis [77,78]. Graphene oxide (GO), an oxidized version of graphite, has received considerable attention during the past decade. GO, on the other hand, can be dispersed in water, which makes it a good candidate to investigate in a biological system. On the other hand, graphenes need to be surface functionalized to make them dispersible in water and safer for the biological environment [79,80]. Their synthesis and surface tailoring entail hazardous and toxic reagents, traces of which may remain with the material to demonstrate further toxicity in vitro/in vivo systems due to GO sheets having intrinsic toxicity. By delivering medications and energy in two different locations at the same time, a hybrid system of nano dimension may be able to lessen the adverse effects of cancer treatment and improve the distinctive properties required for precision medicine. The hybrid carrier systems are frequently eliminated from blood circulation very quickly, and piled-up tumors at the periphery close to the blood arteries. It has a short elimination half-life in blood and high tumor penetration. The membrane of a red blood cell (RBC) was given the appearance of a sponge by being composed of carbon composite. When it is exposed to light, it functions as both a "stealth agent" and a "photolytic carrier". It means that it transfers "tumor-penetrating agents" (such as graphene QDs and docetaxel) as well as heat. When compared to the nanosponge, the RBC-membrane-enveloped nanosponge demonstrates an eight-fold increase in accumulation in tumor tissue. This is because the RBC-membrane-enveloped nanosponge will be integrated with a specific protein that accumulates in tumor spheroids through high lateral bilayer fluidity [81]. The delivery of graphene QDs to tumor areas is accomplished by passing near-infrared light through a structure that is only one atom thick. This makes it much simpler for its therapeutic utilization to penetrate the cancerous tissue and improves the prognosis of cancer therapy utilizing the theranostic approach.

The pathway of accumulation and removal of QDs and hybrid QDs in in vivo systems are depicted in **Figure 3**.

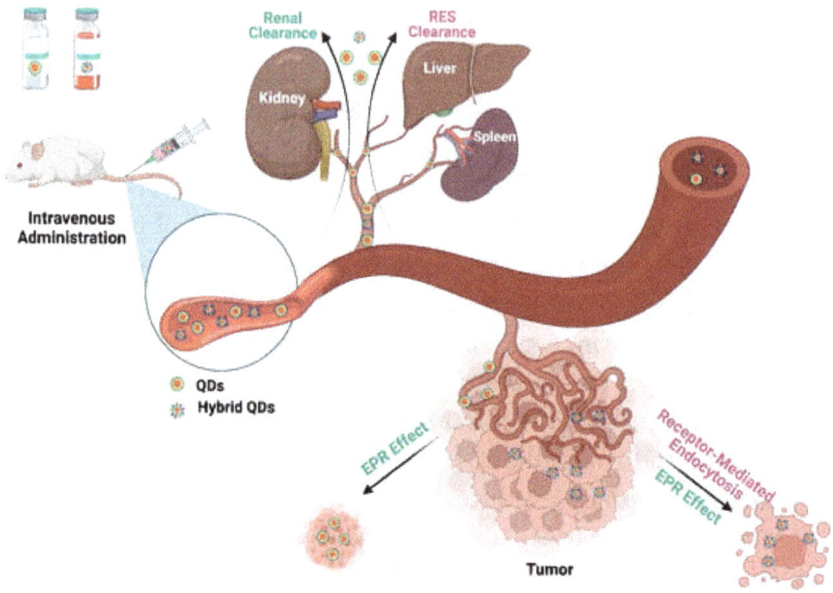

Figure 3. Schematic illustration highlights the accumulation and removal of QDs versus hybrid QDs. QDs are more prone to RES clearance and renal clearance compared to hybrid QDs of a stealth nature. The more targeted delivery of hybrid QDs compared to QDs resulted in major accumulation in tumor tissues due to receptor-mediated endocytosis and the EPR effect, leading to improved therapeutic outcomes. "Image created with BioRender.com".

Different types of hybrid QD-based theranostic systems were explored for cancer applications to improve the biopharmaceutical attributes of this nanoparticulate system (such as aqueous solubility, tumor penetrability, and stability of loaded therapeutics in the tumor microenvironment) and delivery of drug specifically to tumor tissues. Recent contemporary research conducted in this field is discussed in the subsequent section.

3. HYBRID QUANTUM DOT AS CANCER THERANOSTICS: CONTEMPORARY RESEARCH

3.1. Diagnostic Application

Pei et al. developed fluorescent hyper-cross-linked-cyclodextrin–carbon quantum dot (CD-CQD) hybrid nanosponges with outstanding biocompatibility and intense bright blue fluorescence excited at 365 nm with a PLQY of 38.0% [82]. These hybrid QDs systems were generated by simple condensation polymerization of carbon quantum dots (CQDs) with cyclodextrin (CD) at a 1:5 feeding ratio for theranostic applications, specifically in malignancies. In another investigation, Fateh et al. made a hybrid nanostructure of graphene quantum dots (GQDs) and magnetic nanoparticles (MNP) by using hydrophobic interactions between long carbon chains on the surface of GQDs around the edges and MNP in the middle [**83**]. Pyrolysis was used to create GQDs, which were then changed using cetyl alcohol (CA) to produce surfactant-modified GQDs (CA-GQDs). Moreover, an oleate-iron complex has been utilized to make iron-oxide nanoparticles (IONP) as MNP. After that, CA-GQDs and IONP are mixed to make a structure with IONP in the middle and CA-GQDs all around it (IONP@CA-GQD). IONP@CA-GQD possesses both fluorescence and magnetic characteristics. At room temperature, IONP and IONP@CA-GQD have been tested for magnetization hysteresis loops in a moving magnetic field. There have been no observations of coercivity or remanence, indicating superparamagnetism. The computed MS values for IONP and IONP@CA-GQD are 34.1 emu/g and 37.8 emu/g, respectively. Because of GQDs are fluorescent in nature, this hybrid structure could also be used for bioimaging [84,85].

A stable compound of graphene oxide (GO) and graphene quantum dots (GQD) was created by Kumavat et al. by electrostatic layer-by-layer assembly using a polyethylene imine bridge (GO-PEI-GQDs) [86]. In addition, various applications of the mono-equivalents of the GO-PEI-GQDs complex were compared, including cell imaging (diagnostics), photothermal, and oxidative stress responses in MDA-MB-231 breast cancer cells. When exposed to an 808 nm laser for 5 min at a concentration of up to 50 µg/mL, GO-PEI-GQDs displayed an outstanding photothermal response (44–49 °C). According to the study, GO-PEI-GQDs had synergistic effects on cancer cells. It has stable fluorescence imaging, improved photothermal effects, and cytotoxic actions. Composite materials made of GO and GQDs combine many different properties, which makes it possible to improve certain therapeutic systems, such as cancer theranostics [86].

Hyaluronic acid and QDs together have been proven to be useful tools for improving intracellular transport into liver cells. This is accomplished by interacting with CD44-receptors, which allows for in vivo real-time imaging [87,88]. The anionic polysaccharide chondroitin sulfate was employed to coat the positively charged oily core of the cadmium telluride (CdTe) QDs as cancer theranostic nanocapsules [89], which were also loaded with rapamycin and celecoxib as anticancer therapeutics [90]. Chondroitin sulfate nanocapsules have an exterior coating of cationic gelatin-coupled QDs placed on them to prevent non-specific uptake by healthy cells. Matrix metalloproteinase (MMPs) dissolved the gelatin at the tumor location, releasing therapeutic nanocapsules and QDs into cancer cells for therapeutic and imaging action as a cancer therapeutics. An ON–OFF effect, where the fluorescence of QDs was first quenched by energy transfer and then restored after bond cleavage in tumor cells, was seen in a study that substituted lactoferrin for gelatin [90]. Thus, using QDs fluorescence, the in vitro and in vivo localization of nanocapsules into breast tumors was observed.

Recent research related to hybrid QDs utilized for their diagnostic/imaging applicability in different types of cancers is summarized in **Table 1.**

Table 1. Summary of contemporary research carried out utilizing hybrid quantum dots for diagnostic/imaging applications in cancer.

Type of QDs	Type of Cancer	Diagnostic / Imaging Technique	Outcome	Refs.
Lactoferrin QDs	Breast cancer	Fluorescence imaging	Intracellular uptake of QDs showed fluorescence fluorescent due to mercaptopropionic acid-capped cadmium telluride and was successfully used as theranostic	**[89]** (P:2018)
Gelatin/ chondroitin QDs	Breast cancer	Fluorescence imaging	Matrix metalloproteinase layer enabled tracing their internalization into cancer cells and strong non-immunogenic response used as diagnostic	**[90]** (P:2018)
Magnetic graphene-QDs	Cancer cells	Electrochemical detection imaging	Images show high fluorescence in HeLa cells	**[91]** (P:2018)
Graphene-QDs	Cancer cells	MRI and fluorescence imaging	MRI and fluorescence imaging of living Hela cells and monitored intracellular drug release	**[92]** (P:2017)
Carbon-QDs	Tumor cells	Photoluminescence and photoacoustic imaging	Accumulation of C-QDs around the cancer cells via passive targeting with no active targeting species with fluorescence imaging	**[93]** (P:2018)
Carbon-QDs	Cervical cancer	Fluorescence imaging	TAT functionalization enhanced cell labeling and uptake, and that folate selectively tagged tumor cells	**[94]** (P:2013)
Carbon-QDs doped with Fluorine and Nitrogen	Squamous cell carcinoma	Near-infrared fluorescence (NIRF) and PET imaging	Carbon-QDs rapidly uptake by the tumor when administered subcutaneously as compared to intramuscular and intravenous	**[95]** (P:2013)

Type of QDs	Type of Cancer	Diagnostic / Imaging Technique	Outcome	Refs.
Carbon QDs doped with polyethy leneimine	Hepatocellular carcinoma	Bioimaging	Internalized QDs exhibit fluorescent emission authenticating their potential application for gene delivery and bioimaging	**[96]** (P:2012)
Magneton-fluorescence carbon-QDs conjugated with cDNAaptamer	Cervical cancer	Fluorescence and magnetic resonance (MR) imaging	DNA aptamer, which specifically recognizes the receptor tyrosine-protein kinase-like 7 (also known as colon carcinoma kinase 4, CCK4) for targeted dual mode fluorescence/ magnetic resonance (MR) imaging	**[97]** (P:2018)
QDs-conjugated streptavidinprobe	Breast cancer	Diagnosis	QDs-based immunohistochemistry demonstrates the prognostic value of EGFR area in the HER2-positive and lymph node-positive subtype of invasive breast cancer	**[98]** (P:2011)
Carboxyl-modified CdTe-QDs	HeLa and MCF-7 cells	Bioimaging	Sensing probes for cancer- biosensors was the detection of miRNA-21 on lysates of HeLa and MCF-7 cells and other biomarkers.	**[99]** (P:2022)
CdTe-QDs functionalized with single-stranded DNA	Non-specific cells	Fluorescence Diagnosis	QDs detect miRNA-122 within 40 min with enhanced intensity in proportion with miRNA-122 concentrations range 0.16–4.80 nM and has a low detection limit of 9.4 pM	**[100]** (P:2017)
Au-SiO_2-QDs	Breast cancer	Imaging	Photothermal effect provides real-time imaging capability, which makes it appealing as a potential theranostic tool for cancer treatment.	**[101]** (P:2018)

Type of QDs	Type of Cancer	Diagnostic / Imaging Technique	Outcome	Refs.
Graphene-QDs doped nitrogen	Skin cancer	Imaging and diagnosis	Fluorescence intensity of N-GQDs was quenched by the static quenching of UV-damaged DNA through the formation of an N-GQD/UV-damaged DNA complex	**[102]** (P:2022)
Iron selenide-QDs	Skin Cancer	Bioimaging	Synthesized QDs exhibit two bands of photon excitation property and high quantum yield which are suitable for second-window imaging	**[103]** (P:2019)

3.2. Therapeutic Application

Pei et al. formulated doxorubicin (DOX) loaded-fluorescent hyper-cross-linked-cyclodextrin–carbon quantum dot (CD-CQD) hybrid nanosponges (DOX-β-CD-CQD) with a size of around 300 nm with a DOX loading capacity of 39.5% through host-guest complexation [82]. This is because of the supramolecular complexation of DOX with the CD units in the CD-CQD nanosponges. The developed DOX-CD-CQD nanosponges demonstrated pH-responsive controlled release in the simulated tumor microenvironment. The loaded DOX molecules in the surface layer of the DOX-CD-CQD were released in the first 30 h, similar to the pH 7.4 medium. Due to the higher solubility of DOX in acidic media attributed to its protonation, the supramolecular complexation of DOX with β-CD units had a lower inclusion constant and a greater release ratio than in pH 7.4 conditions. Due to the high formation constant, they took longer to get out of the loaded DOX inner layer. Protonated DOX diffusion was prevented by hydrophobic DOX-complexed CD. After 12 h of DOX release, with an accumulative release of approximately 50%, hydrophilic outer shells formed, facilitating protonated DOX diffusion out of the theranostic system. After 24 h of incubation, the DOX concentration gradient climbed to 1.7 μg/mL with the DOX-β-CD-CQD theranostic system concentration of 20 μg/mL. Cell viability (29%) was comparable to free DOX at 10 μg/mL. In terms of antitumor efficacy, the DOX-CD-CQD outperformed free DOX. The DOX-β-CD-CQD had an IC_{50} of 5.00 μg/mL (equivalent to 0.425 μg/mL), compared to 2.26 μg/mL for free DOX. DOX-CD-CQD was internalized by HepG2 cells and accumulated in their nuclei, exhibiting better anticancer activity than the free drug [82]. In another investigation, Fateh et al. developed a hybrid nanostructure of cetyl alcohol-modified graphene quantum dots (CA-GQDs) and conjugated them with iron-oxide nanoparticles (IONP@CA-GQD) [44]. The study indicated no effect on the normal architecture of the liver and cardiac tissues after administration of these hybrid

QDs at a dose of 3 mg/kg for 7 days in mice. Hence, IONP@CA-GQD can be offered as a potential drug delivery system for cancer theranostics. Moreover, IONP@CA-GQD was found to be more toxic for the tumor cells as compared to normal cells in the study [39,83]. It is due to the hydrophobic nature of the carbon chains of cetyl alcohol and oleic acid in the middle of IONP@CA-GQD, hydrophobic drugs can be loaded in this space [83,104]. Furthermore, the magnetic properties of IONP@ CA-GQD would make cancer targeting feasible through this theranostic system. Similarly, Kim et al. examined QD-labeled hyaluronic acid (HA) derivatives for liver-targeted intracellular drug delivery. EDC activation of HA's carboxyl group and conjugation to ADH's amine group produced HA-ADH conjugates. After EDC and sufo-NHS activation of QD carboxyl groups, HA-ADH conjugates were tagged with QDs via amide bond formation. HA binds CD44 via its three carboxyl groups [88]. HA-QD conjugates were endocytosed via HA receptor-mediated endocytosis, as seen in the confocal microscopic images of B16F1 cells. HA receptors such as CD44 are significantly expressed in B16F1 cells. In the case of HEK293 cells without HA receptors, the cellular uptake of HA conjugates and QDs conjugates was noticeably reduced (as shown in **Figure 4**).

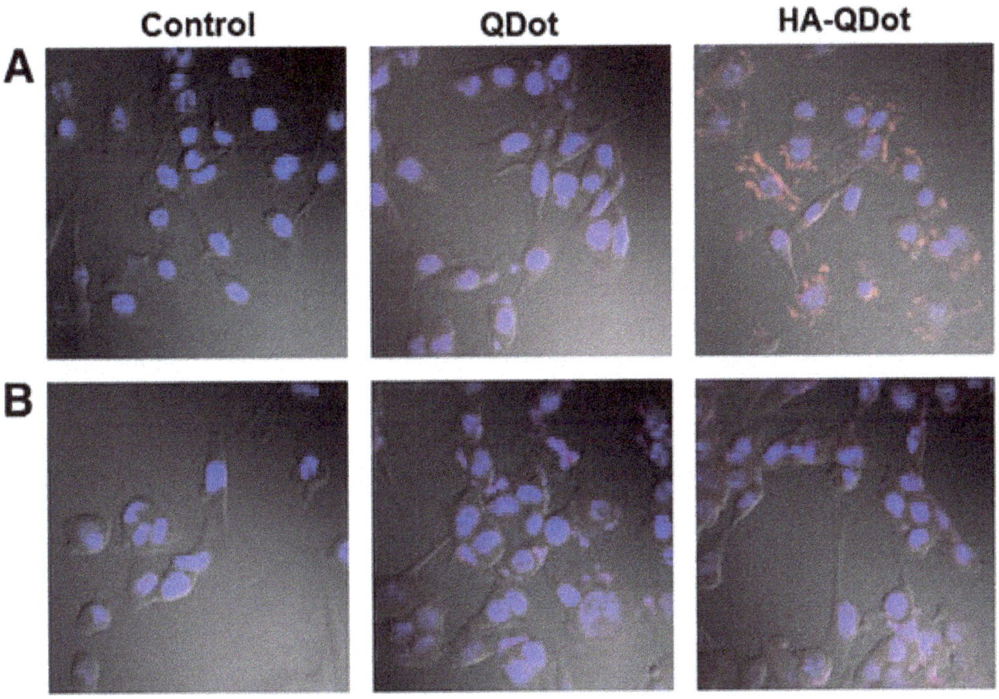

Figure 4. Illustration shows the cellular uptake of hyaluronic acid (HA) conjugated hybrid QDs after 2 h incubation in B16F1 cells—a high expression of HA receptors (A); HEK293 cells—without HA receptors (B). Confocal microscopic images reveal low cellular uptake of developed hybrid QDs system in HEK293 cells. Reproduced from Kim et al. [88], Elsevier, 2012.

In order to deliver targeted anticancer drugs, Chen et al. created a core-shell structured multifunctional nanocarrier system (ZnO-Au-PLA-GPPS-FA) consisting of ZnO-quantum dots-conjugated AuNPs as the core and folic acid (FA)-conjugated amphiphilic hyperbranched block copolymers as the shell. ZnO-quantum dots-conjugated AuNPs could be employed for photothermal therapy to kill tumor cells and fluorescent labeling, respectively. The outer hydrophilic block (GPPS-FA) and the inner hydrophobic block (PLA) were both biocompatible and biodegradable in in vivo system. The cancer cells may be targeted by an FA-conjugated multifunctional nanocarrier system, which may then be absorbed by the target cell through receptor-mediated endocytosis. Additionally, the presence of GPPS on the surface of multifunctional nanocarrier systems resulted in some of their anticancer effects [105]. In another investigation, Sung et al. made a targeted RBC-membrane-encased nanosponge (RBC-NS) that combines stealth and huge payloads of functional molecules to avoid the low EPR effect and the different types of tumors. This biocompatible, light-sensitive, carbon-based, porous particle looks like red blood cells (RBCs) and targets and gets into tumors well [39]. The targeted nanosponge, made of protein/RBC membranes (targeting and stealth properties), porous carbon/silica (hydrophobic, therapeutic agent transport), and graphene QDs (GQDs)/drug (photoresponsive, tumor-penetrating), were injected into a mouse model to deliver docetaxel (DTX) and GQDs to tumors (**Figure 5**). Moreover, Cetuximab (Ct), which can target tumors, is attached to the RBC layer to make it easier for particles to gather around tumors. The nanosponge delivers high amounts of GQD/DTX to the tumor as a photo-penetrative and photolytic agent. The Ct-RBC-GQD/NS-treated tumor can be heated to 68 °C for thermal tumor ablation. Ct-RBC-NS and Ct-NS elevated tumor temperatures to 62 and 53 °C, respectively. Irradiated saline-treated mice showed no temperature increase. Ct-RBC-GQD/stronger NS's improved photothermal conversion may be explained by the accumulation and photothermal combination effects. The localized heat of the NS releases GQD/DTX during NIR exposure, damaging the tumor and improving therapeutic drug penetration (as shown in **Figure 5**).

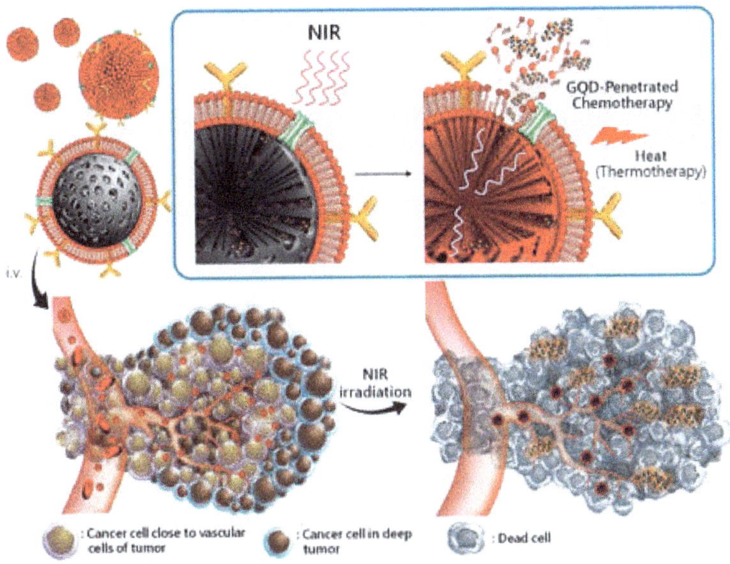

(A)

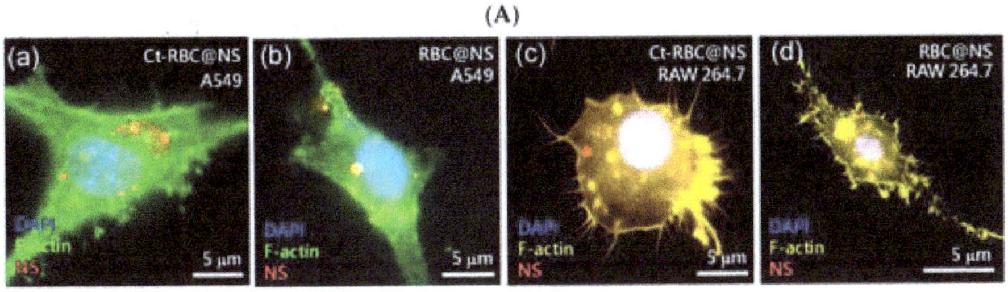

(B)

Figure 5. Schematic illustration highlights the penetration and accumulation of hybrid QDs (graphene quantum dots—GQDs) for RBC-membrane enveloped nanosponge-mediated targeted delivery in a tumor: (A) After the application of near-infrared (NIR) irradiation, generated heat leads to the penetration and accumulation of developed theranostic systems (GQDs with DTX) to deep tumors and the release of drug (DTX) into tumor cells ultimately causes cancer cell death. (B) Cellular uptake of RBC-membrane enveloped nanosponge with cetuximab (Ct-RBC@NS) and without cetuximab conjugation (RBC@NS) upon incubation for 2 h in A549 cancer cells and control RAW 264.7 cells. (a) Developed system conjugated with cetuximab (Ct-RBC@NS) in A549 cancer cells. It is monitored in the cytoplasm (green) and nuclei (blue). (b) Developed system without conjugation of cetuximab (RBC@NS) in A549 cancer cells. It is monitored in the cytoplasm (green) and nuclei (blue). (c) Developed system conjugated with cetuximab (Ct-RBC@NS) in control RAW 264.7 cells. (d) Developed system without conjugation of cetuximab (RBC@NS) in control RAW 264.7 cells. Reprinted with permission from Sung et al. [39], Copyright 2018, American Chemical Society.

Contemporary research related to hybrid QDs utilized to improve the therapeutic performance of loaded drugs for anticancer activity in different types of cancers are summarized in **Table 2.**

Table 2. Summary of contemporary research carried out utilizing hybrid quantum dots for therapeutic/drug delivery applications in cancer.

Type of QDs	Type of Cancer	In Vitro/In Vivo Model	Outcome	Refs.
Lactoferrin-QDs	Breast cancer	In vitro cancer cell line and in vivo tumor model	Enhanced cytotoxicity of breast cancer cells and in vivo antitumor efficacy	[89] (P:2018)
Gelatin/chondroitin-QDs	Breast cancer	In vitro cell line and in vivo model	Targeted internalization into cancer cells and enhanced cytotoxicity against breast cancer cells were demonstrated	[90] (P:2018)
Magnetic graphene-QDs	Cancerous cells	In vitro Hela cell line	G-QD susceptibility of cancerous HeLa cells to DOX is 13% higher and a promising material for cancer cell detection and targeted Dox	[91] (P:2018)
Graphene-QDs	Cancerous cells	In vitro Hela cell line	Cell viability study demonstrated the high cytotoxicity	[92] (P:2017)
Carbon-QDs doped withnitrogen and oxygen	Tumor cells	In vivo antitumor model	Nitrogen and oxygen co-doped C-QDs (N–O-CQDs) with strong absorbance in the NIR region leading to photothermal-based destruction of cancerous cells	[93] (P:2018)
Carbon QDs doped with polyethyleneimine	Hepatocellular carcinoma	In vitro COS-7 cells and HepG2 cells	Facilitate gene transfection in COS-7 and HepG2 cells with lower cytotoxicity	[96] (P:2012)
Magneton-fluorescence carbon-QDs conjugated with cDNA aptamer	Cervical cancer	In vitro cell line and In vivo tumor model	Targeted synergistic killing of lung cancer cells via PDT, PTT, and rapid release of DOX under simultaneous NIR laser irradiation	[97] (P:2018)
Au-SiO_2-QDs	Breast cancer	MCF-7 human breast cancer cells	A targeted synergistic anticancer effect that induced by DOX delivery and efficient heat generation by exploiting the photothermal effect of QDs-gold NPs.	[101] (P:2018)

Type of QDs	Type of Cancer	In Vitro/In Vivo Model	Outcome	Refs.
ZnO-QDs	Cancerous cells	Hela cells	Studies showed that cytotoxicity by both blank and drug-loaded QDs provided high anticancer activity against Hela cells with folate targeting	**[105]** (P:2018)
Graphene-QDs on the surface of hollow Cu_2S NPs	Breast cancer	MDA-MB-231 cells line	Flow cytometry showed a significant level of NIR-triggered Dox release inside MDA-MB-231 cells	**[106]** (P:2020)
Carbon-QDs with nuclear localization signal peptide	Lung cancer	Human lung carcinoma cells	Nucleus-targeted drug delivery of therapeutics functionalized with nuclear signal peptide to improve its antitumor activity	**[107]** (P:2016)
ZnO-QDs	Liver cancer	In vitro HepG2 cells	QDs significantly upregulated mRNA expressions, whereas the anti-apoptotic gene (Bcl-2) was down-regulated	**[108]** (P:2015)
CdSe-QDs	Hepatocellular carcinoma	In vitro HepG2 cancer cell	QDs successfully induced shrinkage and rupture of the membrane, and expression of an apoptotic gene (Bcl2) was positively comparing the untreated HepG2 cell line.	**[109]** (P:2021)
Fe_3O_4-Ag_2O QDs/ Cellulose fibers nanocomposites	Skin Cancer	In vitro cell line study	Magnetic QDs showed that the targeted cytotoxicity of the drug was increased when loaded on nanocomposites, compared to pure Fe_3O4-Ag_2O quantum dots/cellulose fibers nanocomposites	**[110]** (P:2017)
CdTe-QDs and CdSe-QDs	Melanoma tumors	In vivo antitumor model	Result indicated CdTe and CdSe QDs irradiation-induced photothermal therapy shared great potential in the treatment of cancer	**[111]** (P:2012)

Type of QDs	Type of Cancer	In Vitro/In Vivo Model	Outcome	Refs.
Graphene quantum dotmesoporous silica nanohybrids	Breast cancer	4T1 cancer cell line; 4T1 tumor in Balb/c mice	Results indicate that developed hybrid QDs as powerful cancer theranostic for deep tumor localization and regression	[112] (P:2021)
Peptide-based graphene QDs	Breast cancer	HUVEC Cell line; 4T1 tumor-bearing Balb/c mice	Successfully demonstrated multifunctional theranostic peptideticles for targeted drug delivery and tracking in αv integrin overexpressed tumor model	[113] (P:2022)
Tryptophan-sorbitol-based carbon QDs	Liver cancer	Huh7 cell line; Huh7 cells bearing Balb/c mice	Promising cancer nanotheranostic system utilized for diagnosis, targeting, and PDT therapy in hepatocellular carcinoma	[114] (P:2022)
Mn-doped ZnS QDs	Breast cancer	4T1 cancer cell line; 4T1 tumor in Balb/c mice	Theranostic system for image-guided therapy in breast tumor utilizing NIR-II fluorescence and magnetic resonance imaging	[115] (P:2022)

4. CONCISE SUMMARY

The review finds that hybrid quantum dots (QDs) have the potential to be customized for the treatment of many types of malignancies, and their therapeutic efficacy can be tracked in real-time. The quantum dots (QDs) were utilized in conjunction with several nanoparticulate systems, including polymeric, lipid, and inorganic nanoparticles, to create a theranostic system for cancer. The primary objective was to enhance the effectiveness of therapy in multidrug-resistant (MDR) cancer cases. Moreover, this analysis determined that there has been substantial research on carbon-based and graphene-based quantum dots (QDs) in order to combine them with other bio-molecules, with the aim of addressing the difficulties associated with traditional QDs. Multiple preclinical studies have demonstrated the successful utilization of hybrid quantum dots (QDs) as a theranostic system in cancer, hence advancing their potential for clinical investigation. Nevertheless, the literature assessment indicates that the therapeutic efficacy of hybrid quantum dots (QDs) as cancer theranostics has not been thoroughly investigated thus far. Moreover, future investigations should comprehensively address the safety concerns about the hybrid QDs in cancer. This is because there is a possibility that these QDs may accumulate in healthy tissues if they fail to specifically target tumors, which might potentially increase the risks of adverse outcomes. Despite the extensive in vivo studies conducted on the distribution, accumulation, excretion, and toxic effects of QDs, there is still no agreement or consensus among researchers. Furthermore, the intricate nature of in vivo models makes it challenging to replicate pharmacokinetics. Nevertheless, certain in-vitro investigations have facilitated our fundamental comprehension of the mechanics and potential negative consequences of different quantum dots (QDs). The specific category of quantum dots (QDs) significantly influences their spatial arrangement within cells and the rate at which they are eliminated, which directly affects their potential to cause harm to cells (cytotoxicity). The cytotoxic potential of QDs can be predicted based on their local accumulation and biological half-life. Therefore, it is crucial to prioritize the thorough examination of the safety and effectiveness of hybrid QDs.

At a Glance

Cancer is a prominent contributor to global mortality. Over the past few decades, significant progress has been made in cancer treatment. However, the issue of multidrug resistance (MDR) in cancer remains a major challenge with low rates of survival. Significant research is necessary to provide a precise diagnosis and successful treatment. The future of cancer research may involve the use of theranostic methods and tailored delivery of chemotherapy drugs using nanoparticle systems. In recent times, researchers have shown significant interest in carbon-based and graphene-based quantum dots. This is primarily because these quantum dots offer improved biocompatibility and are easier to modify with biological functions compared to traditional heavy metal quantum dots. Furthermore, these quantum dots possess a range of intriguing applications, such as bioimaging, biosensing, drug administration facilitated by quantum dots, and their involvement in photodynamic therapy (PDT) and photothermal therapy (PTT). The present review emphasized the practicality of hybrid quantum dots as a theranostic system in diverse types of cancer. It also examined the different hybrid quantum dots that have been studied for their diagnostic and therapeutic potential in cancer, namely those conjugated with distinct bio-molecules. The impact of linking various biomolecules, such as folic acid and PEG, to hybrid quantum dots on their biopharmaceutical properties, including solubility in water, ability to penetrate tumors, stability of loaded drugs in the tumor microenvironment, targeted drug delivery to tumor tissues, and therapeutic efficacy in different types of cancer, has also been examined.

REFERENCES

1. Kemp, J.A.; Kwon, Y.J. Cancer nanotechnology: Current status and perspectives. *Nano Converg.* **2021**, *8*, 34. **[Google Scholar] [CrossRef]**
2. Singh, R.; Deshmukh, R. Carbon nanotube as an emerging theranostic tool for oncology. *J. Drug Deliv. Sci. Technol.* **2022**, *74*, 103586. **[Google Scholar] [CrossRef]**
3. Siddique, S.; Chow, J.C. Recent advances in functionalized nanoparticles in cancer theranostics. *Nanomaterials* **2022**, *12*, 2826. **[Google Scholar] [CrossRef]**
4. Bernard, V.; Zobač, O.; Sopoušek, J.; Mornstein, V. AgCu bimetallic nanoparticles under effect of low intensity ultrasound: The cell viability study in vitro. *J. Cancer Res.* **2014**, *2014*, 971769. **[Google Scholar] [CrossRef] [Green Version]**
5. Zhao, Y.; Zhu, Y.; Fu, J.; Wang, L. Effective Cancer Cell Killing by Hydrophobic Nanovoid-Enhanced Cavitation under Safe Low-Energy Ultrasound. *Chem. Asian J.* **2014**, *9*, 790–796. **[Google Scholar] [CrossRef] [PubMed]**
6. Miller, D.L.; Smith, N.B.; Bailey, M.R.; Czarnota, G.J.; Hynynen, K.; Makin, I.R.S. Bioeffects Committee of the American Institute of Ultrasound in Medicine. (2012). Overview of therapeutic ultrasound applications and safety considerations. *J. Ultrasound Med.* **2012**, *31*, 623–634. **[Google Scholar] [CrossRef] [PubMed] [Green Version]**
7. Ashikbayeva, Z.; Aitkulov, A.; Atabaev, T.S.; Blanc, W.; Inglezakis, V.J.; Tosi, D. Green-Synthesized Silver Nanoparticle–Assisted Radiofrequency Ablation for Improved Thermal Treatment Distribution. *Nanomaterials* **2022**, *12*, 426. **[Google Scholar] [CrossRef]**
8. Glazer, E.S.; Curley, S.A. Non-invasive radiofrequency ablation of malignancies mediated by quantum dots, gold nanoparticles and carbon nanotubes. *Ther. Deliv.* **2011**, *2*, 1325–1330. **[Google Scholar] [CrossRef] [Green Version]**
9. Nguyen, D.T.; Tzou, W.S.; Zheng, L.; Barham, W.; Schuller, J.L.; Shillinglaw, B.; Sauer, W.H. Enhanced radiofrequency ablation with magnetically directed metallic nanoparticles. *Circ. Arrhythm. Electrophysiol.* **2016**, *9*, e003820. **[Google Scholar] [CrossRef]**
10. Ranoo, S.; Lahiri, B.B.; Nandy, M.; Philip, J. Enhanced magnetic heating efficiency at acidic pH for magnetic nanoemulsions stabilized with a weak polyelectrolyte. *J. Colloid Interface Sci.* **2020**, *579*, 582–597. **[Google Scholar] [CrossRef]**
11. Ranoo, S.; Lahiri, B.B.; Damodaran, S.P.; Philip, J. Tuning magnetic heating efficiency of colloidal dispersions of iron oxide nano-clusters by varying the surfactant concentration during solvothermal synthesis. *J. Mol. Liq.* **2022**, *360*, 119444. **[Google Scholar] [CrossRef]**

12. Dutz, S.; Buske, N.; Landers, J.; Gräfe, C.; Wende, H.; Clement, J.H. Biocompatible magnetic fluids of co-doped iron oxide nanoparticles with tunable magnetic properties. *Nanomaterials* **2020**, *10*, 1019. **[Google Scholar] [CrossRef]**
13. Du, Y.; Liu, X.; Liang, Q.; Liang, X.J.; Tian, J. Optimization and design of magnetic ferrite nanoparticles with uniform tumor distribution for highly sensitive MRI/MPI performance and improved magnetic hyperthermia therapy. *Nano Lett.* **2019**, *19*, 3618–3626. **[Google Scholar] [CrossRef] [PubMed]**
14. Dadfar, S.M.; Camozzi, D.; Darguzyte, M.; Roemhild, K.; Varvarà, P.; Metselaar, J.; Lammers, T. Size-isolation of superparamagnetic iron oxide nanoparticles improves MRI, MPI and hyperthermia performance. *J. Nanobiotechnol.* **2020**, *18*, 22. **[Google Scholar] [CrossRef]**
15. Sharma, P.; Brown, S.; Walter, G.; Santra, S.; Moudgil, B. Nanoparticles for bioimaging. *Adv. Colloid Interface Sci.* **2006**, *123*, 471–485. **[Google Scholar] [CrossRef]**
16. Couvreur, P. Nanoparticles in drug delivery: Past, present and future. *Adv. Drug Deliv. Rev.* **2013**, *65*, 21–23. **[Google Scholar] [CrossRef]**
17. Lucky, S.S.; Soo, K.C.; Zhang, Y. Nanoparticles in photodynamic therapy. *Chem. Rev.* **2015**, *115*, 1990–2042. **[Google Scholar] [CrossRef]**
18. Jaque, D.; Maestro, L.M.; del Rosal, B.; Haro-Gonzalez, P.; Benayas, A.; Plaza, J.L.; Martin Rodriguez, E.; Solé, J.G. Nanoparticles for photothermal therapies. *Nanoscale* **2014**, *6*, 9494–9530. **[Google Scholar] [CrossRef]**
19. Calabrese, G.; Petralia, S.; Franco, D.; Nocito, G.; Fabbi, C.; Forte, L.; Guglielmino, S.; Squarzoni, S.; Traina, F.; Conoci, S. A new Ag-nanostructured hydroxyapatite porous scaffold: Antibacterial effect and cytotoxicity study. *Mater. Sci. Eng. C* **2021**, *118*, 111394. **[Google Scholar] [CrossRef] [PubMed]**
20. Mele, E. Introduction: Smart Materials in Biomedicine. In *Smart Nanoparticles for Biomedicine*; Elsevier: Amsterdam, The Netherlands, 2008; pp. 1–13. **[Google Scholar]**
21. Dreaden, E.C.; Alkilany, A.M.; Huang, X.; Murphy, C.J.; El-Sayed, M.A. The golden age: Gold nanoparticles for biomedicine. *Chem. Soc. Rev.* **2012**, *41*, 2740–2779. **[Google Scholar] [CrossRef]**
22. Eckhardt, S.; Brunetto, P.S.; Gagnon, J.; Priebe, M.; Giese, B.; Fromm, K.M. Nanobio Silver: Its Interactions with Peptides and Bacteria, and Its Uses in Medicine. *Chem. Rev.* **2013**, *113*, 4708–4754. **[Google Scholar] [CrossRef] [Green Version]**
23. Jeyaraj, M.; Gurunathan, S.; Qasim, M.; Kang, M.H.; Kim, J.H. A Comprehensive Review on the Synthesis, Characterization, and Biomedical Application of Platinum Nanoparticles. *Nanomaterials* **2019**, *9*, 1719. **[Google Scholar] [CrossRef] [PubMed] [Green Version]**

24. Phan, T.T.V.; Huynh, T.C.; Manivasagan, P.; Mondal, S.; Oh, J. An Up-To-Date Review on Biomedical Applications of Palladium Nanoparticles. *Nanomaterials* **2020**, *10*, 66. **[Google Scholar] [CrossRef] [Green Version]**
25. Jain, A.; Tiwari, A.; Verma, A.; Saraf, S.; Jain, S.K. Combination cancer therapy using multifunctional liposomes. *Crit. Rev.™ Ther. Drug Carr. Syst.* **2020**, *37*, 105–134. **[Google Scholar] [CrossRef] [PubMed]**
26. Abdellatif, A.A.; Younis, M.A.; Alsharidah, M.; Al Rugaie, O.; Tawfeek, H.M. Biomedical applications of quantum dots: Overview, challenges, and clinical potential. *Int. J. Nanomed.* **2022**, *17*, 1951–1970. **[Google Scholar] [CrossRef]**
27. Wagner, A.M.; Knipe, J.M.; Orive, G.; Peppas, N.A. Quantum dots in biomedical applications. *Acta Biomater.* **2019**, *94*, 44–63. **[Google Scholar] [CrossRef]**
28. Bajwa, N.; Mehra, N.K.; Jain, K.; Jain, N.K. Pharmaceutical and biomedical applications of quantum dots. *Artif. Cells Nanomed. Biotechnol.* **2016**, *44*, 758–768. **[Google Scholar] [CrossRef]**
29. Liu, M.; Yazdani, N.; Yarema, M.; Jansen, M.; Wood, V.; Sargent, E.H. Colloidal quantum dot electronics. *Nat. Electron.* **2021**, *4*, 548–558. **[Google Scholar] [CrossRef]**
30. Smith, A.M.; Duan, H.; Mohs, A.M.; Nie, S. Bioconjugated quantum dots for in vivo molecular and cellular imaging. *Adv. Drug Deliv. Rev.* **2008**, *60*, 1226–1240. **[Google Scholar] [CrossRef] [Green Version]**
31. Xu, Q.; Gao, J.; Wang, S.; Wang, Y.; Liu, D.; Wang, J. Quantum dots in cell imaging and their safety issues. *J. Mater. Chem. B* **2021**, *9*, 5765–5779. **[Google Scholar] [CrossRef]**
32. Devi, S.; Kumar, M.; Tiwari, A.; Tiwari, V.; Kaushik, D.; Verma, R.; Batiha, G.E.S. Quantum dots: An emerging approach for cancer therapy. *Front. Mater.* **2022**, *8*, 585. **[Google Scholar] [CrossRef]**
33. Rahman, M.A.; Abul Barkat, H.; Harwansh, R.K.; Deshmukh, R. Carbon-based Nanomaterials: Carbon Nanotubes, Graphene, and Fullerenes for the Control of Burn Infections and Wound Healing. *Curr. Pharm. Biotechnol.* **2022**, *23*, 1483–1496. **[Google Scholar] [CrossRef] [PubMed]**
34. Kanthi Gudimella, K.; Gedda, G.; Kumar, P.S.; Babu, B.K.; Yamajala, B.; Rao, B.V.; Singh, P.P.; Kumar, D.; Sharma, A. Novel synthesis of fluorescent carbon dots from bio-based Carica Papaya Leaves: Optical and structural properties with antioxidant and anti-inflammatory activities. *Environ. Res.* **2022**, *204*, 111854. **[Google Scholar] [CrossRef]**
35. Zhang, H.; Jin, Y.; Chi, C.; Han, G.; Jiang, W.; Wang, Z.; Cheng, H.; Zhang, C.; Wang, G.; Sun, C.; et al. Sponge particulates for biomedical applications:

Biofunctionalization, multi-drug shielding, and theranostic applications. *Biomaterials* **2021**, *273*, 120824. **[Google Scholar]** **[CrossRef]**

36. Ghaffarkhah, A.; Hosseini, E.; Kamkar, M.; Sehat, A.A.; Dordanihaghighi, S.; Allahbakhsh, A.; van der Kuur, C.I.; Arjmand, M. Synthesis, applications, and prospects of graphene quantum dots: A comprehensive review. *Small* **2022**, *18*, 2102683. **[Google Scholar]** **[CrossRef]**
37. Paduraru, D.N.; Niculescu, A.G.; Bolocan, A.; Andronic, O.; Grumezescu, A.M.; Bîrlă, R. An Updated overview of cyclodextrin-based drug delivery systems for cancer therapy. *Pharmaceutics* **2022**, *14*, 1748. **[Google Scholar]** **[CrossRef]** **[PubMed]**
38. Sung, S.Y.; Su, Y.L.; Cheng, W.; Hu, P.F.; Chiang, C.S.; Chen, W.T.; Hu, S.H. Graphene quantum dots-mediated theranostic penetrative delivery of drug and photolytics in deep tumors by targeted biomimetic nanosponges. *Nano Lett.* **2019**, *19*, 69–81. **[Google Scholar]** **[CrossRef]**
39. Yang, K.; Wan, J.; Zhang, S.; Zhang, Y.; Lee, S.-T.; Liu, Z. In Vivo Pharmacokinetics, Long-Term Biodistribution, and Toxicology of PEGylated Graphene in Mice. *ACS Nano* **2011**, *5*, 516–522. **[Google Scholar]** **[CrossRef]** **[PubMed]**
40. Zhang, S.; Yang, K.; Feng, L.; Liu, Z. In vitro and in vivo behaviors of dextran functionalized graphene. *Carbon* **2011**, *49*, 4040–4049. **[Google Scholar]** **[CrossRef]**
41. Lv, C.; Lin, Y.; Liu, A.A.; Hong, Z.Y.; Wen, L.; Zhang, Z.; Pang, D.W. Labeling viral envelope lipids with quantum dots by harnessing the biotinylated lipid-self-inserted cellular membrane. *Biomaterials* **2016**, *106*, 69–77. **[Google Scholar]** **[CrossRef]**
42. Hashemkhani, M.; Muti, A.; Sennaroğlu, A.; Acar, H.Y. Multimodal image-guided folic acid targeted Ag-based quantum dots for the combination of selective methotrexate delivery and photothermal therapy. *J. Photochem. Photobiol. B Biol.* **2020**, *213*, 112082. **[Google Scholar]** **[CrossRef]** **[PubMed]**
43. Singh, G.; Kumar, M.; Soni, U.; Arora, V.; Bansal, V.; Gupta, D.; Singh, H. Cancer cell targeting using folic acid/anti-HER2 antibody conjugated fluorescent CdSe/CdS/ZnS-Mercaptopropionic acid and CdTe-Mercaptosuccinic acid quantum dots. *J. Nanosci. Nanotechnol.* **2016**, *16*, 130–143. **[Google Scholar]** **[CrossRef]** **[PubMed]**
44. Sahoo, S.L.; Liu, C.H.; Kumari, M.; Wu, W.C.; Wang, C.C. Biocompatible quantum dot-antibody conjugate for cell imaging, targeting and fluorometric immunoassay: Crosslinking, characterization and applications. *RSC Adv.* **2019**, *9*, 32791–32803. **[Google Scholar]** **[CrossRef]** **[Green Version]**

45. Huang, H.; Bai, Y.L.; Yang, K.; Tang, H.; Wang, Y.W. Optical imaging of head and neck squamous cell carcinoma in vivo using arginine-glycine-aspartic acid peptide conjugated near-infrared quantum dots. *OncoTargets Ther.* **2013**, *6*, 1779–1787. **[Google Scholar]**

46. Lu, J.; Liong, M.; Li, Z.; Zink, J.I.; Tamanoi, F. Biocompatibility, biodistribution, and drug-delivery efficiency of mesoporous silica nanoparticles for cancer therapy in animals. *Small* **2010**, *6*, 1794–1805. **[Google Scholar] [CrossRef] [PubMed] [Green Version]**

47. Li, C.; Zou, Z.; Liu, H.; Jin, Y.; Li, G.; Yuan, C.; Jin, M. Synthesis of polystyrene-based fluorescent quantum dots nanolabel and its performance in H5N1 virus and SARS-CoV-2 antibody sensing. *Talanta* **2021**, *225*, 122064. **[Google Scholar] [CrossRef]**

48. Park, J.H.; Gu, L.; Von Maltzahn, G.; Ruoslahti, E.; Bhatia, S.N.; Sailor, M.J. Biodegradable luminescent porous silicon nanoparticles for in vivo applications. *Nat. Mater.* **2009**, *8*, 331–336. **[Google Scholar] [CrossRef]**

49. Zayed, D.G.; AbdElhamid, A.S.; Freag, M.S.; Elzoghby, A.O. Hybrid quantum dot-based theranostic nanomedicines for tumor-targeted drug delivery and cancer imaging. *Nanomedicine* **2019**, *14*, 225–228. **[Google Scholar] [CrossRef] [PubMed] [Green Version]**

50. Liu, W.; Li, C.; Ren, Y.; Sun, X.; Pan, W.; Li, Y.; Wang, J.; Wang, W. Carbon dots: Surface engineering and applications. *J. Mater. Chem. B* **2016**, *4*, 5772–5788. **[Google Scholar] [CrossRef] [PubMed]**

51. Li, B.; Lin, L.; Lin, H.; Wilson, B.C. Photosensitized singlet oxygen generation and detection: Recent advances and future perspectives in cancer photodynamic therapy. *J. Biophotonics* **2016**, *9*, 1314–1325. **[Google Scholar] [CrossRef] [PubMed] [Green Version]**

52. Das, R.K.; Panda, S.; Bhol, C.S.; Bhutia, S.K.; Mohapatra, S. N-doped carbon quantum dot (NCQD)-Deposited carbon capsules for synergistic fluorescence imaging and photothermal therapy of oral cancer. *Langmuir* **2019**, *35*, 15320–15329. **[Google Scholar] [CrossRef] [PubMed]**

53. Sekar, R.; Basavegowda, N.; Jena, S.; Jayakodi, S.; Elumalai, P.; Chaitanyakumar, A.; Baek, K.H. Recent Developments in Heteroatom/Metal-Doped Carbon Dot-Based Image-Guided Photodynamic Therapy for Cancer. *Pharmaceutics* **2022**, *14*, 1869. **[Google Scholar] [CrossRef]**

54. Proskurnin, M.A.; Khabibullin, V.R.; Usoltseva, L.O.; Vyrko, E.A.; Mikheev, I.V.; Volkov, D.S. Photothermal and optoacoustic spectroscopy: State of the art and prospects. *Phys.-Uspekhi* **2022**, *65*, 270. **[Google Scholar] [CrossRef]**

55. Walter, M.; Schubert, L.; Heberle, J.; Schlesinger, R.; Losi, A. Time-resolved photoacoustics of channelrhodopsins: Early energetics and light-driven volume changes. *Photochem. Photobiol. Sci.* **2022**, 1–10. **[Google Scholar] [CrossRef]**
56. He, Z.; Zhang, C.Y.; Lei, Y.; Song, G.; Yao, Y. Plasmonic nanomaterials: A versatile phototheranostic platform of cancers. *Materials Today* **2022. [Google Scholar] [CrossRef]**
57. Gellini, C.; Feis, A. Optothermal properties of plasmonic inorganic nanoparticles for photoacoustic applications. *Photoacoustics* **2021**, *23*, 100281. **[Google Scholar] [CrossRef]**
58. Song, Z.; Quan, F.; Xu, Y.; Liu, M.; Cui, L.; Liu, J. Multifunctional N, S co-doped carbon quantum dots with pH-and thermo-dependent switchable fluorescent properties and highly selective detection of glutathione. *Carbon* **2016**, *104*, 169–178. **[Google Scholar] [CrossRef]**
59. Liu, B.; Wei, S.; Liu, E.; Zhang, H.; Lu, P.; Wang, J.; Sun, G. Nitrogen-doped carbon dots as a fluorescent probe for folic acid detection and live cell imaging. *Spectrochim. Acta Part A Mol. Biomol. Spectrosc.* **2022**, *268*, 120661. **[Google Scholar] [CrossRef]**
60. Lee, C.H.; Rajendran, R.; Jeong, M.S.; Ko, H.Y.; Joo, J.Y.; Cho, S.; Chang, Y.W.; Kim, S. Bioimaging of targeting cancers using aptamer-conjugated carbon nanodots. *Chem. Commun.* **2013**, *49*, 6543–6545. **[Google Scholar] [CrossRef] [PubMed]**
61. Bacon, M.; Bradley, S.J.; Nann, T. Graphene quantum dots. *Part. Part. Syst. Charact.* **2014**, *31*, 415–428. **[Google Scholar] [CrossRef]**
62. Liu, H.; Li, C.; Qian, Y.; Hu, L.; Fang, J.; Tong, W.; Wang, H. Magnetic-induced graphene quantum dots for imaging-guided photothermal therapy in the second near-infrared window. *Biomaterials* **2020**, *232*, 119700. **[Google Scholar] [CrossRef]**
63. Sun, B.; Luo, C.; Yu, H.; Zhang, X.; Chen, Q.; Yang, W.; Wang, M.; Kan, Q.; Zhang, H.; Wang, Y.; et al. Disulfide bond-driven oxidation-and reduction-responsive prodrug nanoassemblies for cancer therapy. *Nano Lett.* **2018**, *18*, 3643–3650. **[Google Scholar] [CrossRef]**
64. Zhu, L.; Zhao, H.; Zhou, Z.; Xia, Y.; Wang, Z.; Ran, H.; Li, P.; Ren, J. Peptide-functionalized phase-transformation nanoparticles for low intensity focused ultrasound-assisted tumor imaging and therapy. *Nano Lett.* **2018**, *18*, 1831–1841. **[Google Scholar] [CrossRef] [PubMed]**
65. Chen, H.; Zhang, W.; Zhu, G.; Xie, J.; Chen, X. Rethinking cancer nanotheranostics. *Nat. Rev. Mater.* **2017**, *2*, 17024. **[Google Scholar] [CrossRef] [PubMed] [Green Version]**

66. Avasthi, A.; Caro, C.; Pozo-Torres, E.; Leal, M.P.; García-Martín, M.L. Magnetic nanoparticles as MRI contrast agents. *Top. Curr. Chem.* **2020**, *378*, 40. **[Google Scholar] [CrossRef]**

67. Pourmadadi, M.; Rahmani, E.; Shamsabadipour, A.; Mahtabian, S.; Ahmadi, M.; Rahdar, A.; Díez-Pascual, A.M. Role of Iron Oxide (Fe_2O_3) Nanocomposites in Advanced Biomedical Applications: A State-of-the-Art Review. *Nanomaterials* **2022**, *12*, 3873. **[Google Scholar] [CrossRef] [PubMed]**

68. Jana, P.; Dev, A. Carbon quantum dots: A promising nanocarrier for bioimaging and drug delivery in cancer. *Mater. Today Commun.* **2022**, *32*, 104068. **[Google Scholar] [CrossRef]**

69. Sasidharan, S.; Bahadur, D.; Srivastava, R. Protein-poly (amino acid) nanocore–shell mediated synthesis of branched gold nanostructures for computed tomographic imaging and photothermal therapy of cancer. *ACS Appl. Mater. Interfaces* **2016**, *8*, 15889–15903. **[Google Scholar] [CrossRef]**

70. Zhu, Y.; Murali, S.; Cai, W.; Li, X.; Suk, J.W.; Potts, J.R.; Ruoff, R.S. Graphene and graphene oxide: Synthesis, properties, and applications. *Adv. Mater.* **2010**, *22*, 3906–3924. **[Google Scholar] [CrossRef] [PubMed]**

71. Navya, P.N.; Kaphle, A.; Srinivas, S.P.; Bhargava, S.K.; Rotello, V.M.; Daima, H.K. Current trends and challenges in cancer management and therapy using designer nanomaterials. *Nano Converg.* **2019**, *6*, 23. **[Google Scholar] [CrossRef] [Green Version]**

72. Luo, G.; Long, J.; Zhang, B.; Liu, C.; Ji, S.; Xu, J.; Ni, Q. Quantum dots in cancer therapy. *Expert Opin. Drug Deliv.* **2012**, *9*, 47–58. **[Google Scholar] [CrossRef] [PubMed]**

73. Kumar, A.; Singh, K.R.; Ghate, M.D.; Lalhlenmawia, H.; Kumar, D.; Singh, J. Bioinspired quantum dots for cancer therapy: A mini-review. *Mater. Lett.* **2022**, *313*, 131742. **[Google Scholar] [CrossRef]**

74. Zhang, H.; Yee, D.; Wang, C. Quantum dots for cancer diagnosis and therapy: Biological and clinical perspectives. *Nanomedicine* **2008**, *3*, 83–91. **[Google Scholar] [CrossRef] [PubMed]**

75. Taghavi, S.; Abnous, K.; Taghdisi, S.M.; Ramezani, M.; Alibolandi, M. Hybrid carbon-based materials for gene delivery in cancer therapy. *J. Control Release* **2020**, *318*, 158–175. **[Google Scholar] [CrossRef]**

76. Robertson, C.A.; Evans, D.H.; Abrahamse, H. Photodynamic therapy (PDT): A short review on cellular mechanisms and cancer research applications for PDT. *J. Photochem. Photobiol. B Biol.* **2009**, *96*, 1–8. **[Google Scholar] [CrossRef]**

77. Thomsen, H.; Marino, N.; Conoci, S.; Sortino, S.; Ericson, M.B. Confined photo-release of nitric oxide with simultaneous two-photon fluorescence tracking in a cellular system. *Sci. Rep.* **2018**, *8*, 9753. **[Google Scholar] [CrossRef] [Green Version]**

78. Zhao, X.; Wei, Z.; Zhao, Z.; Miao, Y.; Qiu, Y.; Yang, W.; Jia, X.; Liu, Z.; Hou, H. Design and development of graphene oxide nanoparticle/chitosan hybrids showing pH-sensitive surface charge-reversible ability for efficient intracellular doxorubicin delivery. *ACS Appl. Mater. Interfaces* **2018**, *10*, 6608–6617. **[Google Scholar] [CrossRef]**

79. Zhuang, W.; He, L.; Wang, K.; Ma, B.; Ge, L.; Wang, Z.; Huang, J.; Wu, J.; Zhang, Q.; Ying, H. Combined adsorption and covalent linking of paclitaxel on functionalized nano-graphene oxide for inhibiting cancer cells. *ACS Omega* **2018**, *3*, 2396–2405. **[Google Scholar] [CrossRef]**

80. Zhao, C.; Song, X.; Liu, Y.; Fu, Y.; Ye, L.; Wang, N.; Wang, F.; Li, L.; Mohammadniaei, M.; Zhang, M.; et al. Synthesis of graphene quantum dots and their applications in drug delivery. *J. Nanobiotechnol.* **2020**, *18*, 1–32. **[Google Scholar] [CrossRef]**

81. Pei, M.; Pai, J.Y.; Du, P.; Liu, P. Facile synthesis of fluorescent hyper-cross-linked β-cyclodextrin-carbon quantum dot hybrid nanosponges for tumor theranostic application with enhanced antitumor efficacy. *Mol. Pharm.* **2018**, *15*, 4084–4091. **[Google Scholar] [CrossRef]**

82. Fateh, S.T.; Kamalabadi, M.A.; Aliakbarniya, A.; Jafarinejad-Farsangi, S.; Koohi, M.; Jafari, E.; Karam, Z.M.; Keyhanfar, F.; Dezfuli, A.S. Hydrophobic@amphiphilic hybrid nanostructure of iron-oxide and graphene quantum dot surfactant as a theranostic platform. *OpenNano* **2022**, *6*, 100037. **[Google Scholar] [CrossRef]**

83. Schroeder, K.L.; Goreham, R.V.; Nann, T. Graphene quantum dots for theranostics and bioimaging. *Pharm. Res.* **2016**, *33*, 2337–2357. **[Google Scholar] [CrossRef]**

84. Dezfuli, A.S.; Kohan, E.; Fateh, S.T.; Alimirzaei, N.; Arzaghi, H.; Hamblin, M.R. Organic dots (O-dots) for theranostic applications: Preparation and surface engineering. *RSC Adv.* **2021**, *11*, 2253–2291. **[Google Scholar] [CrossRef]**

85. Kumawat, M.K.; Thakur, M.; Bahadur, R.; Kaku, T.; Prabhuraj, R.S.; Ninawe, A.; Srivastava, R. Preparation of gra-phene oxide-graphene quantum dots hybrid and its application in cancer theranostics. *Mater. Sci. Eng. C* **2019**, *103*, 109774. **[Google Scholar] [CrossRef] [PubMed]**

86. Kim, K.S.; Hur, W.; Park, S.J.; Hong, S.W.; Choi, J.E.; Goh, E.J.; Yoon, S.K.; Hahn, S.K. Bioimaging for targeted deliv-ery of hyaluronic acid derivatives to the livers in cirrhotic mice using quantum dots. *ACS Nano* **2010**, *4*, 3005–3014. **[Google Scholar] [CrossRef]**

87. Kim, K.S.; Kim, S.; Beack, S.; Yang, J.A.; Yun, S.H.; Hahn, S.K. In vivo real-time confocal microscopy for tar-get-specific delivery of hyaluronic acid-quantum dot conjugates. *Nanomed. Nanotechnol. Biol. Med.* **2012**, *8*, 1070–1073. **[Google Scholar] [CrossRef]**

88. AbdElhamid, A.S.; Helmy, M.W.; Ebrahim, S.M.; Bahey-El-Din, M.; Zayed, D.G.; Zein El Dein, E.A.; El-Gizawy, S.A.; El-zoghby, A.O. Layer-by-layer gelatin/chondroitin quantum dots-based nanotheranostics: Combined rapamy-cin/celecoxib delivery and cancer imaging. *Nanomedicine* **2018**, *13*, 1707–1730. **[Google Scholar] [CrossRef] [PubMed]**

89. AbdElhamid, A.S.; Zayed, D.G.; Helmy, M.W.; Ebrahim, S.M.; Bahey-El-Din, M.; Zein-El-Dein, E.A.; El-Gizawy, S.A.; El-zoghby, A.O. Lactoferrin-tagged quantum dots-based theranostic nanocapsules for combined COX-2 inhibi-tor/herbal therapy of breast cancer. *Nanomedicine* **2018**, *13*, 2637–2656. **[Google Scholar] [CrossRef] [PubMed]**

90. Chowdhury, A.D.; Ganganboina, A.B.; Tsai, Y.C.; Chiu, H.C.; Doong, R.A. Multifunctional GQDs-Concanavalin A@ Fe_3O_4 nanocomposites for cancer cells detection and targeted drug delivery. *Anal. Chim. Acta* **2018**, *1027*, 109–120. **[Google Scholar] [CrossRef]**

91. Su, X.; Chan, C.; Shi, J.; Tsang, M.K.; Pan, Y.; Cheng, C.; Gerile, O.; Yang, M. A graphene quantum dot@ Fe_3O_4@ SiO_2 based nanoprobe for drug delivery sensing and dual-modal fluorescence and MRI imaging in cancer cells. *Biosens. Bi-Oelectron.* **2017**, *92*, 489–495. **[Google Scholar] [CrossRef]**

92. Bao, X.; Yuan, Y.; Chen, J.; Zhang, B.; Li, D.; Zhou, D.; Jing, P.; Xu, G.; Wang, Y.; Holá, K.; et al. In vivo theranostics with near-infrared-emitting carbon dots-Highly efficient photothermal therapy based on passive targeting after intravenous administration. *Light Sci. Appl.* **2018**, *7*, 91. **[Google Scholar] [CrossRef] [PubMed]**

93. Bhunia, S.K.; Saha, A.; Maity, A.R.; Ray, S.C.; Jana, N.R. Carbon nanoparticle-based fluorescent bioimaging probes. *Sci. Rep.* **2013**, *3*, 1473. **[Google Scholar] [CrossRef] [Green Version]**

94. Huang, X.; Zhang, F.; Zhu, L.; Choi, K.Y.; Guo, N.; Guo, J.; Tackett, K.; Anilkumar, P.; Liu, G.; Quan, Q.; et al. Effect of injection routes on the biodistribution, clearance, and tumor uptake of carbon dots. *ACS Nano* **2013**, *7*, 5684–5693. **[Google Scholar] [CrossRef] [Green Version]**

95. Liu, C.; Zhang, P.; Zhai, X.; Tian, F.; Li, W.; Yang, J.; Liu, Y.; Wang, H.; Wang, W.; Liu, W. Nano-carrier for gene de-livery and bioimaging based on carbon dots with PEI-passivation enhanced fluorescence. *Biomaterials* **2012**, *33*, 3604–3613. **[Google Scholar] [CrossRef]**

96. Zhang,M.;Wang,W.;Cui,Y.;Chu,X.;Sun,B.;Zhou,N.;Shen,J.Magnetofluorescent Fe3O4/carbon quantum dots coated single-walled carbon nanotubes as dual-modal targeted imaging and chemo/photodynamic/photothermal tri-ple-modal therapeutic agents. *Chem. Eng. J.* **2018**, *338*, 526–538. **[Google Scholar] [CrossRef]**

97. Yang, X.Q.; Chen, C.; Peng, C.W.; Hou, J.X.; Liu, S.P.; Qi, C.B.; Gong, Y.P.; Zhu, X.B.; Pang, D.W.; Li, Y. Quantum dot-based quantitative immunofluorescence detection and spectrum analysis of epidermal growth factor receptor in breast cancer tissue arrays. *Int. J. Nanomed.* **2011**, *6*, 2265. **[Google Scholar]**

98. Martins, C.S.; LaGrow, A.P.; Prior, J.A. Quantum Dots for Cancer-Related miRNA Monitoring. *ACS Sens.* **2022**, *7*, 1269–1299. **[Google Scholar] [CrossRef]**

99. Lv, S.; Chen, F.; Chen, C.; Chen, X.; Gong, H.; Cai, C. A novel CdTe quantum dots probe amplified resonance light scattering signals to detect microRNA-122. *Talanta* **2017**, *165*, 659–663. **[Google Scholar] [CrossRef] [PubMed]**

100. Volsi, A.L.; Fiorica, C.; D'Amico, M.; Scialabba, C.; Palumbo, F.S.; Giammona, G.; Licciardi, M. Hybrid Gold/Silica/Quantum-Dots supramolecular-nanostructures encapsulated in polymeric micelles as potential theranostic tool for targeted cancer therapy. *Eur. Polym. J.* **2018**, *105*, 38–47. **[Google Scholar] [CrossRef]**

101. Jabeen, G.; Ahmad, M.H.; Aslam, M.; Riaz, S.; Hayat, A.; Nawaz, M.H. N-Doped graphene quantum dots (N-GQDs) as fluorescent probes for detection of UV induced DNA damage. *RSC Adv.* **2022**, *12*, 22458–22464. **[Google Scholar] [CrossRef]**

102. Kwon, J.; Jun, S.W.; Choi, S.I.; Mao, X.; Kim, J.; Koh, E.K.; Kim, Y.H.; Kim, S.K.; Hwang, D.Y.; Kim, C.S.; et al. FeSe quantum dots for in vivo multiphoton biomedical imaging. *Sci. Adv.* **2019**, *5*, eaay0044. **[Google Scholar] [CrossRef] [PubMed] [Green Version]**

103. Hsiao, M.H.; Mu, Q.; Stephen, Z.R.; Fang, C.; Zhang, M. Hexanoyl-chitosan-PEG copolymer coated iron oxide nano-particles for hydrophobic drug delivery. *ACS Macro Lett.* **2015**, *4*, 403–407. **[Google Scholar] [CrossRef]**

104. Chen, T.; Zhao, T.; Wei, D.; Wei, Y.; Li, Y.; Zhang, H. Core–shell nanocarriers with ZnO quantum dots-conjugated Au nanoparticle for tumor-targeted drug delivery. *Carbohydr. Polym.* **2013**, *92*, 1124–1132. **[Google Scholar] [CrossRef] [PubMed]**

105. Zheng, S.; Jin, Z.; Han, C.; Li, J.; Xu, H.; Park, S.; Park, J.O.; Choi, E.; Xu, K. Graphene quantum dots-decorated hol-low copper sulfide nanoparticles for controlled intracellular drug release and enhanced photothermal-chemotherapy. *J. Mater. Sci.* **2020**, *55*, 1184–1197. **[Google Scholar] [CrossRef]**

106. Yang, L.; Wang, Z.; Wang, J.; Jiang, W.; Jiang, X.; Bai, Z.; He, Y.; Jiang, J.; Wang, D.; Yang, L. Doxorubicin conjugated functionalizable carbon dots for nucleus targeted delivery and enhanced therapeutic efficacy. *Nanoscale* **2016**, *8*, 6801–6809. **[Google Scholar] [CrossRef] [Green Version]**

107. Ahmad, J.; Wahab, R.; Siddiqui, M.A.; Musarrat, J.; Al-Khedhairy, A.A. Zinc oxide quantum dots: A potential candi-date to detain liver cancer cells. *Bioprocess Biosyst. Eng.* **2015**, *38*, 155–163. **[Google Scholar] [CrossRef] [PubMed]**

108. Rahman, M.M.; Opo, F.A.; Asiri, A.M. Cytotoxicity Study of Cadmium-Selenium Quantum Dots (Cdse QDs) for De-stroying the Human HepG2 Liver Cancer Cell. *J. Biomed. Nanotechnol.* **2021**, *17*, 2153–2164. **[Google Scholar] [CrossRef]**

109. Fakhri, A.; Tahami, S.; Nejad, P.A. Preparation and characterization of Fe3O4-Ag2O quantum dots decorated cellu-lose nanofibers as a carrier of anticancer drugs for skin cancer. *J. Photochem. Photobiol. B Biol.* **2017**, *175*, 83–88. **[Google Scholar] [CrossRef] [PubMed]**

110. Chu, M.; Pan, X.; Zhang, D.; Wu, Q.; Peng, J.; Hai, W. The therapeutic efficacy of CdTe and CdSe quantum dots for photothermal cancer therapy. *Biomaterials* **2012**, *33*, 7071–7083. **[Google Scholar] [CrossRef]**

111. Prasad, R.; Jain, N.K.; Yadav, A.S.; Jadhav, M.; Radharani, N.N.V.; Gorain, M.; Srivastava, R. Ultrahigh Penetration and Retention of Graphene Quantum Dot Mesoporous Silica Nanohybrids for Image Guided Tumor Regression. *ACS Appl. Bio Mater.* **2021**, *4*, 1693–1703. **[Google Scholar] [CrossRef]**

112. Ghafary, S.M.; Rahimjazi, E.; Hamzehil, H.; Mousavi, S.M.M.; Nikkhah, M.; Hosseinkhani, S. Design and preparation of a theranostic peptideticle for targeted cancer therapy: Peptide-based codelivery of doxorubicin/curcumin and graphene quantum dots. *Nanomed. Nanotechnol. Biol. Med.* **2022**, *42*, 102544. **[Google Scholar] [CrossRef] [PubMed]**

113. Wang, Y.; Chen, J.; Tian, J.; Wang, G.; Luo, W.; Huang, Z.; Fan, X. Tryptophan-sorbitol based carbon quantum dots for theranostics against hepatocellular carcinoma. *J. Nanobiotechnol.* **2022**, *20*, 78. **[Google Scholar] [CrossRef] [PubMed]**

114. Li, Y.; Zhang, P.; Tang, W.; McHugh, K.J.; Kershaw, S.V.; Jiao, M.; Han, B. Bright, magnetic NIR-II quantum dot probe for sensitive dual-modality imaging and intensive combination therapy of cancer. *ACS Nano* **2022**, *16*, 8076–8094. **[Google Scholar] [CrossRef] [PubMed]**

15 Chapter

RECENT PROGRESS IN THE IDENTIFICATION AND CHARACTERIZATION OF MONO-TO-MULTI-LAYER GRAPHENE AND ITS APPLICATIONS IN THE FIELD OF BIOMEDICINE

1. INTRODUCTION: AN OVERVIEW

The sixth element, carbon, is highly intriguing because of its ability to exist in several allotropic forms [1,2]. The allotropes of this substance exhibit a softness comparable to graphite [3,4] and a hardness comparable to diamond [5]. Graphite is a three-dimensional structure composed of numerous layers, each only one atom thick. These layers are made up of strong carbon atoms that are grouped in a hexagonal lattice by sp2 hybridization [6,7]. Graphene is a two-dimensional crystal structure made up of a single atomic sheet of graphite [8]. In recent times, graphene, which is a variant form of carbon, has garnered significant attention in scientific study circles owing to its favorable physico-chemical characteristics [9,10]. When arranged in layers, graphene can take on several shapes, including monolayer graphene [11], few-layer graphene [12], and multilayer graphene [13]. The properties of graphene vary depending on the number of stacked layers. Monolayer graphene exhibits superior qualities compared to few-layer graphene and multi-layer graphene [14,15]. Single layer graphene was theoretically postulated by scientists over 80 years ago [16]. Subsequently, the physical possibility of the practical existence of two-dimensional graphene was deemed implausible [17]. Nevertheless, in 2004, Geim et al. successfully separated a single layer of graphene using the scotch tape technique and conducted laboratory tests to verify its properties [18]. Researchers originally discovered the extraordinary features of graphene during this instance [19]. Subsequently, there has been a significant increase in research on materials derived from graphene to explore its diverse features and potential uses, including in the field of biomedicine [20,21]. By devoting specialized attention to graphene-based materials, researchers were able to successfully manufacture high-performance materials. Nevertheless, attaining these exceptional levels of performance poses numerous obstacles, particularly when it comes to monolayer graphene [22,23]. One of the hurdles involves synthesizing monolayer or few-layer graphene in large quantities with excellent purity, a highly complex procedure [24]. Additional difficulties arise from the need to rearrange monolayer graphene into few-layer and multi-layer graphene structures [25,26]. The act of restacking leads to a reduction in the performance of devices that are based on graphene. Hence, it is crucial to have control over the lateral size and aggregation states, as well as the oxidation process of graphene in graphene oxide, in order to construct high-performance devices based on graphene [27,28]. Following the successful synthesis of monolayer graphene, which has promising properties for the development of graphene-based devices, researchers were interested in studying the stacking order of graphene [29,30]. The characterisation techniques employed are atomic force microscopy (AFM) [31], Raman spectroscopy [32], Raman mapping [33], and Transmission electron microscopy (TEM) [34]. Raman spectra and TEM are commonly employed techniques to ascertain the composition of produced materials, specifically if they consist of monolayer graphene, few-layer graphene, or multi-layer graphene [32,34]. Following the process of characterisation, the graphene material can be utilized for various purposes, including biosensors [35], tissue engineering [36], drug carriers [37], and other biological applications. This review focuses on the distinct variations of graphene and their specific benefits in biological applications, which are determined by the number of layers.

2. PRODUCTION OF MONO-, FEW-, AND MULTI-LAYER GRAPHENE

Graphene, particularly monolayer graphene, has garnered significant interest since 2004 because of its favorable mechanical, electrical, and thermal characteristics [18]. Nevertheless, the process of creating a single layer of graphene is highly challenging and costly [38,39]. Hence, the synthesis of few-layer graphene or multi-layer graphene is also a topic of interest [40]. Figure 1 illustrates a generic framework depicting the different layers of graphene, including monolayer, few-layer, and multi-layer graphene. Mono-, few-, and multi-layer graphene can be produced by many techniques, such as micromechanical exfoliation [41], chemical vapor deposition [42], and chemical processes like oxidizing graphene to graphene oxide [43] and subsequently reducing it chemically [44] or thermally [45]. Several techniques offer high-quality, large-scale production of few-layer and multi-layer graphene, but they yield only limited quantities of monolayer graphene, such as the chemical method. However, the level of purity and the number of defects in the graphene produced using this method are still areas of concern [46,47]. Chemical vapour deposition can be used to create large-scale monolayer graphene. However, the issue of purity remains a worry [48].

The various methods for producing mono-, few-, and multi-layer graphene are as follows.

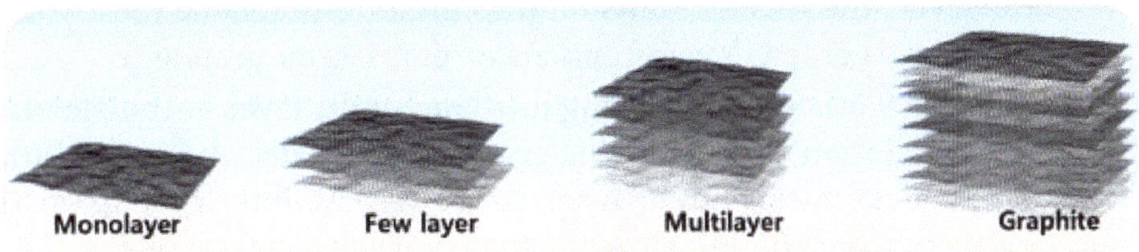

Figure 1. Schematic overview of monolayer, few-layer and multi-layer graphene.

2.1. Synthesis of Few- to Multi-Layer Graphene

2.1.1. Exfoliation of Graphite

Graphite is an abundant material and well known for its lubricating properties [**49,50**]. However, it has poor mechanical, electrical, and thermal properties. Graphite can be exfoliated into different types of graphene, but we must first overcome the weak van der Waals forces which hold the different types of graphene into graphite [**51**]. A general scheme of the process for obtaining different layers of graphene is shown in **Figure 2**.

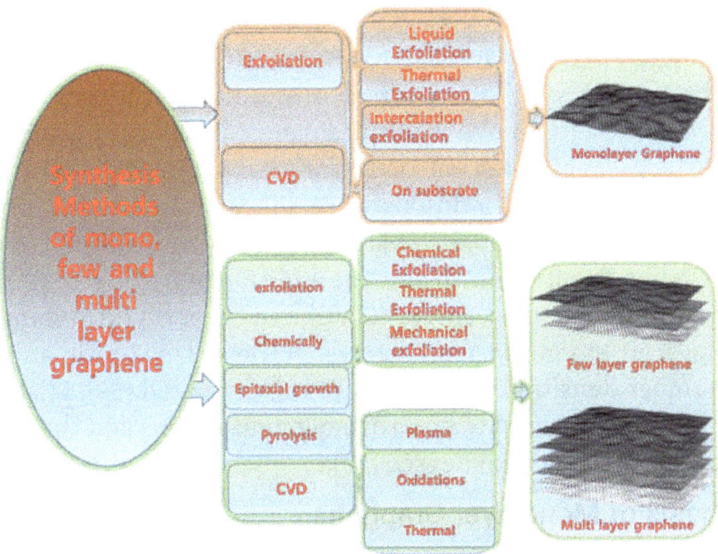

Figure 2. Schematic overview of the process for obtaining different layers of graphene.

a. There are several ways to overcome these weak interactions, and the most promising among them is the sonication of graphite in different solvents; however, the yields of multi-layer and monolayer graphene [**52**] are very poor in this process.
b. Another promising strategy to obtain graphene from graphite involves

oxidizing graphite by various methods, such as Hummer's method, and reducing it chemically or thermally to obtain a large-scale yield [53]. However, the redox results of graphite into graphene mostly provide high yields of few-layer graphene or multi-layer graphene.

c. Another method of producing few- and multi-layer graphene involves the exfoliation of graphite via graphite intercalation [54,55]. Different types of chemicals can be inserted to graphite interlayer space, thereby increasing the interlayer distance of adjacent graphene sheets in graphite. This phenomenon also changes the properties of graphene, since the increase in interlayer spacing affects electronic coupling between adjacent graphene sheets in graphite [56].

d. Another way to exfoliate graphite into few- and multi-layer graphene is via ball milling [57]. This is a way of exfoliating graphite via mechanical exfoliation. Ball milling has been extensively used in the past to reduce the particle size of a material [58]. Scientists thus propose ball milling as a way to mechanically exfoliate graphite in small-size nano-graphite, increasing the mixing time to obtain few- or multi-layer graphene. Thus, ball milling is a promising technique for exfoliating graphite into graphene. The advantage of using ball milling to produce graphene is its low production cost, its easy handling, and its ability to produce graphene at large scale.

e. The plasma synthesis method is another significant way to produce graphene with few-to-multiple layers. Microwave plasmas produced by surface waves at a stimulation frequency of 2.45 GHz and under atmospheric pressure conditions were successfully used to produce highly structured and stable self-standing graphene sheets [16]. There were also investigations into how the addition of hydrogen affects the density of the carbon precursor (C2, C) and the structural soundness of synthetic graphene sheets. Changes in the sp3/sp2 ratio and the C2 and C number densities were shown to be correlated [59]. Microwave-driven plasmas were used to control oxygen functions and the sp2/sp3 carbon ratio (~15) to a high degree [60].

2.1.2. Synthesis of Monolayer Graphene

a. In 2004, for the first time, Geim and Novoselov developed a method of synthesizing graphene using micromechanical cleavage as "scotch tape" via mechanical exfoliation [18]. This was for the first time in history that any scientist experimentally synthesized monolayer graphene. After synthesizing the monolayer graphene, these scientists further demonstrated its outstanding properties [61]. However, due to the uneven thickness of the graphene flakes and its high production costs, the mechanical exfoliation method was not suitable for the mass production of graphene that might be used to study graphene-based devices.

b. Another method of producing monolayer graphene is the chemical vapour deposition method [62]. Monolayer graphene can be grown epitaxially on a silicon carbide substrate, and can be used for various applications, such as transistors. The size of the monolayer graphene grown depends on the size of the silicon wafer. The surface of the silicon wafer also influences the properties of the synthesized monolayer graphene.

3. PROPERTIES OF MONO-, FEW-, AND MULTI-LAYER GRAPHENE

The properties of graphene depend on a number of internal factors (such as the number of graphene layers stacked in a graphene flake) and external factors (such as temperature); these factors affect the final performance of devices based on graphene flakes. It is well known that monolayer graphene exhibits extremely good mechanical, thermal, and electrical properties, and that these properties decrease as the number of layers increases—such as from monolayer graphene flakes to few-layer and multi-layer graphene flakes. The sections below describe the effects of these factors on the mechanical, thermal, and optical properties of mono-, few-, and multi-layer graphene.

3.1. Mechanical Properties of Mono-, Few-, and Multi-Layer Graphene

Figure 3 shows the effect of the number of layers in graphene flake on mechanical properties (such as fracture stress and fracture strain). As expected, it was found that the mechanical properties decrease as the number of layer increases from monolayer to few-layer and then to multi-layer graphene. However, these changes in mechanical properties are not significant, as shown by Zhang et al. [14]. It is clear that the fracture strain and tensile strength are not significantly affected by the transition from few-layer graphene to multi-layer graphene. This is due to the dominance of the "nano effect" of graphite as the number of layers increases from few-layer to multi-layer graphene. Zhang et al. [14] further report that the mechanical properties are significantly affected by temperature, which is an external factor. The studies show that, as the temperature increases from 300 K to 2000 K, the tensile strength decreases significantly from 125 GPa to around 43 GPa. This fall in tensile strength is due to the softening of the structures. In addition, it is well known, and was demonstrated by Zhang et at [14], that the atoms of graphene undergo severe movement at higher temperatures, leading to such reductions in the mechanical properties. Similarly, fracture strain (67%) and Young's modulus (23%) drop significantly as the temperature increases from 300 K to 2000 K [14].

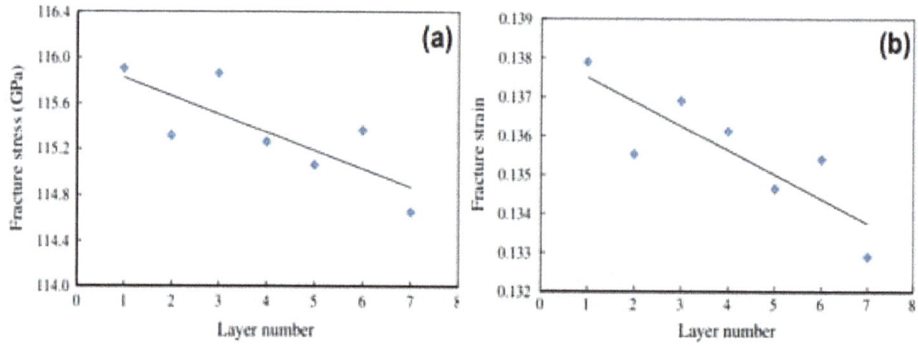

Figure 3. (a) The fracture stress and (b) fracture strain of mono- to few- to multi-layer graphene. Reproduced from [14] with the permission of Elsevier.

3.2. Thermal Properties of Mono-, Few-, and Multi-Layer Graphene

The coefficient of thermal expansion (CTE) was studied as a function of temperature and the number of layers stacked in graphene flakes. It was found and demonstrated by Mag-isa et al. [15] that thermal properties such as CTE are highly dependent on internal factors (the number of layers in the graphene flake) and external factors (temperature) (**Figure 4**). For example, it was noted that, as the temperature increases from 20 °C to 140 °C, the COE is affected. This was expected, due to increase in the thermal kinetics of the processes as the temperature increases. Similarly, the thermal properties are affected by increasing the number of layers, as the thermal properties of monolayer graphene differ from those of few- and multi-layer graphene [12,15]. It is interesting to note that the results for few-layer graphene are placed between those of monolayer graphene and multi-layer graphene, as expected. Moreover, all of the values of CTE studies reported by Mag-isa et al. [15] and other researchers are negative. A negative CTE is generally regarded as a non-close-packed system with directional interactions, such as ice or graphene.

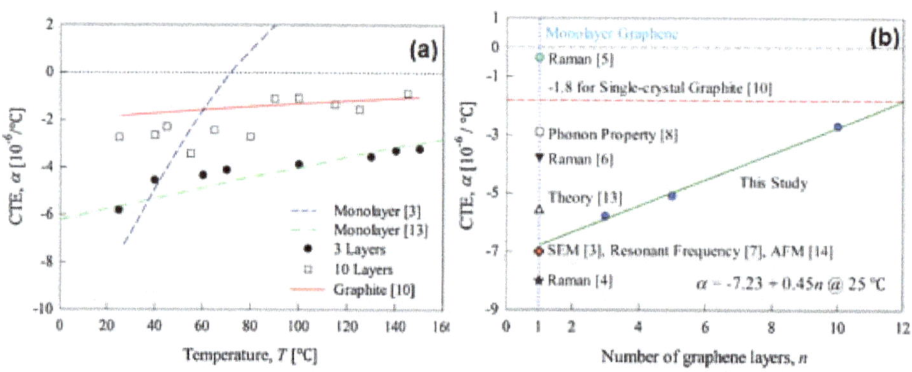

Figure 4. The thermal properties of the graphene-based flakes: (a) coefficient of thermal expansion against temperature; (b) coefficient of thermal expansion against mono-, few-, and multi-layer graphene. Reproduced from [15] with the permission of Elsevier.

3.3. Optical Properties of Mono-, Few-, and Multi-Layer Graphene

By taking into account the surface conductivity (σ), the refractive index (*n*) and the extinction coefficient (*k*) of graphene can be determined. Knowledge of the frequency-dependent surface conductivity sG, which can be determined from either a microscopic model or from experiments, is necessary for understanding the optical characteristics of graphene [63]. The surface conductivity (σ_G) of graphene is dictated by the high-frequency expression acquired from the Kubo model [64].

4. DETERMINATION OF MONO-, FEW-, AND MULTI-LAYER GRAPHENE

The determination of the number of graphene layers and defects is critical for tailoring its properties for intended applications, and involves microscopic as well as spectroscopic measurements.

4.1. Atomic Force Microscopy (AFM)

AFM is defined as high resolution microscopy with a resolution at the atomic scale. AFM can determine various features of a material, such as thickness, grain height, topological features, phase diagrams, and roughness. These features are then correlated with the properties and target applications of the material, such as graphene. Xu et al. [65] studied and employed the AFM technique to determine thickness of graphene flake. The corresponding histogram shows that the flake thickness ranges from 1.1 to 1.6 nm. By considering an interlayer distance 0.33 nm, a total of 4–5 layers can be estimated to be present in the graphene flake. It therefore falls into the category of few-layer graphene [65]. Chen et al. [66] demonstrated the use of AFM for determining the roughness of graphene flakes. The authors found that the pristine polymer forms a homogenous film with a mean roughness in the range of 2.2 nm. As the graphene is added to the polymer, the composites made of polymer-graphene also exhibit good film-formation properties, with a mean roughness of 3.1 nm [66]. Zhang et al. [67] also studied the early stages of atomic layer deposition (ALD) on epitaxial graphene (EG) via AFM (**Figure 5**). It can be seen that the EG-based sample shows flat topological features. The small surface roughness and very narrow distribution of height values can also be noticed. Fitted with a single, the Gaussian peak shows a full-width half-maximum (FWHM) of 0.2 nm. Similarly, different topological images of the specimen were shown as a function of different ALD cycles, increasing from 10 to 80 [68]. No significant differences can be seen in the topology images; the height values and Gaussian fit of FWHM remained practically the same as the number of ALD cycles increases. Gao et al. [69] also studied the thickness of graphene flakes with the help of the AFM technique. The authors state that the thickness of the graphene flake used in the work was in the range of 0.4–1 nm. This shows that the graphene flake was monolayer-to-few-layer graphene.

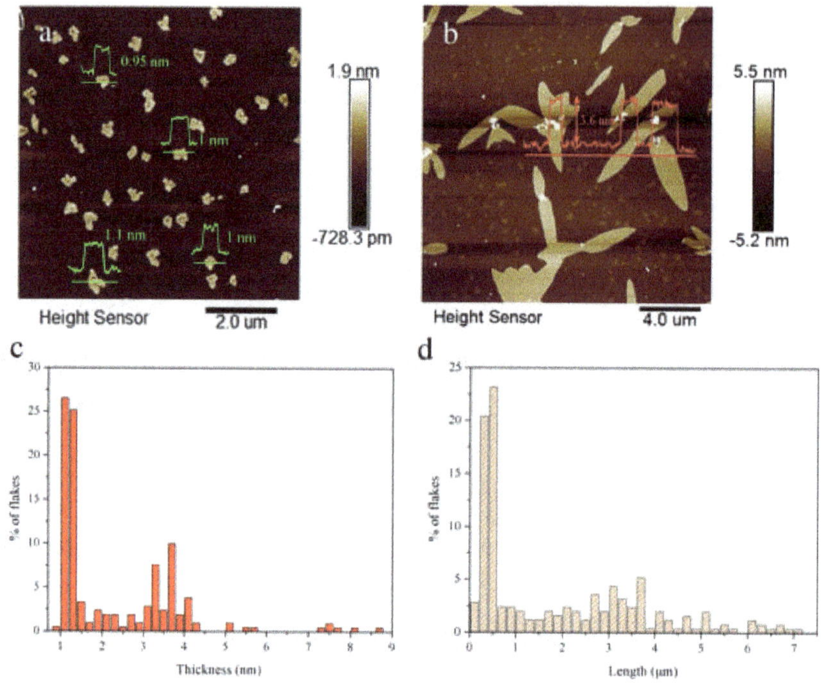

Figure 5. (a) AFM images of monolayer graphene flakes with adjacent height profiles; (b) different types of few-layer graphene flakes with adjacent height profiles; (c) the adjacent thickness of the flakes; (d) the length distribution of different flakes analyzed through AFM microscopy. Reproduced from [67]. [CC-BY].

Yuan et al. [70] investigated the thickness of the graphene flake with the help of the AFM tool. The authors demonstrated the sheet-like topological feature of the graphene flake. They further investigated the flake thickness and its lateral dimension. It was found from the histogram that the graphene flake consisted mostly of monolayer graphene, while a low percentage of few-layer graphene and multi-layer graphene was witnessed. The lateral dimension was in the range of 2 to 10 μm, and therefore had very high shape anisotropy [70]. Finally, Bhuiyan et al. [71] investigated the graphene flake thickness with the help of AFM microscopy. From this study, the authors demonstrate the smooth, sheet-like topology of the graphene flake, which had a thickness of around 1 nm; this shows that the graphene flake used in the work is few-layer graphene [71].

4.2. Transmission Electron Microscopy (TEM)

The TEM technique involves the transmission of a beam of electrons through a sample that is usually 100 nm thick and suspended on a grid. This technique gives an image of atomic-scale resolution and can be used in determining the characteristics of mono-, few-, and multi-layer graphene.

Araby et al. [72] demonstrated the use of TEM in determining the graphene thickness via high resolution TEM images (**Figure 6**). The images show that the graphene flake is few-layer graphene, the type of graphene obtained using an indium catalyst at 150 °C. On the other hand, Long et al. [73] examined the morphology of graphene flakes via TEM and HRTEM. The TEM images show the wrinkled morphology of the graphene flakes, indicating a stable structure. The HRTEM images of the graphene flakes show the stacking morphology of borophene on the reduced graphene oxide. Moreover, the investigation shows a good interaction between borophene and the reduced graphene oxide [73]. Navik et al. [74] investigated the morphology of few-to-multi-layer graphene via HRTEM. This technique shows that, in the graphene flakes, the distribution of few-layer graphene is around 70% and the remaining 30% belongs to multi-layer graphene. The HRTEM images provided by the authors show exfoliated graphene sheets in which the exfoliation process produces large transparent graphene flakes, which mostly belong to few-layer graphene [74]. Ding et al. [75] studied the number of graphene layers per graphene flake via HRTEM. The authors demonstrated that the graphene flake used in the work contains mono- and bi-layer graphene. Thus, a high degree of exfoliation was achieved in the work and the remarkable properties of the related materials were studied.

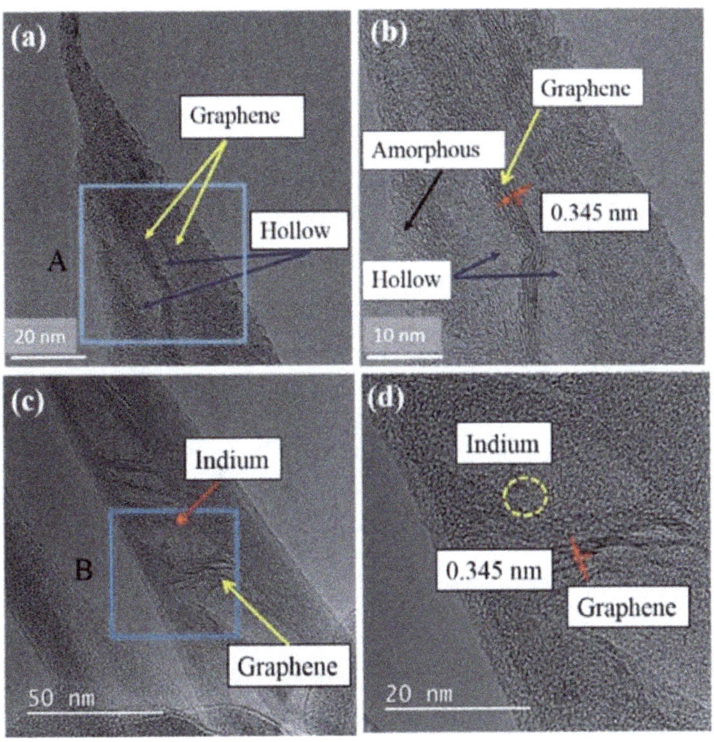

Figure 6. (a-d) Typical TEMs of graphene nano flakes in different forms, synthesized by an indium catalyst at 150 °C. Reproduced from [72] (CC BY).

4.3. Raman Spectra

Raman spectra (named after Sir C.V. Raman from India) is a famous spectroscopy technique that provides information about the vibrational modes of molecules, etc. It is most prominently used in determining the number of graphene layers in a graphene flake. The graphene flake generally shows three characteristic bands, namely, the D-band, G-band, and 2D band [76]. The position of the 2D band is different for different types of graphene flake. It is around 2702 cm^{-1} for tri-layer graphene, and increases with the increase in the number of layers to as much as 2720 cm^{-1} for multi-layer graphene and 2725 cm^{-1} for bulk graphite [14]. Campanelli et al. [77] used Raman spectra to demonstrate the interaction of Yb^{3+} with monolayer and bi-layer graphene. Their paper demonstrates that there is a good interaction between Yb^{3+} and graphene flakes, as confirmed using Raman spectra. A shift of ~5 cm^{-1} to ~7 cm^{-1} was observed at different Raman bands in the presence of Yb^{3+}, indicating a good interaction [77]. Wang et al. [78] demonstrated the use of the Raman spectra technique for identifying monolayer and bi-layer graphene. The authors reported on systematic and statistical tests of the signatures of the so-called buffer layer and its coupling with the epitaxial graphene layer via Raman spectroscopy.

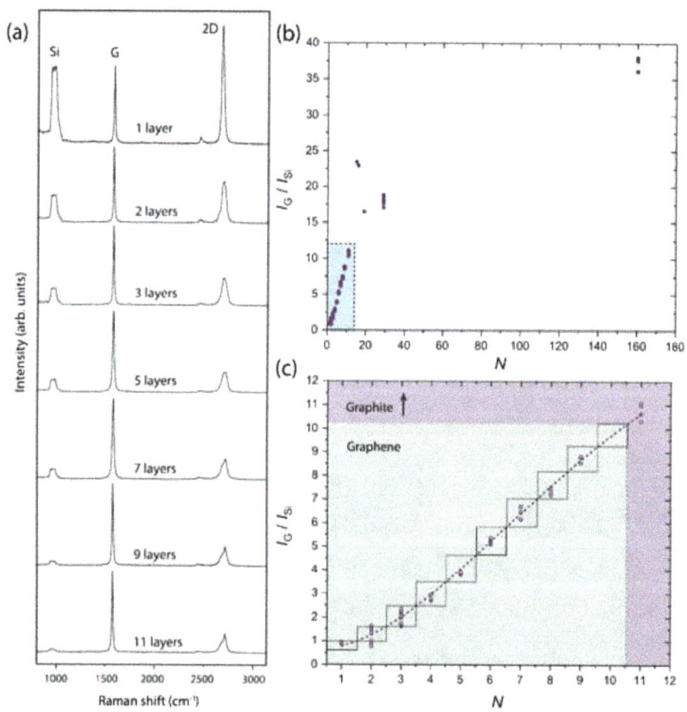

Figure 7. (a) Representative Raman spectra of different graphene flakes varying from monolayer to few-layer to multi-layer graphene; (b) a plot of the experimental values obtained from the ratio intensity of the G-band to Raman signals from the Si substrate; (c) a magnified view of the (b) graph from monolayer to multi-layer graphene [79].

The findings show a coupling between graphene and buffer layer [78]. Silva et al. [79] demonstrated the different number of graphene flakes with mono-, few-, and multi-layer graphene via Raman spectra (**Figure 7**). A statistical analysis of the different types of graphene flake with mono-, few-, and multi-layer graphene was presented. The protocol is based on the position of the 2D band, which is different for different type of graphene layers, as discussed above. The authors demonstrated that the Raman spectra is suitable for the statistical analysis of mono-, few-, and multi-layer graphene with relatively low costs [79].

4.4. Raman Mapping

Raman mapping is known as a laser-based microscopic technique which is obtained via Raman spectra. Recently, Raman mapping has played a significant role in determining the homogeneity of the graphene flakes used for different applications. Huang et al. [80] studied monolayer graphene through Raman mapping at different temperatures of over 150 °C and 250 °C. The Raman signatures indicate the formation of defects in graphene, even at low temperatures. The Raman maps show that the ratio of intensity of the D to G bands was 1.85 and 4.01 for 150 °C and 250 °C, respectively, indicating higher defect density for the 250 °C sample [80]. Bouhafs et al. showed [81] the synthesis and characterization of large-area few-layer graphene by chemical vapor deposition. The Raman map used in this study shows clear ABA and ABC stacking in the studied few-layer graphene. Bleu et al. [82] studied few-layer graphene via Raman spectra and Raman mapping. The most significant peaks for the few-layer graphene studied through Raman spectra are the D, G, and 2D bands that are located around 1350, 1580, and 2700 cm^{-1}, respectively (**Figure 8**). In addition, the intensity ratio of the D to G band can be used to estimate the defect density, and the intensity ratio of 2D to G reveals the number of graphene layers stacked. The histogram was estimated via Raman maps and few-layer graphene was found to be dominant in the sample [82]. Moreover, the Raman maps show the good homogeneity of the synthesized few-layer graphene, with low defect density.

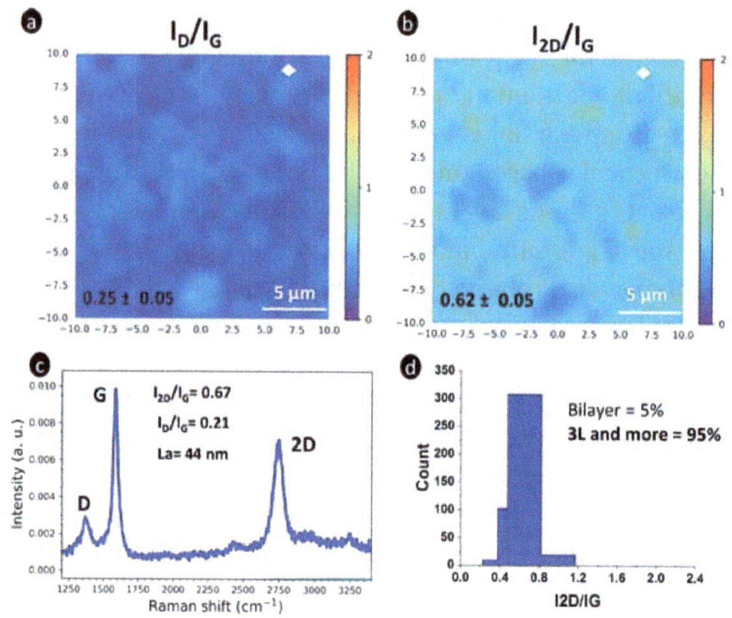

Figure 8. Raman analysis of the synthesized few-layer graphene; (a) a Raman map of the ratio of intensity of the D-band to the G-band; (b) a Raman map of ratio of the intensity of the 2D band to the G-band; (c) Raman spectra of few-layer graphene; (d) the distribution of the number of graphene layers in few-layer graphene. Reproduced from [82] with the permission of Elsevier.

5. BIOMEDICAL DEVICES AND OTHER BIOMEDICAL APPLICATIONS

5.1. Use as a Biomedical Device

The anatomical characteristics of different layers of graphene are different. Mono-to-multi-layer graphene is highly sensitive and is used in sensor applications, including biomolecules, elements, gas, pressure, and electrochemical detection [83,84]. Due to its large surface area, electrical conductivity, biocompatibility, easy operation, tunable structural transformation, and excellent mechanical strength, layered graphene is used as a biomedical device for regenerative drugs [85,86], bio-mining [87] detection [88], and cancer cures [89].

On the other hand, there is a basic restriction on the use of monolayer graphene in device applications: that is, the light–graphene interaction is weak due to the atomic thickness of graphene. In particular, graphene exhibits a wavelength-independent absorption of 2.3% in the visible and near-infrared regions, which limit its further application in areas such as sensors and photodetectors [90]. In recent years, several approaches have been proposed to solve these problems, such as the angle-insensitive broadband absorption enhancement of graphene [91], and electrically modulating and switching the infrared absorption of monolayer graphene [92].

5.2. Use as a Sensor Device

Additionally, biomolecules with hydrophobic domains or systems, such as DNA strands or proteins, have a tendency to naturally adsorb on graphene [93]. Another simple way to modify graphene for the targeted immobilization of bio-receptor units is to functionalize it [94].

5.2.1. Nucleic Acids Sensor Device

Compared with devices made using bilayer and few-layer graphene, sensor devices made with functionalized monolayer graphene exhibit superior sensitivity, better repeatability, and stronger construction [95]. Monolayer-graphene-based FET devices are used as nucleic acid sensors with high selectivity and low detection limits [96,97]. In this device, the fixed graphene layers are determined by using 1-pyrenebutanoic acid succinimidyl ester (PBASE) as a link chemical and acquired a 10 pM detection limit (**Figure 9**). To recognize pesticides, more GFET detectors based on microfluidics were created [98].

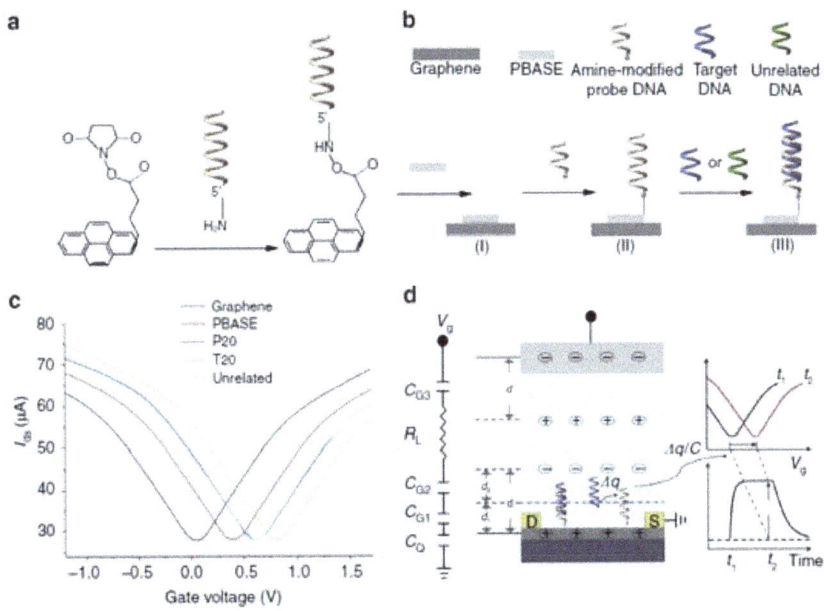

Figure 9. Characteristics of G-FETs. (a) The relationship between PBASE's molecular structure and the probe DNA. (b) PBASE is used to functionalize the graphene surface. A probe DNA (shown in orange) is immobilized by interacting with the PBASE. The probe DNA is then hybridized with the target DNA (shown in blue) and unrelated control DNAs (in green). (c) Graphs of the source drain current versus the constant source drain voltage (0.1 V) for different conditions of graphene. (d) A schematic showing the equivalent circuit made up of four parallel plate capacitors and a resistance (RL) linked in series and the sensing model of a G-FET. Reproduced from [97] (CC BY).

5.2.2. Mammalian Cell Sensor Device

For cell-sensing applications, layered graphene is used as an immobilizer for detecting cells or various pathogens. Recently, scientists developed few-layer graphene-based FETs with HER2-specific aptamers and the ability to detect very low concentrations of HER2-cells [98]. Moreover, the devices exhibited highly sensitive detection against breast cancer cells and SK-BK-3 cells at the single-cell level.

5.2.3. Monolayer Graphene and Immune Sensor Devices

The surface modifications of functionalized specific polymers improve the interfacial interactions between them when used as high-performing medical devices [99,100]. As evidence, when the monolayers of graphene are mixed with a bioactive immobilizer (biotinylated cholera toxin), they showed a remarkable response to the anti-CT test [101]. Additionally, pyrene-NTA-conjugated graphene-monolayer-based sensor devices can detect up to 4 pL^{-1} of anti-CT, which is also a very significant result for anti-CT approaches [102]. A high-sensitivity surface plasmon resonance (SPR) biosensor made of a monolayer of graphene, a four-layer MoS2, and a layer of gold was studied [103].

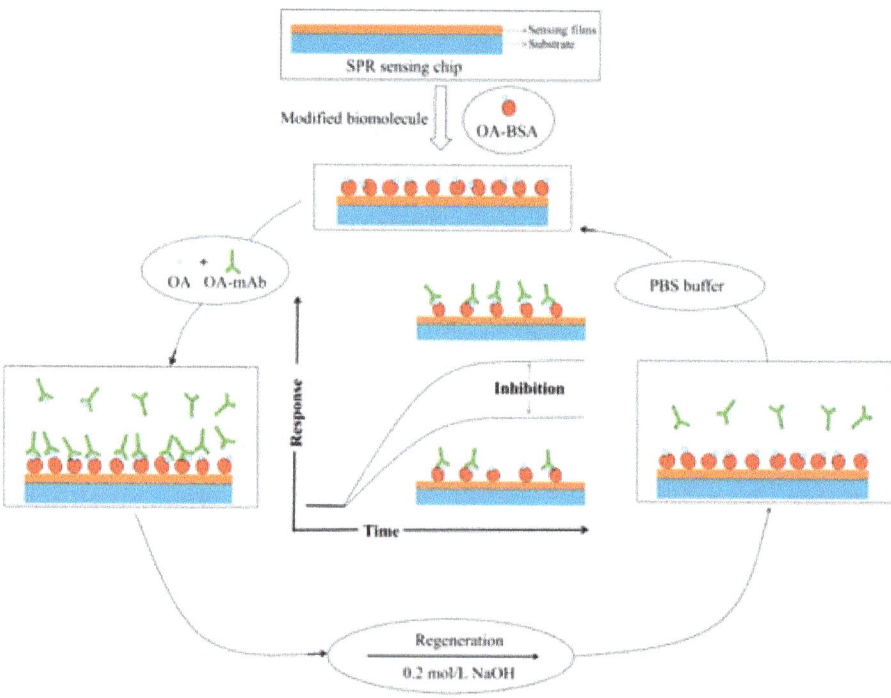

Figure 10. The flowchart for the method of detecting okadaic acid (OA). Reproduced from [103] (CC BY).

The maximal sensitivity of the sensor, 282º/RIU, was shown by computational models using the transfer matrix method (TMM), which was about two times more sensitive than the typical Au-based SPR sensor. Additionally, the Kretschmann-configured manufactured sensors were employed to detect okadaic acid (OA). To create a competitive inhibitory immunoassay, the okadaic acid-bovine serum albumin bioconjugate (OA-BSA) was mounted on the graphene layer of the sensors. According to the findings, the sensor had a very low sensitivity limit (LOD) for OA of 1.18 ng/mL, which is roughly 22.6 times lower than that of a traditional Au immunoassay. **Figure 10** depicts the schematic diagram of the OA detection process (**Figure 10**).

5.2.4. The Development of Imaging Devices for Biomedical Purposes

Bio-imaging is an important aspect of regenerative medicine [104]. Imaging techniques use different layers of graphene (a single layer to multiple layers) for imaging contrast agents such as fluorescence/confocal imaging [105,106], Forster resonance energy transfer (FRET) imaging [106], coherent anti-Stokes Raman scattering imaging (CARS) [107,108,109], magnetic resonance imaging (MRI) [110], surface-enhanced Raman scattering (SERS) [111], ultrasound imaging [112,113], photoacoustic imaging [114], electron paramagnetic resonance imaging (EPRI) [115,116], and positron-emission tomography (PET) [117,118]. Among these techniques, a few, such as computer tomography (CT), PET, MRI, and ultrasound imaging, are well-established and for use in humans. These imaging techniques used to better understand and monitor the affected site of the body for therapeutic purposes. Additionally, there are two fundamental necessities for bio-imaging: (1) rapid, selective, and sensitive instruments, and (2) good, non-toxic, biocompatible, biodegradable imaging contrast agents that have good bio clearance and are capable of crossing the physiological body membrane barrier.

Currently, the most developed EPRI technique is used to detect and quantify various physiological parameters in a cancer-effected microenvironment in in vivo models, such as pO2 [119,120,121], pH [116,122], and oxidation situations [116,123]. Normally, for this method, TAM (triphenylmethyl radical) [124] or other compound chemicals [125] are used. In EPRI for localized oxygenation sensing, carbon-based ink is used [126]. CARS imaging technology is highly sensitive in providing signals to process images in in vivo, ex vivo, and in cell cultures, up to micrometer-level penetrations [127,128].

Mono-to-multi-layered graphene-based materials are used in the biomedical field as contrast agents [129,130]. The key benefit of this is that the processes are biocompatible, less toxic, and metal-free [131]. Mono-to-multi-layered graphene-based materials are mainly used in fluorescence imaging [132] where they are used in modified form, display layer, and surface-chemistry-dependent fluorescence [133]. To enhance the biocompatibility of mono-to-multi-layered graphene, a few common polymers are used, such as polyethylene mine [134], PEG (polyethylene glycol) [135,136,137,138], polystyrene [139], and polypeptides [140]. Recently, scientists shortened the varied imaging techniques of different layers of graphene used, and determined that various imaging techniques can be performed on one platform [141]. Therefore, we assume that, in the coming decade, different layers of graphene-based materials and their diverse morphology will play a significant role in biomedical applications and in determining the future of therapeutic approaches.

5.3. Use in Other Biomedical Applications

5.3.1. As a Delivery Carrier and in Treatment

Ligand-Based Drug Delivery

Most interferences with drug-delivery systems are due to the presence of many proteolytic enzymes in the cytoplasm of a cell. Graphene and its derivatives are used as drug carriers or carriers that bypass those interfering enzymes. Its properties make it a promising carrier for biomedical applications; for example, COOH and OH are the main group ligands of GO that are acceptable for binding to various organic molecules, such as proteins, DNA, and many polymers. This ligand makes it biocompatible, cell selectable, and efficient. Some reported polymer binding ligands are used to deliver anti-cancer drugs such as doxorubicin (DOX) [142]. Other reports noted that the drug Ibuprofen has been delivered using chitosan-containing GO [138].

Figure 11. Cartoon illustrating the applications of monolayer, few-layer, and multi-layer graphene.

Graphene's unmodified basal plane site with open surface π electrons is hydrophobic, creating π-π interactions for loading drugs [143,144]. GO has been transformed into a carrier for the supply of water-soluble cancer drugs. The solubility of physiological and aqueous solutions can be increased by functional NGO with PEG [145]. **Figure 11** shows a scheme of monolayer, few-layer, and multi-layer graphene applications.

Stimuli Response Delivery

Some drug delivery occurs due to changes in temperature, pH, light, and salt concentrations, whereby the polymer ligand can detect the particular physiological change and deliver the drug. A few reports show that the biopolymers bound with graphene materials are capable of pH-responsive drug delivery or of being used as transporters [146]. For example, DOX is an anticancer drug that is easily released from the GO-Dox complex in a low pH body environment due to its high solubility at a low pH or in acidic conditions [147]. On the other hand, pH-sensitive delivery has been shown to be effective in the treatment of tumors using folic acid with DOX and camptothecin [148]. A photochemically regulated gene-delivery carrier was also reported where the low molecular weight PEI and rGO were combined with polyethylene glycol (PEG) and delivered the plasmid DNA by physiochemical assays [149].

Gene Delivery

Gene delivery is a good alternative way to cure many genetic diseases, as genes can be delivered to a cell through a vehicle or carrier. Many scientific works have reported gene delivery through GO or GO-based materials. To prepare for gene delivery, polyehylenimine (PEI) was adhered to the surface of the GO sheet via covalent coupling and electrostatic contact with stacking plasmid DNA (PDNA) [150]. The long chain [138] and the branching chain are parallel to the GO [151]. In contrast to other complex compounds, PEI has high transfection effectiveness and low cytotoxicity. Additionally, chitosan complex GO has been used for the delivery of anti-cancer medication drugs and plasmid DNA [152].

5.3.2. Therapeutic Applications

Many results report that the different layers of graphene or graphene oxide can inhibit the growth of tumor spheres in various cell lines, including breast, lung, ovarian, pancreatic cancers, and glioblastoma [153]. Few-layer GO not only inhibits tumor development when injected into mice as an immunocompetent containing CT26 colon cancer cells [154], but also inhibits autophagy [155], immunity, and cell death [156]. In xenograft models, the conjugate PEG-GO exhibits malignancy and thus has a high therapeutic value [157]. An example of this is an rGO/iron oxide NP coated with PEG (rGO-IONP-PEG), which can be used for photothermal treatment (PTT) in vivo and as a triple-mode sensor for fluorescence, photoacoustic tomography (PAT), and magnetic resonance imaging (MRI) [158,159]. Another example is fluorinated GO, which can be used as a magnetically sensitive drug carrier and can be imaged via MRI and photoacoustic tomography [160].

5.3.3. Tissue Engineering for Cell Growth

Monolayer Graphene and Scaffold Formation for Fibroblasts

In a recent study, monolayer graphene's cytotoxicity and usefulness as a scaffold for murine fibroblast L929 cell lines were assessed. The authors demonstrated the impact of monolayer graphene toxicity through tests on cell viability, morphology, cytoskeleton architecture (microfilaments and microtubules), cell adhesion, and migration. They discovered that the fibroblasts grown on a monolayer of graphene showed modifications in cell attachment, motility, and cytoskeleton organization [161]. Within 24 h of culture, monolayer graphene was discovered to show no cytotoxicity toward L929 fibroblasts and to improve cell growth and adhesion. Additionally, as shown by the results, the monolayer graphene aided in the migration of cells. In the end, it was shown that monolayer graphene is not hazardous to murine subcutaneous connective tissue fibroblasts, and may even help injured tissues heal [161].

Few-Layer and Multi-Layer Graphene and Scaffold Formation

Graphene with different numbers of layers has different effects on different cell line cultures. A mammalian cell culture was tested using a few-layer graphene sheet platform. They described an investigation on the behavior of NIH-3T3 fibroblasts on several carbon nano-materials layers, including carbon nanotubes, RGO, and GEO. Layers treated with carbon nanomaterials demonstrated excellent compatibility and improved gene transfection efficiency [162]. When compared to conventional polystyrene tissue culture plates, graphene-chitosan hybrid films have demonstrated positive outcomes in tissue engineering to repair and enhance tissue function [163].

Differentiation of Stem Cells

Stem cells are important for living organisms due to their continuous regeneration powers and ability to convert into any kind of cell type where required. In regenerative medicine, stem cells are used as a therapeutic agent. Different layers of graphene are used in stem cell cultures for different purposes. The different layers of graphene showed various types of toxicity effects, electrical conductivity behaviors, biocompatibility, etc. A recent study showed that human mesenchymal stem cells (hMSCs) could differentiate into neurons very well on a surface made of few-layer graphene because the graphene served as an electrically coupled cell-adhesion layer for the hMSCs [164]. In another study, multi-layer graphene also improved the osteogenic differentiation to a greater extent than the general growth factor [165]. A comparative study found that osteogenic differentiation was stimulated but adipogenic differentiation was impeded on graphene as compared to the GO. This occurred due to the different surface characteristics of different layers of graphene or graphene derivatives which come from chemical functionalization [166,167].

5.3.4. Anti-Microbial Effects

The different layers of graphene and their derivatives showed various diversified antimicrobial effects on both Gram-positive and Gram-negative bacteria [168,169]. Few-layer graphene oxide and reduced graphene oxide both showed antibacterial effects on their surfaces [168]. Meanwhile, some reports showed that the bacterial growth was intensified, rather than inhibited, on the monolayer graphene surface [170]. The monolayer GO is more effective than the few-layer and multi-layer GOs because monolayer GO is more efficient in charge transfer with bacterial cells [169]. This happens due to oxidative stress induced by membrane disruption [170,171].

5.3.5. Digital and Analog Devices and Equipment

Graphene is frequently employed by researchers in the field of digital and analog device and equipment. A novel electro-optical encoder based on a graphene–Al_2O_3 multi-layer stack has been designed. It was demonstrated that employing various chemical opportunities for graphene yields the desired encoding operation for the intended structure [172]. In a different study, a switch made of graphene was proposed as a solution to power and footprint issues. The structure may be a good contender for PIC uses, since adding graphene monolayers improves the interaction between light and the material, and lowers the cutoff intensity for the switching function [173]. By altering the chemical prospects of graphene, a graphene-based plasmonic waveguide was created and studied for switching functions at terahertz frequencies [174]. A graphene layer and a silicon ridge were added above and below the channel, respectively, to create a new plasmonic channel. Low energy consumption and the opportunity for on-chip integration are benefits of the sub-wavelength scale [175]. The graphene waveguides are encased between two SnO_2 layers in a small, graphene-based plasmonic D flip-flop, which is shown. The flip-flop's compact area is a crucial component for use in optical integrated circuits [176].

6. CONCISE SUMMARY

Graphene has garnered significant interest from scientists worldwide since its extraordinary capabilities were initially demonstrated in 2004. The properties of monolayer graphene undergo changes as it is rearranged to produce few-layer graphene, and the properties of few-layer graphene are further altered when it is layered to form multi-layer graphene. Nevertheless, the exceptional characteristics are only observed in monolayer graphene, whereas few-layer and multi-layer graphene only exhibit the "nano effect" of graphite. Nevertheless, the widespread application of monolayer graphene is hindered by the difficulties associated with its large-scale synthesis and the high costs involved in its processing. Scientists have devised various methods to produce graphene, including the exfoliation of graphite to obtain graphene and the synthesis of graphene by chemical vapor deposition. Chemical vapor deposition is the optimal method for producing monolayer graphene, although graphite exfoliation can be employed to produce few-layer and multi-layer graphene. Scientists have seen a notable progress in the characterization of graphene flakes, and they have put forward several approaches for characterizing graphene. Out of these, AFM, TEM, and Raman spectra have been identified as the most favorable options. Graphene has a promising future. Over the past twenty years, scientists worldwide have shown significant enthusiasm for graphene. The growing abundance of papers on graphene serves as a testament to the significant focus and interest directed towards this material. Monolayer graphene is particularly intriguing because of its exceptional mechanical, electrical, and thermal characteristics. The process of producing large quantities of monolayer graphene is still a significant obstacle, and researchers will investigate novel approaches of synthesize graphene in the next ten years. The exceptional and uncomplicated modification capabilities, such as surface modifications with any desired ligands, make differentially layered 2D graphene materials highly intriguing for biomedical applications. Monolayer graphene and few-layer graphene exhibit no toxicity, possess greater biocompatibility, and demonstrate exceptional UV absorption, surface-enhanced Raman scattering (SERS), fluorescence, and fluorescence quenching. These properties render them highly potent nanomaterials for applications in biosensors, therapeutics, tissue engineering, and biomedical electronic devices. However, there are still certain obstacles to overcome, despite the encouraging outcomes observed in the functionalization and optimization of graphene by fractionation depending on factors such as the number of layers, size, shape, and chemical functionalities. Furthermore, there are numerous unparalleled advantages and ample prospects to investigate the intriguing attributes of these materials and their potential applications. We anticipate that collaborative endeavors involving chemistry, biology, and engineering will enhance the automated detection of biological platforms based on graphene layers, resulting in a range of successful and groundbreaking applications. Furthermore, we deduce that the many strata of graphene-derived substances and their varied structure will play a crucial role in the coming decade in the field of biomedical applications, and will determine the trajectory of therapeutic advancements.

At a Glance

Monolayer graphene possesses extraordinary mechanical, electrical, and thermal properties, rendering it a fascinating material. The properties of graphene are influenced as the number of layers rises, specifically in the case of few-layer graphene (with a maximum of 5 layers) and multi-layer graphene (with a maximum of 10 layers). To acquire the essential characteristics, it is vital to control the quantity of layers in the graphene flake. Hence, in our present analysis, we examine the diverse methods employed to manufacture single-layer and few-/multi-layer graphene. The attributes of mono-/few-/multi-layer graphene, including its mechanical, thermal, and optical properties, are subsequently evaluated. Graphene exhibits distinctive electrical characteristics, including high carrier mobility, ambipolar behavior, and a distinct energy band structure. These properties make it suitable for use in field effect transistors (FETs) and many applications such as radio frequency (RF) circuits, sensors, memory, and more. This review focuses on the incorporation of graphene into devices used for the detection of biomolecules, as well as its applications in the field of biomedicine. This text examines the benefits of utilizing graphene in many scenarios, and it also provides an analysis and evaluation of the latest advancements in biomedical devices and other uses.

REFERENCES

1. Dong, H.; Zhang, Z.; Feng, Z.; Kang, J.; Wu, D.; Wang, Q.; Li, J.; Su, R. Origins of low lattice thermal conductivity in 2D carbon allotropes. *J. Mater. Res. Technol.* **2021**, *11*, 1982–1990. **[Google Scholar] [CrossRef]**

2. Liu, W.-D.; Yu, Y.; Dargusch, M.; Liu, Q.; Chen, Z.-G. Carbon allotrope hybrids advance thermoelectric development and applications. *Renew. Sustain. Energy Rev.* **2021**, *141*, 110800. **[Google Scholar] [CrossRef]**

3. Mailian, A.; Panosyan, Z.; Yengibaryan, Y.; Mailian, M. Identification of Carbon Allotropes in Tribolayers Obtained by Rubbing of Graphite. *Mater. Today Proc.* **2017**, *4*, 6842–6848. **[Google Scholar] [CrossRef]**

4. Kabir, H.; Zhu, H.; May, J.; Hamal, K.; Kan, Y.; Williams, T.; Echeverria, E.; McIlroy, D.N.; Estrada, D.; Davis, P.H.; et al. The sp2-sp3 carbon hybridization content of nanocrystalline graphite from pyrolyzed vegetable oil, comparison of electrochemistry and physical properties with other carbon forms and allotropes. *Carbon* **2019**, *144*, 831–840. **[Google Scholar] [CrossRef]**

5. Liu, Y.; Jiang, X.; Fu, J.; Zhao, J. New metallic carbon: Three dimensionally carbon allotropes comprising ultrathin diamond nanostripes. *Carbon* **2018**, *126*, 601–610. **[Google Scholar] [CrossRef]**

6. Zhang, X.; Schneider, R.; Müller, E.; Gerthsen, D. Practical aspects of the quantification of sp2-hybridized carbon atoms in diamond-like carbon by electron energy loss spectroscopy. *Carbon* **2016**, *102*, 198–207. **[Google Scholar]** **[CrossRef]**

7. Theye, M.-L.; Paret, V. Spatial organization of the sp2-hybridized carbon atoms and electronic density of states of hydrogenated amorphous carbon films. *Carbon* **2002**, *40*, 1153–1166. **[Google Scholar]** **[CrossRef]**

8. Mandal, T.K.; Hou, Y.; Gao, Z.Y.; Ning, H.; Yang, W.S.; Gao, M.Y. Graphene oxide-based sensor for ultrasensitive visual detection of fluoride. *Adv. Sci.* **2016**, *3*, 1600217. **[Google Scholar]** **[CrossRef]** **[Green Version]**

9. Bin Hamid, M.A.; Chan, K.T.; Raymond Ooi, C.H.; Zainuddin, H.; Mohd Shah, N.; Shahrol Nidzam, N.N. Structural stability and electronic properties of graphene/germanene heterobilayer. *Results Phys.* **2021**, *28*, 104545. **[Google Scholar]** **[CrossRef]**

10. Zhang, H.; Zhang, B.; Gao, Q.; Song, J.; Han, G. A review on microstructures and properties of graphene-reinforced aluminum matrix composites fabricated by friction stir processing. *J. Manuf. Process.* **2021**, *68*, 126–135. **[Google Scholar]** **[CrossRef]**

11. Wu, Y.-C.; Shao, J.-L.; Zhan, H. Damage and self-healing characteristics of monolayer graphene enhanced Cu under ballistic impact. *Mech. Mater.* **2021**, *155*, 103736. **[Google Scholar]** **[CrossRef]**

12. Graf, D.; Molitor, F.; Ensslin, K.; Stampfer, C.; Jungen, A.; Hierold, C.; Wirtz, L. Spatially Resolved Raman Spectroscopy of Single- and Few-Layer Graphene. *Nano Lett.* **2007**, *7*, 238–242. **[Google Scholar]** **[CrossRef]** **[PubMed]** **[Green Version]**

13. Liu, B.; Cao, S.; Gao, N.; Cheng, L.; Liu, Y.; Zhang, Y.; Feng, D. Thermosetting CFRP interlaminar toughening with multi-layers graphene and MWCNTs under mode I fracture. *Compos. Sci. Technol.* **2019**, *183*, 107829. **[Google Scholar]** **[CrossRef]**

14. Zhang, Y.Y.; Gu, Y.T. Mechanical properties of graphene: Effects of layer number, temperature and isotope. *Comput. Mater. Sci.* **2013**, *71*, 197–200. **[Google Scholar]** **[CrossRef]** **[Green Version]**

15. Mag-isa, A.E.; Kim, S.-M.; Kim, J.-H.; Oh, C.-S. Variation of thermal expansion coefficient of freestanding multilayer pristine graphene with temperature and number of layers. *Mater. Today Commun.* **2020**, *25*, 101387. **[Google Scholar]** **[CrossRef]**

16. Tsyganov, D.; Bundaleska, N.; Henriques, J.; Felizardo, E.; Dias, A.; Abrashev, M.; Kissovski, J.; Botelho do Rego, A.M.; Ferraria, A.M.; Tatarova, E. Simultaneous Synthesis and Nitrogen Doping of Free-Standing Graphene Applying Microwave Plasma. *Materials* **2020**, *13*, 4213. **[Google Scholar] [CrossRef]**
17. Mukri, M.-R.; Elias, M.S.; Aziz, M.; Tanemura, M.; Mohd Yusop, M.Z. Structural Modification of Pristine Graphene Network Towards Nanoporous Graphene Membrane: A Review. *J. Appl. Membr. Sci. Technol.* **2018**, *22*. **[Google Scholar] [CrossRef]**
18. Novoselov, K.S.; Geim, A.K.; Morozov, S.V.; Jiang, D.; Zhang, Y.; Dubonos, S.V.; Grigorieva, I.V.; Firsov, A.A. Electric field effect in atomically thin carbon films. *Science* **2004**, *306*, 666–669. **[Google Scholar] [CrossRef] [Green Version]**
19. Geim, A.K.; Novoselov, K.S. The rise of graphene. In *Nanoscience and Technology*; Macmillan Publishers Ltd.: London, UK, 2009; pp. 11–19. **[Google Scholar]**
20. Shende, P.; Pathan, N. Potential of carbohydrate-conjugated graphene assemblies in biomedical applications. *Carbohydr. Polym.* **2021**, *255*, 117385. **[Google Scholar] [CrossRef]**
21. Song, S.; Shen, H.; Wang, Y.; Chu, X.; Xie, J.; Zhou, N.; Shen, J. Biomedical application of graphene: From drug delivery, tumor therapy, to theranostics. *Colloids Surf. B Biointerfaces* **2020**, *185*, 110596. **[Google Scholar] [CrossRef]**
22. Lin, L.; Peng, H.; Liu, Z. Synthesis challenges for graphene industry. *Nat. Mater.* **2019**, *18*, 520–524. **[Google Scholar] [CrossRef]**
23. Urban, K.W. The challenges of graphene. *Nat. Mater.* **2011**, *10*, 165–166. **[Google Scholar] [CrossRef]**
24. Reina, G.; González-Domínguez, J.M.; Criado, A.; Vázquez, E.; Bianco, A.; Prato, M. Promises, facts and challenges for graphene in biomedical applications. *Chem. Soc. Rev.* **2017**, *46*, 4400–4416. **[Google Scholar] [CrossRef] [PubMed] [Green Version]**
25. Jeong, J.H.; Kim, Y.H.; Roh, K.C.; Kim, K.-B. Effect of thermally decomposable spacers on graphene microsphere structure and restacking of graphene sheets during electrode fabrication. *Carbon* **2019**, *150*, 128–135. **[Google Scholar] [CrossRef]**
26. Xie, Q.; Zhang, Y.; Zhao, P. Facile fabrication of honeycomb-like restacking-inhibited graphene architecture with superior electrochemical performance for energy storage. *Mater. Lett.* **2018**, *225*, 93–96. **[Google Scholar] [CrossRef]**
27. Kota, M.; Park, H.S. Restacking-inhibited nitrogen-incorporated mesoporous reduced graphene oxides for high energy supercapacitors. *Ceram. Int.* **2018**, *44*, 3195–3200. **[Google Scholar] [CrossRef]**

28. Xu, G.; Yuan, J.; Geng, X.; Dou, H.; Chen, L.; Yan, X.; Zhu, H. Caterpillar-like graphene confining sulfur by restacking effect for high performance lithium sulfur batteries. *Chem. Eng. J.* **2017**, *322*, 454–462. **[Google Scholar] [CrossRef]**

29. Tambe, P. Synthesis and characterization of acid treated reduced graphene oxide. *Mater. Today Proc.* **2022**, *49*, 1294–1297. **[Google Scholar] [CrossRef]**

30. Fu, H.; Gao, B.; Liu, Z.; Liu, W.; Wang, Z.; Wang, M.; Li, J.; Feng, Z.; Reza Kamali, A. Electrochemical performance of honeycomb graphene prepared from acidic graphene oxide via a chemical expansion method. *J. Electroanal. Chem.* **2022**, *920*, 116545. **[Google Scholar] [CrossRef]**

31. Hauquier, F.; Alamarguy, D.; Viel, P.; Noël, S.; Filoramo, A.; Huc, V.; Houzé, F.; Palacin, S. Conductive-probe AFM characterization of graphene sheets bonded to gold surfaces. *Appl. Surf. Sci.* **2012**, *258*, 2920–2926. **[Google Scholar] [CrossRef]**

32. Agarwal, P.B.; Paulchowdhury, P.; Mukherjee, A.; Lohani, P.; Thakur, N.K. Optimization of oxygen plasma based etching of single layered graphene through Raman and FESEM characterization. *Mater. Today Proc.* **2022**, *48*, 616–618. **[Google Scholar] [CrossRef]**

33. Shtepliuk, I.; Ivanov, I.G.; Pliatsikas, N.; Iakimov, T.; Jamnig, A.; Sarakinos, K.; Yakimova, R. Probing the uniformity of silver-doped epitaxial graphene by micro-Raman mapping. *Phys. B Condens. Matter* **2020**, *580*, 411751. **[Google Scholar] [CrossRef]**

34. Pelaez-Fernandez, M.; Bermejo, A.; Benito, A.M.; Maser, W.K.; Arenal, R. Detailed thermal reduction analyses of graphene oxide via in-situ TEM/EELS studies. *Carbon* **2021**, *178*, 477–487. **[Google Scholar] [CrossRef]**

35. Wang, Z.; Yu, H.; Zhao, Z. Silk fibroin hydrogel encapsulated graphene filed-effect transistors as enzyme-based biosensors. *Microchem. J.* **2021**, *169*, 106585. **[Google Scholar] [CrossRef]**

36. GV, Y.D.; Prabhu, A.; Anil, S.; Venkatesan, J. Preparation and characterization of dexamethasone loaded sodium alginate-graphene oxide microspheres for bone tissue engineering. *J. Drug Deliv. Sci. Technol.* **2021**, *64*, 102624. **[Google Scholar] [CrossRef]**

37. Karimi, S.; Namazi, H. Fe_3O_4@PEG-coated dendrimer modified graphene oxide nanocomposite as a pH-sensitive drug carrier for targeted delivery of doxorubicin. *J. Alloys Compd.* **2021**, *879*, 160426. **[Google Scholar] [CrossRef]**

38. Sharif, S.; Ahmad, K.S.; Rehman, F.; Bhatti, Z.; Thebo, K.H. Two-dimensional graphene oxide based membranes for ionic and molecular separation: Current status and challenges. *J. Environ. Chem. Eng.* **2021**, *9*, 105605. **[Google Scholar] [CrossRef]**

39. Ikram, R.; Jan, B.M.; Ahmad, W. Advances in synthesis of graphene derivatives using industrial wastes precursors; prospects and challenges. *J. Mater. Res. Technol.* **2020**, *9*, 15924–15951. **[Google Scholar]** **[CrossRef]**

40. Rotte, N.K.; Naresh, V.; Muduli, S.; Reddy, V.; Srikanth, V.V.S.; Martha, S.K. Microwave aided scalable synthesis of sulfur, nitrogen co-doped few-layered graphene material for high-performance supercapacitors. *Electrochim. Acta* **2020**, *363*, 137209. **[Google Scholar]** **[CrossRef]**

41. Salussolia, G.; Barbieri, E.; Pugno, N.M.; Botto, L. Micromechanics of liquid-phase exfoliation of a layered 2D material: A hydrodynamic peeling model. *J. Mech. Phys. Solids* **2020**, *134*, 103764. **[Google Scholar]** **[CrossRef]**

42. Hwang, G.; Kim, T.; Shin, J.; Shin, N.; Hwang, S. Machine learnings for CVD graphene analysis: From measurement to simulation of SEM images. *J. Ind. Eng. Chem.* **2021**, *101*, 430–444. **[Google Scholar]** **[CrossRef]**

43. Yildiz, G.; Bolton-Warberg, M.; Awaja, F. Graphene and graphene oxide for bio-sensing: General properties and the effects of graphene ripples. *Acta Biomater.* **2021**, *131*, 62–79. **[Google Scholar]** **[CrossRef]**

44. Guo, J.; Mao, B.; Li, J.; Wang, X.; Yang, X. Rethinking the reaction pathways of chemical reduction of graphene oxide. *Carbon* **2021**, *171*, 963–967. **[Google Scholar]** **[CrossRef]**

45. Barkauskas, J.; Gaidukevič, J.; Niaura, G. Thermal reduction of graphite oxide in the presence of nitrogen-containing dyes. *Carbon Lett.* **2021**, *31*, 1097–1110. **[Google Scholar]** **[CrossRef]**

46. Wadekar, P.H.; Ghosh, A.; Khose, R.V.; Pethsangave, D.A.; Mitra, S.; Some, S. A novel chemical reduction/co-precipitation method to prepare sulfur functionalized reduced graphene oxide for lithium-sulfur batteries. *Electrochim. Acta* **2020**, *344*, 136147. **[Google Scholar]** **[CrossRef]**

47. Vázquez-Sánchez, P.; Rodríguez-Escudero, M.A.; Burgos, F.J.; Llorente, I.; Caballero-Calero, O.; González, M.M.; Fernández, R.; García-Alonso, M.C. Synthesis of Cu/rGO composites by chemical and thermal reduction of graphene oxide. *J. Alloys Compd.* **2019**, *800*, 379–391. **[Google Scholar]** **[CrossRef]**

48. Demetriou, G.; Biancalana, F.; Abraham, E.; Ji, W.; Wang, Y.; Kar, A.K. Direct observation of an irradiance dependent nonlinear refraction in CVD single layer graphene. *Opt. Commun.* **2021**, *481*, 126535. **[Google Scholar]** **[CrossRef]**

49. Jiang, X.; Song, J.; Chen, S.; Su, Y.; Fan, H.; Zhang, Y.; Hu, L. In-situ fabricated bulk metallic glass/graphite composites with a 3D lubricating layer: Tribological properties under dry sliding and in seawater. *Tribol. Int.* **2020**, *148*, 106301. **[Google Scholar]** **[CrossRef]**

50. Liu, Q.; Pang, M.; Chen, J.; Liu, G.; Zhang, L. Microstructure and properties characterization of Ti-containing Ni60/Graphite self-lubricating composite coatings applied on 300 M ultra-high strength steel by laser cladding. *Mater. Chem. Phys.* **2021**, *266*, 124554. **[Google Scholar] [CrossRef]**

51. Ferreira, M.P.; da Nova Mussel, W.; Dutra, P.R.; Ângela de Barros Correia Menezes, M.; Pedrosa, T.A. Physical-chemical exfoliation of pristine graphite flakes. *Radiat. Phys. Chem.* **2021**, *188*, 109652. **[Google Scholar] [CrossRef]**

52. Tene, T.; Guevara, M.; Viteri, E.; Maldonado, A.; Pisarra, M.; Sindona, A.; Vacacela Gomez, C.; Bellucci, S. Calibration of Fermi Velocity to Explore the Plasmonic Character of Graphene Nanoribbon Arrays by a Semi-Analytical Model. *Nanomaterials* **2022**, *12*, 2028. **[Google Scholar] [CrossRef] [PubMed]**

53. Aixart, J.; Díaz, F.; Llorca, J.; Rosell-Llompart, J. Increasing reaction time in Hummers' method towards well exfoliated graphene oxide of low oxidation degree. *Ceram. Int.* **2021**, *47*, 22130–22137. **[Google Scholar] [CrossRef]**

54. Rozmanowski, T.; Krawczyk, P. Methanol electrooxidation at $NiCl_2$–$FeCl_3$-graphite intercalation compound affected by ozone treatment. *J. Phys. Chem. Solids* **2021**, *157*, 110223. **[Google Scholar] [CrossRef]**

55. Yamamoto, H.; Matsumoto, K.; Hagiwara, R. Stage-number dependence of intercalated species for fluorosilicate graphite intercalation compounds: Pentafluorosilicate vs. hexafluorosilicate. *J. Fluor. Chem.* **2021**, *242*, 109714. **[Google Scholar] [CrossRef]**

56. Cahen, S.; Vangelisti, R. Chemical vapor transport for intercalation reactions: Synthesis of a 1st stage DyCl3 graphite intercalation compound. *J. Solid State Chem.* **2021**, *299*, 122185. **[Google Scholar] [CrossRef]**

57. Suvarna, K.S.; Binitha, N.N. Graphene preparation by jaggery assisted ball-milling of graphite for the adsorption of Cr(VI). *Mater. Today Proc.* **2020**, *25*, 236–240. **[Google Scholar] [CrossRef]**

58. Tenorio Gonzalez, F.N.; Barajas Rosales, I.R.; Vera Serna, P.; Sánchez de Jesus, F.; Bolarin Miró, A.M.; Garrido Hernández, A.; Kusý, M. Reducing the crystallite and particle size of SrFe12O19 with PVA by high energy ball milling. *J. Alloys Compd.* **2019**, *771*, 464–470. **[Google Scholar] [CrossRef]**

59. Tsyganov, D.; Bundaleska, N.; Tatarova, E.; Dias, A.; Henriques, J.; Rego, A.; Ferraria, A.; Abrashev, M.V.; Dias, F.M.; Luhrs, C.C.; et al. On the plasma-based growth of 'flowing' graphene sheets at atmospheric pressure conditions. *Plasma Sources Sci. Technol.* **2016**, *25*, 015013. **[Google Scholar] [CrossRef] [Green Version]**

60. Tatarova, E.; Dias, A.; Henriques, J.; Abrashev, M.; Bundaleska, N.; Kovacevic, E.; Bundaleski, N.; Cvelbar, U.; Valcheva, E.; Arnaudov, B.; et al. Towards large-scale in free-standing graphene and N-graphene sheets. *Sci. Rep.* **2017**, *7*, 10175. **[Google Scholar] [CrossRef] [Green Version]**

61. Zhao, C.; Hong, Y.; Chu, X.; Dong, Y.; Hu, Z.; Sun, X.; Yan, S. Enhanced ferroelectric properties of P(VDF-TrFE) thin film on single-layer graphene simply adjusted by crystallization condition. *Mater. Today Energy* **2021**, *20*, 100678. **[Google Scholar] [CrossRef]**

62. Wu, Y.; Wang, S.; Komvopoulos, K. A review of graphene synthesis by indirect and direct deposition methods. *J. Mater. Res.* **2020**, *35*, 76–89. **[Google Scholar] [CrossRef]**

63. Depine, R.A. *Graphene Optics: Electromagnetic Solution of Canonical Problems*; IOP Publishing: Bristol, UK, 2016; ISBN 978-1-6817-4309-7. **[Google Scholar]**

64. Hanson, G.W. Dyadic Green's functions and guided surface waves for a surface conductivity model of graphene. *J. Appl. Phys.* **2008**, *103*, 064302. **[Google Scholar] [CrossRef] [Green Version]**

65. Xu, X.; Sun, S.; Luo, J.; Ma, R.; Lin, J.; Fang, L.; Zhang, P.; Chen, Y. Few-layer graphene prepared via microwave irradiation of black sesame for supercapacitor applications. *Chem. Eng. J.* **2021**, *425*, 130664. **[Google Scholar] [CrossRef]**

66. Chen, J.; Liu, R.; Zhu, L.; Chen, W.; Dong, C.; Wan, Z.; Cao, W.; Zhang, X.; Peng, R.; Wang, M. Sb2S3-based bulk/nano planar heterojunction film solar cells with graphene/polymer composite layer as hole extracting interface. *Mater. Lett.* **2021**, *300*, 130190. **[Google Scholar] [CrossRef]**

67. Zhang, Z.; Jin, H.; Wu, C.; Ji, J. Efficient Production of High-Quality Few-Layer Graphene Using a Simple Hydrodynamic-Assisted Exfoliation Method. *Nanoscale Res. Lett.* **2018**, *13*, 416. **[Google Scholar] [CrossRef] [PubMed] [Green Version]**

68. Schilirò, E.; Lo Nigro, R.; Panasci, S.E.; Gelardi, F.M.; Agnello, S.; Yakimova, R.; Roccaforte, F.; Giannazzo, F. Aluminum oxide nucleation in the early stages of atomic layer deposition on epitaxial graphene. *Carbon* **2020**, *169*, 172–181. **[Google Scholar] [CrossRef]**

69. Gao, B.; Hu, C.; Fu, H.; Sun, Y.; Li, K.; Hu, L. Preparation of single-layer graphene based on a wet chemical synthesis route and the effect on electrochemical properties by double layering surface functional groups to modify graphene oxide. *Electrochim. Acta* **2020**, *361*, 137053. **[Google Scholar] [CrossRef]**

70. Yuan, S.-J.; Dong, B.; Dai, X.-H. Facile and scalable synthesis of high-quality few-layer graphene from biomass by a universal solvent-free approach. *Appl. Surf. Sci.* **2021**, *562*, 150203. **[Google Scholar] [CrossRef]**

71. Bhuiyan, A.G.; Terai, T.; Katsuzaki, T.; Takeda, N.; Hashimoto, A. Growth of single crystalline Si on graphene using RF-MBE: Orientation control with an AlN interface layer. *Appl. Surf. Sci.* **2021**, *548*, 149295. **[Google Scholar] [CrossRef]**

72. Araby, M.I.; Rosmi, M.S.; Vishwakarma, R.; Sharma, S.; Wakamatsu, Y.; Takahashi, K.; Kalita, G.; Kitazawa, M.; Tanemura, M. Graphene formation at 150 °C using indium as catalyst. *RSC Adv.* **2017**, *7*, 47353–47356. **[Google Scholar] [CrossRef]**

73. Long, C.; Xie, X.; Fu, J.; Wang, Q.; Guo, H.; Zeng, W.; Wei, N.; Wang, S.; Xiong, Y. Supercapacitive brophene-graphene aerogel as elastic-electrochemical dielectric layer for sensitive pressure sensors. *J. Colloid Interface Sci.* **2021**, *601*, 355–364. **[Google Scholar] [CrossRef] [PubMed]**

74. Navik, R.; Gai, Y.; Wang, W.; Zhao, Y. Curcumin-assisted ultrasound exfoliation of graphite to graphene in ethanol. *Ultrason. Sonochem.* **2018**, *48*, 96–102. **[Google Scholar] [CrossRef] [PubMed]**

75. Ding, J.; Zhao, H.; Yu, H. Graphene nanofluids based on one-step exfoliation and edge-functionalization. *Carbon* **2021**, *171*, 29–35. **[Google Scholar] [CrossRef]**

76. Ferrari, A.C.; Meyer, J.C.; Scardaci, V.; Casiraghi, C.; Lazzeri, M.; Mauri, F.; Piscanec, S.; Jiang, D.; Novoselov, K.S.; Roth, S.; et al. Raman Spectrum of Graphene and Graphene Layers. *Phys. Rev. Lett.* **2006**, *97*, 187401. **[Google Scholar] [CrossRef] [Green Version]**

77. Campanelli, D.A.; Menezes, J.W.; Valsecchi, C.; Zegarra, L.B.R.; Jacinto, C.; Armas, L.E.G. Interaction between Yb^{3+} doped glasses substrates and graphene layers by raman spectroscopy. *Thin Solid Films* **2020**, *712*, 138315. **[Google Scholar] [CrossRef]**

78. Wang, T.; Huntzinger, J.-R.; Bayle, M.; Roblin, C.; Decams, J.-M.; Zahab, A.-A.; Contreras, S.; Paillet, M.; Landois, P. Buffer layers inhomogeneity and coupling with epitaxial graphene unravelled by Raman scattering and graphene peeling. *Carbon* **2020**, *163*, 224–233. **[Google Scholar] [CrossRef] [Green Version]**

79. Silva, D.L.; Campos, J.L.E.; Fernandes, T.F.D.; Rocha, J.N.; Machado, L.R.P.; Soares, E.M.; Miquita, D.R.; Miranda, H.; Rabelo, C.; Vilela Neto, O.P.; et al. Raman spectroscopy analysis of number of layers in mass-produced graphene flakes. *Carbon* **2020**, *161*, 181–189. **[Google Scholar] [CrossRef]**

80. Huang, S.; Li, S.; Hsu, K.-J.; Villalobos, L.F.; Agrawal, K.V. Systematic design of millisecond gasification reactor for the incorporation of gas-sieving nanopores in single-layer graphene. *J. Membr. Sci.* **2021**, *637*, 119628. **[Google Scholar] [CrossRef]**

81. Bouhafs, C.; Pezzini, S.; Geisenhof, F.R.; Mishra, N.; Mišeikis, V.; Niu, Y.; Struzzi, C.; Weitz, R.T.; Zakharov, A.A.; Forti, S.; et al. Synthesis of large-area rhombohedral few-layer graphene by chemical vapor deposition on copper. *Carbon* **2021**, *177*, 282–290. **[Google Scholar] [CrossRef]**

82. Bleu, Y.; Bourquard, F.; Michalon, J.-Y.; Lefkir, Y.; Reynaud, S.; Loir, A.-S.; Barnier, V.; Garrelie, F.; Donnet, C. Transfer-free graphene synthesis by nickel catalyst dewetting using rapid thermal annealing. *Appl. Surf. Sci.* **2021**, *555*, 149492. **[Google Scholar] [CrossRef]**

83. Lebioda, M.; Pawlak, R.; Szymański, W.; Kaczorowski, W.; Jeziorna, A. Laser Patterning a Graphene Layer on a Ceramic Substrate for Sensor Applications. *Sensors* **2020**, *20*, 2134. **[Google Scholar] [CrossRef]**

84. Kaur, M.; Kaur, M.; Sharma, V.K. Nitrogen-doped graphene and graphene quantum dots: A review onsynthesis and applications in energy, sensors and environment. *Adv. Colloid Interface Sci.* **2018**, *259*, 44–64. **[Google Scholar] [CrossRef]**

85. Rodriguez-Losada, N.; Aguirre, J. The impact of graphene on neural regenerative medicine. *Neural Regen. Res.* **2017**, *12*, 1071. **[Google Scholar] [CrossRef]**

86. Jahanshahi, M.; Kowsari, E.; Haddadi-Asl, V.; Khoobi, M.; Bazri, B.; Aryafard, M.; Lee, J.H.; Kadumudi, F.B.; Talebian, S.; Kamaly, N.; et al. An innovative and eco-friendly modality for synthesis of highly fluorinated graphene by an acidic ionic liquid: Making of an efficacious vehicle for anti-cancer drug delivery. *Appl. Surf. Sci.* **2020**, *515*, 146071. **[Google Scholar] [CrossRef]**

87. Potsi, G.; Bourlinos, A.B.; Mouselimis, V.; Poláková, K.; Chalmpes, N.; Gournis, D.; Kalytchuk, S.; Tomanec, O.; Błoński, P.; Medve□, M.; et al. Intrinsic photoluminescence of amine-functionalized graphene derivatives for bioimaging applications. *Appl. Mater. Today* **2019**, *17*, 112–122. **[Google Scholar] [CrossRef]**

88. Gu, H.; Tang, H.; Xiong, P.; Zhou, Z. Biomarkers-based Biosensing and Bioimaging with Graphene for Cancer Diagnosis. *Nanomaterials* **2019**, *9*, 130. **[Google Scholar] [CrossRef] [PubMed] [Green Version]**

89. Kovalska, E.; Lesongeur, P.; Hogan, B.T.; Baldycheva, A. Multi-layer graphene as a selective detector for future lung cancer biosensing platforms. *Nanoscale* **2019**, *11*, 2476–2483. **[Google Scholar] [CrossRef] [PubMed] [Green Version]**

90. Nair, R.R.; Blake, P.; Grigorenko, A.N.; Novoselov, K.S.; Booth, T.J.; Stauber, T.; Peres, N.M.R.; Geim, A.K. Fine Structure Constant Defines Visual Transparency of Graphene. *Science* **2008**, *320*, 1308. **[Google Scholar] [CrossRef] [PubMed] [Green Version]**

91. Sang, T.; Gao, J.; Yin, X.; Qi, H.; Wang, L.; Jiao, H. Angle-Insensitive Broadband Absorption Enhancement of Graphene Using a Multi-Grooved Metasurface. *Nanoscale Res. Lett.* **2019**, *14*, 105. **[Google Scholar] [CrossRef] [Green Version]**

92. Chen, J.; Chen, S.; Gu, P.; Yan, Z.; Tang, C.; Xu, Z.; Liu, B.; Liu, Z. Electrically modulating and switching infrared absorption of monolayer graphene in metamaterials. *Carbon* **2020**, *162*, 187–194. **[Google Scholar] [CrossRef]**

93. Divya, J.; Selvendran, S.; Raja, A.S.; Sivasubramanian, A. Graphene-Au-Coated Plasmonic Sensor Based on D-Shaped Bezier Polygonal Hollow Core Photonic Crystal Fiber. *Braz. J. Phys.* **2021**, *51*, 1314–1323. **[Google Scholar] [CrossRef]**

94. Georgakilas, V.; Otyepka, M.; Bourlinos, A.B.; Chandra, V.; Kim, N.; Kemp, K.C.; Hobza, P.; Zboril, R.; Kim, K.S. Functionalization of Graphene: Covalent and Non-Covalent Approaches, Derivatives and Applications. *Chem. Rev.* **2012**, *112*, 6156–6214. **[Google Scholar] [CrossRef]**

95. Serban, B.-C.; Cobianu, C.; Buiu, O.; Bumbac, M.; Dumbravescu, N.; Avramescu, V.; Nicolescu, C.M.; Brezeanu, M.; Radulescu, C.; Craciun, G.; et al. Quaternary Oxidized Carbon Nanohorns—Based Nanohybrid as Sensing Coating for Room Temperature Resistive Humidity Monitoring. *Coatings* **2021**, *11*, 530. **[Google Scholar] [CrossRef]**

96. Kumar Manoharan, A.; Jayavel, R.; Shanmugam, M.; Sengottaiyan, C.; Chinnathambi, S.; Mohankumar, N. A Compact Sensory Platform Based pH Sensor Using Graphene Field Effect Transistor. *J. Nanosci. Nanotechnol.* **2021**, *21*, 3299–3305. **[Google Scholar] [CrossRef]**

97. Xu, S.; Zhan, J.; Man, B.; Jiang, S.; Yue, W.; Gao, S.; Guo, C.; Liu, H.; Li, Z.; Wang, J.; et al. Real-time reliable determination of binding kinetics of DNA hybridization using a multi-channel graphene biosensor. *Nat. Commun.* **2017**, *8*, 14902. **[Google Scholar] [CrossRef] [PubMed] [Green Version]**

98. Ahmed, S.; Jiang, X.; Wang, C.; Kalsoom, U.E.; Wang, B.; Khan, J.; Muhammad, Y.; Duan, Y.; Zhu, H.; Ren, X.; et al. An Insightful Picture of Nonlinear Photonics in 2D Materials and their Applications: Recent Advances and Future Prospects. *Adv. Opt. Mater.* **2021**, *9*, 2001671. **[Google Scholar] [CrossRef]**

99. Hu, Z.; Shao, Q.; Moloney, M.G.; Xu, X.; Zhang, D.; Li, J.; Zhang, C.; Huang, Y. Nondestructive Functionalization of Graphene by Surface-Initiated Atom Transfer Radical Polymerization: An Ideal Nanofiller for Poly(p-phenylene

benzobisoxazole) Fibers. *Macromolecules* **2017**, *50*, 1422–1429. **[Google Scholar]** **[CrossRef]**

100. Devi, N.A.; Swain, B.P. Investigation of Metal-Oxide/Reduced Graphene-Oxide Nanocomposites for Gas Sensor Applications. In *Nanostructured Materials and Their Applications*; Springer: Singapore, 2021; pp. 211–227. **[Google Scholar]**

101. Nurrohman, D.T.; Chiu, N.-F. A Review of Graphene-Based Surface Plasmon Resonance and Surface-Enhanced Raman Scattering Biosensors: Current Status and Future Prospects. *Nanomaterials* **2021**, *11*, 216. **[Google Scholar]** **[CrossRef]**

102. Wenjuan, Y.; Le Goff, A.; Spinelli, N.; Holzinger, M.; Diao, G.-W.; Shan, D.; Defrancq, E.; Cosnier, S. Electrogenerated trisbipyridyl Ru(II)-/nitrilotriacetic-polypyrene copolymer for the easy fabrication of label-free photoelectrochemical immunosensor and aptasensor: Application to the determination of thrombin and anti-cholera toxinantibody. *Biosens. Bioelectron.* **2013**, *42*, 556–562. **[Google Scholar]** **[CrossRef]**

103. Cai, H.; Wang, M.; Wu, Z.; Liu, J.; Wang, X. Performance Enhancement of SPR Biosensor Using Graphene–MoS_2 Hybrid Structure. *Nanomaterials* **2022**, *12*, 2219. **[Google Scholar]** **[CrossRef]** **[PubMed]**

104. Fatima, N.; Qazi, U.Y.; Mansha, A.; Bhatti, I.A.; Javaid, R.; Abbas, Q.; Nadeem, N.; Rehan, Z.A.; Noreen, S.; Zahid, M. Recent developments for antimicrobial applications of graphene-based polymeric composites: A review. *J. Ind. Eng. Chem.* **2021**, *100*, 40–58. **[Google Scholar]** **[CrossRef]**

105. Wychowaniec, J.K.; Iliut, M.; Zhou, M.; Moffat, J.; Elsawy, M.A.; Pinheiro, W.A.; Hoyland, J.A.; Miller, A.F.; Vijayaraghavan, A.; Saiani, A. Designing Peptide/Graphene Hybrid Hydrogels through Fine-Tuning of Molecular Interactions. *Biomacromolecules* **2018**, *19*, 2731–2741. **[Google Scholar]** **[CrossRef]** **[PubMed]** **[Green Version]**

106. Vafaei, S.; Allabush, F.; Tabaei, S.R.; Male, L.; Dafforn, T.R.; Tucker, J.H.R.; Mendes, P.M. Förster Resonance Energy Transfer Nanoplatform Based on Recognition-Induced Fusion/Fission of DNA Mixed Micelles for Nucleic Acid Sensing. *ACS Nano* **2021**, *15*, 8517–8524. **[Google Scholar]** **[CrossRef]** **[PubMed]**

107. Wychowaniec, J.K.; Litowczenko, J.; Tadyszak, K. Fabricating versatile cell supports from nano- and micro-sized graphene oxide flakes. *J. Mech. Behav. Biomed. Mater.* **2020**, *103*, 103594. **[Google Scholar]** **[CrossRef]** **[PubMed]**

108. Koren, K.; Zieger, S.E. Optode Based Chemical Imaging—Possibilities, Challenges, and New Avenues in Multidimensional Optical Sensing. *ACS Sens.* **2021**, *6*, 1671–1680. **[Google Scholar]** **[CrossRef]**

109. Parvin, N.; Mandal, T.; Roy, P. Soluble Nanoparticles Versatile Tools for Plant Tissue Imaging. *J. Bionanosci.* **2013**, *7*, 256–259. **[Google Scholar]** **[CrossRef]**

110. Kim, S.; Moriya, S.; Maruki, S.; Fukaminato, T.; Ogata, T.; Kurihara, S. Adsorption and release on three-dimensional graphene oxide network structures. *R. Soc. Open Sci.* **2021**, *8*, 201585. **[Google Scholar]** **[CrossRef]**

111. Driscoll, J.; Moirangthem, A.; Yan, I.K.; Patel, T. Fabrication and Characterization of a Biomaterial Based on Extracellular-Vesicle Functionalized Graphene Oxide. *Front. Bioeng. Biotechnol.* **2021**, *9*, 686510. **[Google Scholar]** **[CrossRef]**

112. Moody, A.S.; Dayton, P.A.; Zamboni, W.C. Imaging methods to evaluate tumor microenvironment factors affecting nanoparticle drug delivery and antitumor response. *Cancer Drug Resist.* **2021**, *4*, 382. **[Google Scholar]** **[CrossRef]**

113. Jakubovic, R.; Ramjist, J.; Gupta, S.; Guha, D.; Sahgal, A.; Foster, F.S.; Yang, V.X.D. High-Frequency Micro-Ultrasound Imaging and Optical Topographic Imaging for Spinal Surgery: Initial Experiences. *Ultrasound Med. Biol.* **2018**, *44*, 2379–2387. **[Google Scholar]** **[CrossRef]**

114. Phan, T.T.V.; Huynh, T.-C.; Manivasagan, P.; Mondal, S.; Oh, J. An Up-To-Date Review on Biomedical Applications of Palladium Nanoparticles. *Nanomaterials* **2019**, *10*, 66. **[Google Scholar]** **[CrossRef]** **[PubMed]** **[Green Version]**

115. Tadyszak, K.; Wychowaniec, J.; Litowczenko, J. Biomedical Applications of Graphene-Based Structures. *Nanomaterials* **2018**, *8*, 944. **[Google Scholar]** **[CrossRef]** **[PubMed]** **[Green Version]**

116. Parvin, N.; Jin, Q.; Wei, Y.; Yu, R.; Zheng, B.; Huang, L.; Zhang, Y.; Wang, L.; Zhang, H.; Gao, M.; et al. Few-Layer Graphdiyne Nanosheets Applied for Multiplexed Real-Time DNA Detection. *Adv. Mater.* **2017**, *29*, 201606755. **[Google Scholar]** **[CrossRef]**

117. Dolatkhah, M.; Hashemzadeh, N.; Barar, J.; Adibkia, K.; Aghanejad, A.; Barzegar-Jalali, M.; Omidi, Y. Graphene-based multifunctional nanosystems for simultaneous detection and treatment of breast cancer. *Colloids Surf. B Biointerfaces* **2020**, *193*, 111104. **[Google Scholar]** **[CrossRef]**

118. Jang, S.-C.; Kang, S.-M.; Lee, J.Y.; Oh, S.Y.; Vilian, A.E.; Lee, I.; Han, Y.-K.; Park, J.H.; Cho, W.-S.; Roh, C.; et al. Nano-graphene oxide composite for in vivo imaging. *Int. J. Nanomed.* **2018**, *13*, 221–234. **[Google Scholar]** **[CrossRef]** **[Green Version]**

119. Matsumoto, K.; Mitchell, J.B.; Krishna, M.C. Multimodal Functional Imaging for Cancer/Tumor Microenvironments Based on MRI, EPRI, and PET. *Molecules* **2021**, *26*, 1614. **[Google Scholar]** **[CrossRef]**

120. Sheth, D.B.; Gratzl, M. Electrochemical mapping of oxygenation in the three-dimensional multicellular tumour hemi-spheroid. *Proc. R. Soc. A Math. Phys. Eng. Sci.* **2019**, *475*, 20180647. **[Google Scholar] [CrossRef] [Green Version]**

121. Kuppusamy, P.; Shankar, R.A.; Zweier, J.L. In vivo measurement of arterial and venous oxygenation in the rat using 3D spectral-spatial electron paramagnetic resonance imaging. *Phys. Med. Biol.* **1998**, *43*, 1837–1844. **[Google Scholar] [CrossRef]**

122. Khramtsov, V.V.; Bobko, A.A.; Tseytlin, M.; Driesschaert, B. Exchange Phenomena in the Electron Paramagnetic Resonance Spectra of the Nitroxyl and Trityl Radicals: Multifunctional Spectroscopy and Imaging of Local Chemical Microenvironment. *Anal. Chem.* **2017**, *89*, 4758–4771. **[Google Scholar] [CrossRef] [PubMed] [Green Version]**

123. Nguyen, H.V.-T.; Chen, Q.; Paletta, J.T.; Harvey, P.; Jiang, Y.; Zhang, H.; Boska, M.D.; Ottaviani, M.F.; Jasanoff, A.; Rajca, A.; et al. Nitroxide-Based Macromolecular Contrast Agents with Unprecedented Transverse Relaxivity and Stability for Magnetic Resonance Imaging of Tumors. *ACS Cent. Sci.* **2017**, *3*, 800–811. **[Google Scholar] [CrossRef]**

124. Boś-Liedke, A.; Walawender, M.; Woźniak, A.; Flak, D.; Gapiński, J.; Jurga, S.; Kucińska, M.; Plewiński, A.; Murias, M.; Elewa, M.; et al. EPR Oximetry Sensor—Developing a TAM Derivative for In Vivo Studies. *Cell Biochem. Biophys.* **2018**, *76*, 19–28. **[Google Scholar] [CrossRef]**

125. Mrówczyński, R.; Coy, L.E.; Scheibe, B.; Czechowski, T.; Augustyniak-Jabłokow, M.; Jurga, S.; Tadyszak, K. Electron Paramagnetic Resonance Imaging and Spectroscopy of Polydopamine Radicals. *J. Phys. Chem. B* **2015**, *119*, 10341–10347. **[Google Scholar] [CrossRef] [PubMed]**

126. Ferini, G.; Valenti, V.; Tripoli, A.; Illari, S.I.; Molino, L.; Parisi, S.; Cacciola, A.; Lillo, S.; Giuffrida, D.; Pergolizzi, S. Lattice or Oxygen-Guided Radiotherapy: What If They Converge? Possible Future Directions in the Era of Immunotherapy. *Cancers* **2021**, *13*, 3290. **[Google Scholar] [CrossRef]**

127. Hirai, N.; Maeda, Y.; Hashimoto, K.; Andriana, B.B.; Matsuyoshi, H.; Sato, H. Coherent Anti-Stokes Raman Scattering Spectroscopy Using a Double-Wavelength-Emission Electronically Tuned Ti:Sapphire Laser. *Appl. Spectrosc.* **2021**, *75*, 988–993. **[Google Scholar] [CrossRef] [PubMed]**

128. Pence, I.J.; Goss, A.; Fast, A.; Brinkmann, M.; Hellwig, T.; Fallnich, C.; Evans, C.L. Development of a clinical coherent Raman imaging system for in vivo drug monitoring. In *Proceedings of the Biophotonics Congress 2021*; Optica Publishing Group: Washington, DC, USA; 2021; p. NTu3C.4. **[Google Scholar] [CrossRef]**

129. Yeh, P.-C.; Yoon, S.; Kurniawan, D.; Chung, Y.G.; Chiang, W.-H. Unraveling the Fluorescence Quenching of Colloidal Graphene Quantum Dots for Selective Metal Ion Detection. *ACS Appl. Nano Mater.* **2021**, *4*, 5636–5642. **[Google Scholar]** **[CrossRef]**

130. Lin, J.; Huang, Y.; Huang, P. Graphene-Based Nanomaterials in Bioimaging. In *Biomedical Applications of Functionalized Nanomaterials*; Elsevier: Amsterdam, The Netherlands, 2018; pp. 247–287. **[Google Scholar]**

131. Wen, W.; Song, Y.; Yan, X.; Zhu, C.; Du, D.; Wang, S.; Asiri, A.M.; Lin, Y. Recent advances in emerging 2D nanomaterials for biosensing and bioimaging applications. *Mater. Today* **2018**, *21*, 164–177. **[Google Scholar]** **[CrossRef]**

132. Liang, L.; Peng, X.; Sun, F.; Kong, Z.; Shen, J.-W. A review on the cytotoxicity of graphene quantum dots: From experiment to simulation. *Nanoscale Adv.* **2021**, *3*, 904–917. **[Google Scholar]** **[CrossRef]**

133. Samanta, S.; Banerjee, S.L.; Bhattacharya, K.; Singha, N.K. Graphene Quantum Dots-Ornamented Waterborne Epoxy-Based Fluorescent Adhesive via Reversible Addition–Fragmentation Chain Transfer-Mediated Miniemulsion Polymerization: A Potential Material for Art Conservation. *ACS Appl. Mater. Interfaces* **2021**, *13*, 36307–36319. **[Google Scholar]** **[CrossRef]**

134. Bapli, A.; Chatterjee, A.; Gautam, R.K.; Jana, R.; Seth, D. Modulation of the Protein–Ligand Interaction in the Presence of Graphene Oxide: A Detailed Spectroscopic Study. *Langmuir* **2021**, *37*, 5034–5048. **[Google Scholar]** **[CrossRef]**

135. Ghosh, S.; Chatterjee, K. Poly(Ethylene Glycol) Functionalized Graphene Oxide in Tissue Engineering: A Review on Recent Advances. *Int. J. Nanomed.* **2020**, *15*, 5991–6006. **[Google Scholar]** **[CrossRef]** **[PubMed]**

136. Anusuya, T.; Kumar, V.; Kumar, V. Hydrophilic graphene quantum dots as turn-off fluorescent nanoprobes for toxic heavy metal ions detection in aqueous media. *Chemosphere* **2021**, *282*, 131019. **[Google Scholar]** **[CrossRef]**

137. A Ocsoy, M.; Yusufbeyoglu, S.; Ildiz, N.; Ulgen, A.; Ocsoy, I. DNA Aptamer-Conjugated Magnetic Graphene Oxide for Pathogenic Bacteria Aggregation: Selective and Enhanced Photothermal Therapy for Effective and Rapid Killing. *ACS Omega* **2021**, *6*, 20637–20643. **[Google Scholar]** **[CrossRef]** **[PubMed]**

138. Huang, Q.; Cai, Y.; Zhang, X.; Liu, J.; Liu, Z.; Li, B.; Wong, H.; Xu, F.; Sheng, L.; Sun, D.; et al. Aligned Graphene Mesh-Supported Double Network Natural Hydrogel Conduit Loaded with Netrin-1 for Peripheral Nerve Regeneration. *ACS Appl. Mater. Interfaces* **2021**, *13*, 112–122. **[Google Scholar]** **[CrossRef]** **[PubMed]**

139. Oh, S.D.; Kim, J.; Lee, D.H.; Kim, J.H.; Jang, C.W.; Kim, S.; Choi, S.-H. Structural and optical characteristics of graphene quantum dots size-controlled and well-aligned on a large scale by polystyrene-nanosphere lithography. *J. Phys. D Appl. Phys.* **2016**, *49*, 025308. **[Google Scholar] [CrossRef]**

140. Borandeh, S.; Alimardani, V.; Abolmaali, S.S.; Seppälä, J. Graphene Family Nanomaterials in Ocular Applications: Physicochemical Properties and Toxicity. *Chem. Res. Toxicol.* **2021**, *34*, 1386–1402. **[Google Scholar] [CrossRef] [PubMed]**

141. Jampilek, J.; Kralova, K. Advances in Drug Delivery Nanosystems Using Graphene-Based Materials and Carbon Nanotubes. *Materials* **2021**, *14*, 1059. **[Google Scholar] [CrossRef]**

142. Wang, F.; Duo, T.; Wang, Y.; Xiao, Z.; Xu, A.; Liu, R. Novel Polyethyleneimine/k-Carrageenan Composite from Facile One-Step Fabrication for the Removal of Copper Ion from Aqueous Solution. *J. Polym. Environ.* **2022**, *30*, 1001–1011. **[Google Scholar] [CrossRef]**

143. Oh, J.-S.; Park, J.-S.; Lee, E.-J. Enhanced Effect of Polyethyleneimine-Modified Graphene Oxide and Simvastatin on Osteogenic Differentiation of Murine Bone Marrow-Derived Mesenchymal Stem Cells. *Biomedicines* **2021**, *9*, 501. **[Google Scholar] [CrossRef]**

144. Jonoush, Z.A.; Farahani, M.; Bohlouli, M.; Niknam, Z.; Golchin, A.; Hatamie, S.; Rezaei-Tavirani, M.; Omidi, M.; Zali, H. Surface Modification of Graphene and its Derivatives for Drug Delivery Systems. *Mini-Rev. Org. Chem.* **2021**, *18*, 78–92. **[Google Scholar] [CrossRef]**

145. Olson, E.; Plaut, J.S.; Barnhill, S.A.; Sabuncu, S.; Dambacher, C.M.; Speese, S.D.; Ranganathan, S.V.; Branchaud, B.P.; Yildirim, A. Enzyme-Instructed Formation of β-Sheet-Rich Nanoplatelets for Label-Free Protease Sensing. *ACS Appl. Nano Mater.* **2021**, *4*, 7800–7810. **[Google Scholar] [CrossRef]**

146. Pontrelli, G.; Toniolo, G.; McGinty, S.; Peri, D.; Succi, S.; Chatgilialoglu, C. Mathematical modelling of drug delivery from pH-responsive nanocontainers. *Comput. Biol. Med.* **2021**, *131*, 104238. **[Google Scholar] [CrossRef]**

147. Ren, M.-X.; Wang, Y.-Q.; Lei, B.-Y.; Yang, X.-X.; Hou, Y.-L.; Meng, W.-J.; Zhao, D.-L. Magnetite nanoparticles anchored on graphene oxide loaded with doxorubicin hydrochloride for magnetic hyperthermia therapy. *Ceram. Int.* **2021**, *47*, 20686–20692. **[Google Scholar] [CrossRef]**

148. Yoon, J.; Lim, J.; Shin, M.; Lee, S.-N.; Choi, J.-W. Graphene/MoS_2 Nanohybrid for Biosensors. *Materials* **2021**, *14*, 518. **[Google Scholar] [CrossRef]**

149. Lemus, L.M.R.; Azamar-Barrios, J.A.; Ortiz-Vazquez, E.; Quintana-Owen, P.; Freile-Pelegrín, Y.; Perera, F.G.; Madera-Santana, T.J. Development and physical characterization of novel bio-nanocomposite films based on reduced graphene oxide, agar and melipona honey. *Carbohydr. Polym. Technol. Appl.* **2021**, *2*, 100133. **[Google Scholar] [CrossRef]**

150. Mathew, T.; Sree, R.A.; Aishwarya, S.; Kounaina, K.; Patil, A.G.; Satapathy, P.; Hudeda, S.P.; More, S.S.; Muthucheliyan, K.; Kumar, T.N.; et al. Graphene-based functional nanomaterials for biomedical and bioanalysis applications. *FlatChem* **2020**, *23*, 100184. **[Google Scholar] [CrossRef]**

151. Işıklan, N.; Hussien, N.A.; Türk, M. Synthesis and drug delivery performance of gelatin-decorated magnetic graphene oxide nanoplatform. *Colloids Surf. A Physicochem. Eng. Asp.* **2021**, *616*, 126256. **[Google Scholar] [CrossRef]**

152. Kesavan, S.; Meena, K.; Sharmili, S.A.; Govindarajan, M.; Alharbi, N.S.; Kadaikunnan, S.; Khaled, J.M.; Alobaidi, A.S.; Alanzi, K.F.; Vaseeharan, B. Ulvan loaded graphene oxide nanoparticle fabricated with chitosan and d-mannose for targeted anticancer drug delivery. *J. Drug Deliv. Sci. Technol.* **2021**, *65*, 102760. **[Google Scholar] [CrossRef]**

153. Ebrahimi, M.; Teimouri, M.; Pooladi, M. The Synergistic Anticancer Traits of Graphene Oxide Plus Doxorubicin Against BT474 and MCF7 Breast Cancer Stem Cells In Vitro. *Appl. Biochem. Biotechnol.* **2021**, *193*, 3586–3601. **[Google Scholar] [CrossRef]**

154. Zhou, F.; Wang, M.; Luo, T.; Qu, J.; Chen, W.R. Photo-activated chemo-immunotherapy for metastatic cancer using a synergistic graphene nanosystem. *Biomaterials* **2021**, *265*, 120421. **[Google Scholar] [CrossRef]**

155. Mudaliar, P.; Pradeep, P.; Abraham, R.; Sreekumar, E. Targeting cap-dependent translation to inhibit Chikungunya virus replication: Selectivity of p38 MAPK inhibitors to virus-infected cells due to autophagy-mediated down regulation of phospho-ERK. *J. Gen. Virol.* **2021**, *102*, 001629. **[Google Scholar] [CrossRef]**

156. Xie, L.; Chen, F.; Du, H.; Zhang, X.; Wang, X.; Yao, G.; Xu, B. Graphene oxide and indole-3-acetic acid cotreatment regulates the root growth of Brassica napus L. via multiple phytohormone pathways. *BMC Plant Biol.* **2020**, *20*, 101. **[Google Scholar] [CrossRef] [PubMed] [Green Version]**

157. Parvin, N.; Mandal, T.K. Dually emissive P,N-co-doped carbon dots for fluorescent and photoacoustic tissue imaging in living mice. *Microchim. Acta* **2017**, *184*, 1117–1125. **[Google Scholar] [CrossRef]**

158. Zhang, X.; Zhang, X.; Song, L.; Huang, X.; Li, Y.; Qiao, M.; Liu, W.; Zhang, T.; Qi, Y.; Wang, W.; et al. An ultrasensitive, homogeneous fluorescence quenching immunoassay integrating separation and detection of aflatoxin M1 based on

magnetic graphene composites. *Microchim. Acta* **2021**, *188*, 59. **[Google Scholar]** **[CrossRef]**

159. Vasilopoulos, V.; Pitou, M.; Fekas, I.; Papi, R.; Ouranidis, A.; Pavlidou, E.; Patsalas, P.; Choli-Papadopoulou, T. Graphene-Wrapped Copper Nanoparticles: An Antimicrobial and Biocompatible Nanomaterial with Valuable Properties for Medical Uses. *ACS Omega* **2020**, *5*, 26329–26334. **[Google Scholar]** **[CrossRef]**
160. Liu, C.; Chen, F.; Tang, Y.; Huo, P. An environmentally friendly nanocomposite polypyrrole@silver/reduced graphene oxide with high catalytic activity for bacteria and antibiotics. *J. Mater. Sci. Mater. Electron.* **2021**, *32*, 15211–15225. **[Google Scholar]** **[CrossRef]**
161. Mandal, T.K.; Lee, Y.R.; Parvin, N. Red phosphorus Decorating Graphene oxide Nanosheets: Label-free DNA detection. *Biomater. Sci.* **2020**, *8*, 125–131. **[Google Scholar]** **[CrossRef]** **[PubMed]**
162. Ryoo, S.-R.; Kim, Y.-K.; Kim, M.-H.; Min, D.-H. Behaviors of NIH-3T3 Fibroblasts on Graphene/Carbon Nanotubes: Proliferation, Focal Adhesion, and Gene Transfection Studies. *ACS Nano* **2010**, *4*, 6587–6598. **[Google Scholar]** **[CrossRef]** **[PubMed]**
163. Tiwari, S.; Patil, R.; Dubey, S.K.; Bahadur, P. Graphene nanosheets as reinforcement and cell-instructive material in soft tissue scaffolds. *Adv. Colloid Interface Sci.* **2020**, *281*, 102167. **[Google Scholar]** **[CrossRef]** **[PubMed]**
164. Kang, E.-S.; Kim, H.; Han, Y.; Cho, Y.-W.; Son, H.; Luo, Z.; Kim, T.-H. Enhancing osteogenesis of adipose-derived mesenchymal stem cells using gold nanostructure/peptide-nanopatterned graphene oxide. *Colloids Surf. B Biointerfaces* **2021**, *204*, 111807. **[Google Scholar]** **[CrossRef]** **[PubMed]**
165. Long, X.; Duan, L.; Weng, W.; Cheng, K.; Wang, D.; Ouyang, H. Light-induced osteogenic differentiation of BMSCs with graphene/TiO_2 composite coating on Ti implant. *Colloids Surf. B Biointerfaces* **2021**, *207*, 111996. **[Google Scholar]** **[CrossRef]** **[PubMed]**
166. Lee, E.A.; Kwak, S.-Y.; Yang, J.-K.; Lee, Y.-S.; Kim, J.-H.; Kim, H.D.; Hwang, N.S. Graphene oxide film guided skeletal muscle differentiation. *Mater. Sci. Eng. C* **2021**, *126*, 112174. **[Google Scholar]** **[CrossRef]** **[PubMed]**
167. Yang, Y.; Jiang, X.; Li, Y.; Jin, W.; Huang, X. Construction of halogenated graphenes by halogenation of hydrogenated graphene. *Compos. Commun.* **2021**, *25*, 100771. **[Google Scholar]** **[CrossRef]**
168. Feng, X.; Wang, J.; Cai, P.; Yang, Z.; Shen, J.; Zhang, Y.; Zhang, X. Graphene/protamine assembled hybrid paper with antibacterial activity. *Colloids Surf. A Physicochem. Eng. Asp.* **2021**, *625*, 126977. **[Google Scholar]** **[CrossRef]**

169. Du, J.; Wang, T.; Zhou, Q.; Hu, X.; Wu, J.; Li, G.; Li, G.; Hou, F.; Wu, Y. Graphene oxide enters the rice roots and disturbs the endophytic bacterial communities. *Ecotoxicol. Environ. Saf.* **2020**, *192*, 110304. **[Google Scholar] [CrossRef]**

170. Ruiz, O.N.; Fernando, K.A.S.; Wang, B.; Brown, N.A.; Luo, P.G.; McNamara, N.D.; Vangsness, M.; Sun, Y.-P.; Bunker, C.E. Graphene Oxide: A Nonspecific Enhancer of Cellular Growth. *ACS Nano* **2011**, *5*, 8100–8107. **[Google Scholar] [CrossRef]**

171. Abdalla, O.; Wahab, M.A.; Abdala, A. Fabrication of Graphene Oxide-Based Membranes and their Applications in Water Treatment. *Curr. Pharm. Biotechnol.* **2021**, *22*, 1686–1704. **[Google Scholar] [CrossRef]**

172. Haddadan, F.; Soroosh, M.; Alaei-Sheini, N. Designing an electro-optical encoder based on photonic crystals using the graphene–Al_2O_3 stacks. *Appl. Opt.* **2020**, *59*, 2179. **[Google Scholar] [CrossRef]**

173. Jalali Azizpour, M.R.; Soroosh, M.; Dalvand, N.; Seifi-Kavian, Y. All-Optical Ultra-Fast Graphene-Photonic Crystal Switch. *Crystals* **2019**, *9*, 461. **[Google Scholar] [CrossRef] [Green Version]**

174. Haddadan, F.; Soroosh, M.; Alaei-Sheini, N. Cross-talk reduction in a graphene-based ultra-compact plasmonic encoder using an Au nano-ridge on a silicon substrate. *Appl. Opt.* **2022**, *61*, 3209. **[Google Scholar] [CrossRef]**

175. Haddadan, F.; Soroosh, M. Design and simulation of a subwavelength 4-to-2 graphene-based plasmonic priority encoder. *Opt. Laser Technol.* **2023**, *157*, 108680. **[Google Scholar] [CrossRef]**

176. Bagheri, F.; Soroosh, M.; Haddadan, F.; Seifi-Kavian, Y. Design and simulation of a compact graphene-based plasmonic D flip-flop. *Opt. Laser Technol.* **2022**, *155*, 108436. **[Google Scholar] [CrossRef]**

INDEX

www.ingramcontent.com/pod-product-compliance
Lightning Source LLC
LaVergne TN
LVHW062223200926
848484LV00006B/1463
9781787153844